Fluoropolymers 1

Synthesis

TOPICS IN APPLIED CHEMISTRY

Series Editor: **Alan R. Katritzky, FPS**
University of Florida
Gainesville, Florida

Gebran J. Sabongi
3M Company
St. Paul, Minnesota

Current volumes in the series:

ANALYSIS AND DEFORMATION OF POLYMERIC MATERIALS
Paints, Plastics, Adhesives, and Inks
Jan W. Gooch

CHEMISTRY AND APPLICATIONS OF LEUCO DYES
Edited by Ramaiah Muthyala

FLUOROPOLYMERS 1: Synthesis
FLUOROPOLYMERS 2: Properties
Edited by Gareth Hougham, Patrick E. Cassidy, Ken Johns,
and Theodore Davidson

FROM CHEMICAL TOPOLOGY TO THREE-DIMENSIONAL
GEOMETRY
Edited by Alexandru T. Balaban

LEAD-BASED PAINT HANDBOOK
Jan W. Gooch

ORGANIC PHOTOCHROMIC AND THERMOCHROMIC
COMPOUNDS
Volume 1: Main Photochromic Families
Volume 2: Physicochemical Studies, Biological Applications, and
Thermochromism
Edited by John C. Crano and Robert J. Guglielmetti

ORGANOFLUORINE CHEMISTRY
Principles and Commercial Applications
Edited by R. E. Banks, B. E. Smart, and J. C. Tatlow

PHOSPHATE FIBERS
Edward J. Griffith

RESORCINOL
Its Uses and Derivatives
Hans Dressler

A Continuation Order Plan is available for this series. A continuation order will bring delivery of each new volume immediately upon publication. Volumes are billed only upon actual shipment. For further information please contact the publisher.

Fluoropolymers 1
Synthesis

Edited by

Gareth Hougham

IBM T. J. Watson Research Center
Yorktown Heights, New York

Patrick E. Cassidy

Southwest Texas State University
San Marcos, Texas

Ken Johns

Chemical and Polymer
Windlesham, Surrey, England

Theodore Davidson

Princeton, New Jersey

Kluwer Academic / Plenum Publishers
New York • Boston • Dordrecht • London • Moscow

Chemistry Library

Chem

Library of Congress Cataloging-in-Publication Data

Fluoropolymers / edited by Gareth Hougham ... [et al.].
        p.   cm. -- (Topics in applied chemistry)
    Includes bibliographical references and indexes.
    Contents: 1. Synthesis -- 2. Properties.
    ISBN 0-306-46060-2 (v. 1). -- ISBN 0-306-46061-0 (v. 2)
    1. Fluoropolymers.   I. Hougham, Gareth.   II. Series.
QD383.F48F54   1999
547'.70459--dc21   99-23732
                                                              CIP

Cover graphic: Conformational energy surface of 6FDA-PFMB fluorinated polyimide. Used to test correspondence between experimental activation energy of b transition and energy of rotation about phenyl-imide bond. [G. Hougham and T. Jackman, Polymer Preprints **37**(1), 1996.] Graphic by G. Hougham and T. Jackman.

ISBN: 0-306-46060-2

© 1999 Kluwer Academic/Plenum Publishers, New York
233 Spring Street, New York, N.Y. 10013

10 9 8 7 6 5 4 3 2 1

A C.I.P. record of this book is available from the Library of Congress.

Printed in the United States of America

Contributors

Richard C. Allen, Specialty Materials Division, 3M Company, St. Paul, Minnesota 55144-1000

Bruno Ameduri, Ecole Nationale Supérieure de Chimie de Montpellier, F-34296 Montpellier Cédex 5, France

Milton H. Andrus, Jr., Specialty Materials Division, 3M Company, St. Paul, Minnesota 55144-1000

M. Antonietti, Max-Planck Institute of Colloids and Interfaces, D-14513 Teltow-Seehof, Germany

David Babb, Central Research and Development, Organic Product Research, Dow Chemical Company, Freeport, Texas 77541

Georgy Barsamyan, Samsung Corning Co., Ltd., Dong-Suwon, Kyeongii-Do, Korea 442-600. Present address: Solvay S.A. Moscow Office, Moscow 123310, Russia

Bernard Boutevin, Ecole Nationale Supérieure de Chimie de Montpellier, F-34296, Montpellier Cédex 5, France

Maria Bruma, Institute of Macromolecular Chemistry, Iasi, Romania

Leonard J. Buckley, Naval Research Laboratory, Washington, D.C. 20375

Gerardo Caporiccio, Via Filiberto 13, Milan 20149, Italy

P. A. B. Carstens, Atomic Energy Corporation of South Africa, Ltd., Pretoria 0001, South Africa

Patrick E. Cassidy, Polymer Research Group, Department of Chemistry, Southwest Texas State University, San Marcos, Texas 78666

J. A. de Beer, Atomic Energy Corporation of South Africa, Ltd., Pretoria 0001, South Africa

Joseph M. DeSimone, Department of Chemistry, University of North Carolina at Chapel Hill, Chapel Hill, North Carolina 27599-3290

James P. DeYoung, MICELL Technologies Inc., Raleigh, North Carolina 27613

Gilbert L. Eian, Speciality Materials Division, 3M Company, St. Paul, Minnesota 55144-1000

William B. Farnham, DuPont Central Research and Development, Experimental Station, Wilmington, Delaware 19880-0328

Andrew E. Feiring, DuPont Central Research and Development, Experimental Station, Wilmington, Delaware 19880-0328

John W. Fitch, Polymer Research Group, Department of Chemistry, Southwest Texas State University, San Marcos, Texas 78666

Matilda M. Fone, Corporate Research and Development, Raychem Corporation, Menlo Park, California 94025

Andy A. Goodwin, Department of Materials Engineering, Monash University, Clayton, Victoria, Australia

James R. Griffith, Naval Research Laboratory, Washington, D.C. 20375

Francine Guida-Pietrasanta, Ecole Nationale Supérieure de Chimie de Montpellier, F-34296 Montpellier Cédex 5, France

Henry S.-W. Hu, Geo-Centers, Inc., Ft. Washington, Maryland 20744. Present address: Sensors for Medicine and Science, Inc., Germantown, Maryland 20874

Ming-H. Hung, DuPont Fluoroproducts, Wilmington, Delaware 19880-0293

Richard J. Lagow, Department of Chemistry and Biochemistry, University of Texas at Austin, Austin, Texas 78712-1167

Chi-I Lang, Department of Chemistry, Rensselaer Polytechnic Institute, Troy, New York 12180-3590

K. J. Lawson, Cranfield University, Cranfield, Bedford MK43 0AL, United Kingdom

J. P. Le Roux, Atomic Energy Corporation of South Africa, Ltd., Pretoria 0001, South Africa

Abdellatif Manseri, CNRS, Ecole Nationale Supérieure de Chimie de Montpellier, F-34296 Montpellier Cédex 5, France

Martin T. McKenzie, Corporate Research and Development, Raychem Corporation, Menlo Park, California 94025

Frank W. Mercer, Corporate Research and Development, Raychem Corporation, Menlo Park, California 94025

Boris V. Mislavsky, Institute of Chemical Physics, Russian Academy of Science, 117977 Moscow, Russia

James A. Moore, Department of Chemistry, Rensselaer Polytechnic Institute, Troy, New York 12180-3590

Shiego Nakamura, Faculty of Engineering, Kanagawa University, Rokkakubashi Kanagawa-ku, Yokohama 221-8686, Japan

Yuko Nishimoto, Faculty of Science, Kanagawa University Tsuchiya, Hiratsuka 259-1293, Japan

J. R. Nicholls, Cranfield University, Cranfield, Bedford, MK43 0AL, United Kingdom

S. Oestreich, Max-Planck Institute of Colloids and Interfaces, D-14513 Teltow-Seehof, Germany

Robert J. Olsen, Speciality Materials Division, 3M Company, St. Paul, Minnesota 55144-1000

Amédée Ratsimihety, Ecole Nationale Supérieure de Chimie de Montpellier, F-34296 Montpellier Cédex 5, France

V. Sreenivasulu Reddy, Polymer Research Group, Department of Chemistry, Southwest Texas State University, San Marcos, Texas 78666

Timothy J. Romack, MICELL Technologies Inc., Raleigh, North Carolina 276013

Shlomo Rozen, DuPont Central Research and Development, Experimental Station, Willmington, Delaware 19880-0328; Permanent address: Department of Chemistry, Tel-Aviv University, Tel-Aviv 69978, Israel

Diederich Schilo, Acordis Research GmbH, 63784 Obernburg, Germany

Arthur W. Snow, Naval Research Laboratory, Washington, D.C. 20375

Vladimir Sokolov, Russian Research Center, "Kurchatov Institute," Moscow 123182, Russia

Marcus O. Weber, Acordis Research GmbH, 63784 Obernburg, Germany

Han-Chao Wei, Department of Chemistry and Biochemistry, University of Texas at Austin, Austin, Texas 78712-1167

Preface

The fluorine atom, by virtue of its electronegativity, size, and bond strength with carbon, can be used to create compounds with remarkable properties. Small molecules containing fluorine have many positive impacts on everyday life of which blood substitutes, pharmaceuticals, and surface modifiers are only a few examples.

Fluoropolymers, too, while traditionally associated with extreme high-performance applications have found their way into our homes, our clothing, and even our language. A recent American president was often likened to the tribology of PTFE.

Since the serendipitous discovery of Teflon at the Dupont Jackson Laboratory in 1938, fluoropolymers have grown steadily in technological and marketplace importance. New synthetic fluorine chemistry, new processes, and new appreciation of the mechanisms by which fluorine imparts exceptional properties all contribute to accelerating growth in fluoropolymers.

There are many stories of harrowing close calls in the fluorine chemistry lab, especially from the early years, and synthetic challenges at times remain daunting. But, fortunately, modern techniques and facilities have enabled significant strides toward taming both the hazards and synthetic uncertainties.

In contrast to past environmental problems associated with fluorocarbon refrigerants, the exceptional properties of fluorine in polymers have great environmental value. Some fluoropolymers are enabling green technologies such as hydrogen fuel cells for automobiles and oxygen-selective membranes for cleaner diesel combustion.

Curiously, fluorine incorporation can result in property shifts to opposite ends of a performance spectrum. Certainly with reactivity, fluorine compounds occupy two extreme positions, and this is true of some physical properties of fluoropolymers as well. One example depends on the combination of the low electronic polarizability and high dipole moment of the carbon–fluorine bond. At one extreme, some fluoropolymers have the lowest dielectric constants known. At the other, closely related materials are highly capacitive and even piezoelectric.

Much progress has been made in understanding the sometimes confounding properties of fluoropolymers. Computer simulation is now contributing to this with new fluorine force fields and other parameters, bringing realistic prediction within reach of the practicing physical chemist.

These two volumes attempt to bring together in one place the chemistry, physics, and engineering properties of fluoropolymers. The collection was intended to provide balance between breadth and depth, with contributions ranging from the introduction of fluoropolymer structure–property relationships, to reviews of subfields, to more focused topical reports.

GGH

Acknowledgments

Gareth Hougham thanks G. Tesoro, IBM, K. C. Appleby, D. L. Wade, R. H. Henry, and K. Howell.

Patrick Cassidy expresses his appreciation to the Robert A. Welch Foundation, the National Aeronautics and Space Administration, the National Science Foundation, and the Institute for Environmental and Industrial Science at Southwest Texas State University.

Ken Johns thanks Diane Kendall and Catherine Haworth, Senior Librarian, of the Paint Research Association.

Theodore Davidson wishes to acknowledge his students and collaborators who have shared in the work on polytetrafluoroethylene. Sincere thanks go to Professor Bernhard Wunderlich for providing the stimulus for a career in polymer science.

Contents

I. Synthesis

1. Polyacrylates Containing the Hexafluoroisopropylidene Function in the Pendant Groups

V. Sreenivasulu Reddy, Maria Bruma, John Fitch, and Patrick Cassidy

1.1. Introduction	3
1.2. Experimental	4
1.2.1. Materials	4
1.2.2. Synthesis of the Monomers	4
1.2.3. Synthesis of the Polymers	5
1.2.4. Measurements	5
1.3. Results and Discussion	6
1.4. Conclusions	9
1.5. References	9

2. Fluorinated Cyanate Ester Resins

Arthur Snow and Leonard J. Buckley

2.1. Introduction	11
2.2. Background	12
2.3. Monomer Synthesis	12
2.4. Resin Cure	15
2.5. Resin Properties	17
2.6. Dielectric Properties	19
2.7. Processing	21
2.8. Summary	22
2.9. References	23

3. **Polymers from the Thermal (2π + 2π) Cyclodimerization of Fluorinated Olefins**

    David A. Babb

    3.1. Introduction   25
         3.1.1. Cyclodimerization of Fluorinated Olefins   25
         3.1.2. Polymers Containing the Hexafluorocyclobutane Ring   27
         3.1.3. Perfluorocyclobutane Aryl Ether Polymers   29
    3.2. Experimental   29
         3.2.1. Synthesis of 2-Bromotetrafluoroethyl Aryl Ethers   31
         3.2.2. Synthesis of Trifluorovinyloxy Aryl Ether Monomers   34
         3.2.3. Polymerization of Bis(Trifluorovinyloxy)Aryl Ether
                Monomers   36
         3.2.4. Fabrication of Polymer Plaques for Mechanical Testing . . .   38
         3.2.5. Fabrication of Polymer Film Samples for Thermal Stability
                Testing   39
    3.3. Results and Discussion   39
         3.3.1. Monomer Synthesis   39
         3.3.2. Polymerization   42
         3.3.3. Thermal Stability Testing   44
         3.3.4. Thin-Film Processing   46
    3.4. Conclusions   47
    3.5. References   49

4. **Functional Fluoromonomers and Fluoropolymers**

    *Ming-H. Hung, William B. Farnham, Andrew E. Feiring,
    and Shlomo Rozen*

    4.1. Introduction   51
    4.2. Results and Discussion   52
         4.2.1. Linear Polymers   53
         4.2.2. Cyclic Oligomers   55
         4.2.3. Perfluorinated Polymers/Oligomers   56
    4.3. Conclusions   61
    4.4. Appendix   62
         4.4.1. Preparation of 9,9-Dihydro-9-Hydroxyperfluoro-
                (3,6-Dioxa-5-Methyl-1-Nonene   62
         4.4.2. General Procedure for Substituting Hydrogens with Fluorine   64
    4.5. References   65

5. **Use of Original Fluorinated Telomers in the Synthesis of Hybrid Silicones**

   Bruno Ameduri, Bernard Boutevin, Gerardo Caporiccio, Francine Guida-Pietrasanta, Abdellatif Manseri, Amédée Ratsimihety

   5.1. Introduction.. 67
   5.2. Results and Discussion 68
       5.2.1. Synthesis of Fluorinated Precursors................ 68
       5.2.2. Introduction of Fluorinated Groups in Silanes and Silicones 72
   5.3. Conclusions.. 79
   5.4. References... 79

6. **Chlorotrifluoroethylene Suspension Polymerization**

   Milton H. Andrus Jr., Robert J. Olsen, Gilbert L. Eian, and Richard C. Allen

   6.1. Introduction....................................... 81
   6.2. Experimental 82
       6.2.1. Materials 82
       6.2.2. Preparation of Perfluorooctanoyl Peroxide 82
       6.2.3. Polymerization 82
       6.2.4. Molecular Weight Measurements................. 82
       6.2.5. Polymer Testing 83
   6.3. Results and Conclusions 83
       6.3.1. Polymerizations.............................. 83
       6.3.2. Organic Peroxides............................ 85
       6.3.3. Gel Permeation Chromatography................. 86
       6.3.4. Capillary Rheometry 87
   6.4. Summary.. 88
   6.5. References... 90

7. **Fluorinated Polymers with Functional Groups: Synthesis and Applications. Langmuir–Blodgett Films from Functional Fluoropolymers**

   Boris V. Mislavsky

   7.1. Introduction....................................... 91
   7.2. Synthesis of Functional Fluoropolymers 92
       7.2.1. Polymerization of Functional Fluoromonomers and Copolymerization of Functional Monomers with Fluoroolefins................................. 92
       7.2.2. Radiation Grafting of Functional Monomers onto Fluoropolymers.............................. 93

7.3. Applications of Functional Fluoropolymers 96
    7.3.1. Chlorine-Alkali Electrolysis 96
    7.3.2. Fuel Cells 96
    7.3.3. Water Electrolysis 96
    7.3.4. Polymer Catalysts 96
    7.3.5. Permselective Membranes 99
    7.3.6. Selective Adsorbents 99
    7.3.7. Ion Selective Electrodes 100
    7.3.8. Mixed Conductivity Polymers 102
7.4. Langmuir–Blodgett Films from Functional Fluoropolymers 102
    7.4.1. Multilayered Structures with Metal Cations Based on
           LB Films of Perfluorinated Sulfoacid Polymer
           (Nafion-Type) 103
    7.4.2. LB Mono- and Multilayers of Fluorocarbon Amphiphilic
           Polymers and Their Application in Photogalvanic
           Metal–Insulator–Semiconductor Structures 107
7.5. References 108

8. **Synthesis of Fluorinated Poly(Aryl Ether)s Containing
   1,4-Naphthalene Moieties**

   *Frank W. Mercer, Matilda M. Fone, Martin T. McKenzie,
   and Andy A. Goodwin*

8.1. Introduction 111
8.2. Experimental 112
    8.2.1. Starting Materials 112
    8.2.2. Polymerizations 113
    8.2.3. Polymer Films 114
    8.2.4. Measurements 114
8.3. Results and Discussion 115
8.4. Conclusions 123
8.5. References 124

9. **Synthesis and Properties of Fluorine-Containing Aromatic
   Condensation Polymers Obtained from Bisphenol AF and
   Its Derivatives**

   Shigeo Nakamura and Yuko Nishimoto

9.1. Introduction 127
9.2. Poly(Carbonate)s 128
    9.2.1. Synthesis 128
    9.2.2. Properties 131

9.3. Poly(Formal)s. 132
   9.3.1. Synthesis 132
   9.3.2. Properties. 133
   9.3.3. Thermal Degradation. 135
9.4. Poly(Ketone)s 137
   9.4.1. Synthesis 137
   9.4.2. Properties. 139
9.5. Poly(Azomethine)s 140
   9.5.1. Synthesis 140
   9.5.2. Properties. 141
9.6. Poly(Azole)s 143
   9.6.1. Synthesis 143
   9.6.2. Properties. 146
9.7. Poly(Siloxane) 148
   9.7.1. Synthesis 148
   9.7.2. Properties. 148
9.8. Conclusions. 150
9.9. References. 150

10. Novel Fluorinated Block Copolymers: Synthesis and Application

M. Antonietti and S. Oestreich

10.1. Introduction 151
10.2. Polymer Synthesis. 153
10.3. Properties 156
   10.3.1. Micelle Formation 156
   10.3.2. Dispersion Polymerization of Styrene in
         Fluorinated Solvents 156
   10.3.3. Surface Energies of the Block Copolymers 159
   10.3.4. Mesophase Formation. 161
   10.3.5. Gas Permeation Measurements 163
10.4. Conclusions and Outlook. 164
10.5. References. 165

11. Synthesis and Structure–Property Relationships of Low-Dielectric-Constant Fluorinated Polyacrylates

Henry S.-W. Hu and James R. Griffith

11.1. Introduction 167
11.2. Experimental 169
   11.2.1. Materials 169
   11.2.2. Techniques 170

11.3. Results and Discussion 172
    11.3.1. Monomers and Polymers 172
    11.3.2. Dielectric Constant and Structure–Property Relationships 177
    11.3.3. Processability and Applications 178
    11.3.4. Lower Dielectric Constants? 178
11.4. Conclusions 179
11.5. References 179

12. Epoxy Networks from a Fluorodiimidediol

Henry S.-W. Hu and James R. Griffith

12.1. Introduction 181
12.2. Experimental 183
    12.2.1. Methods 183
    12.2.2. Materials 183
    12.2.3. Techniques 184
12.3. Results and Discussion 185
12.4. Conclusions 188
12.5. References 188

13. Synthesis of Fluoropolymers in Liquid and Supercritical Carbon Dioxide Solvent Systems

James P. DeYoung, Timothy J. Romack, and Joseph M. DeSimone

13.1. Introduction 191
    13.1.1. Alternative Solvent Technologies 191
    13.1.2. Solvent Properties of Carbon Dioxide 192
13.2. Fluoroalkyl Acrylate Polymerization in Carbon Dioxide 193
    13.2.1. Homopolymers 193
    13.2.2. Random Copolymers 194
    13.2.3. Applications of Amphiphilic Copolymers 195
13.3. Fluoroolefin Polymerization in Carbon Dioxide 195
    13.3.1. Overview 195
    13.3.2. Melt-Processable Fluoropolymers 196
    13.3.3. Ion-Exchange Resins 197
    13.3.4. β-Scission and Acid End Groups 201
    13.3.5. Other Fluoropolymers 202
13.4. Other Systems of Interest 202
    13.4.1. Photooxidation of Fluoroolefins in Liquid Carbon Dioxide 202
    13.4.2. Hybrid Carbon Dioxide/Aqueous Systems 203
13.5. Conclusions 203
13.6. References 204

II. Direct Fluorination

14. Direct Fluorination of Polymers

Richard J. Lagow and Han-Chao Wei

14.1. Introduction 209
14.2. Poly(Carbon Monofluoride) 211
14.3. Perfluoropolyethers 212
14.4. Perfluorinated Nitrogen-Containing Ladder Polymers 217
14.5. Surface Fluorination of Polymers 219
14.6. Conclusion. 220
14.7. References 220

15. Surface Fluorination of Polymers Using Xenon Difluoride

Georgy Barsamyan and Vladimir B. Sokolov

15.1. Introduction 223
15.2. Fluorination of Substances and Surfaces 224
15.3. Polymer Surface Fluorination with Xenon Difluoride 231
15.4. Experimental 231
15.5. Results and Discussion 232
15.6. Rubber Surface Fluorination with Xenon Difluoride 233
15.7. Industrial Application of Rubber Surface Fluorination
      with Xenon Difluoride 236
15.8. Recent Development of the Xenon Difluoride Method 239
15.9. Conclusions 239
15.10. References 239

16. New Surface-Fluorinated Products

P. A. B. Carstens, J. A. de Beer, and J. P. Le Roux

16.1. Introduction 241
16.2. Permeation-Based Products 242
      16.2.1. Containerization of Crop Protection Chemicals 242
      16.2.2. Fluorinated HDPE Containers for Other Applications ... 243
      16.2.3. Rotational-Molded LLDPE Automotive Fuel Tanks 244
      16.2.4. HDPE Pipes 244
      16.2.5. Benefits of Surface Fluorination for Flexible Plastics ... 245
16.3. Products Based on Improved Adhesion 247
      16.3.1. Introduction 247
      16.3.2. Sheet Cladding 249
      16.3.3. Pipe Products Manufactured with Fluorination
              Technology 251

16.3.4. Adhesive Performance of Surface-Activated Polyolefin
Surfaces with Respect to Reinforcement Resins. 252
16.3.5. Surface-Fluorinated Fibers in Cementitious Mixtures . . . 255
16.4. Current Status and Future Developments 257
16.5. References. 258

17. Modified Surface Properties of Technical Yarns

Marcus O. Weber and Diederich Schilo

17.1. Introduction 261
17.2. Adhesion to Rubber and PVC 264
17.3. Wetting Behavior of Fluorinated Polyester 266
17.4. Influences on Biocompatibility 268
17.5. References. 269

III. Vapor Deposition

18. Vapor Deposition Polymerization as a Route to Fluorinated Polymers

J. A. Moore and Chi-I Lang

18.1. Introduction 273
18.2. Vapor Deposition Polymerization 277
18.2.1. Parylene N 277
18.2.2. Parylene F 279
18.2.3. Poly(Benzocyclobutenes) 284
18.2.4. Polynaphthalenes 294
18.3. Conclusions. 310
18.4. References. 311

19. Ultrathin PTFE, PVDF, and FEP Coatings Deposited Using Plasma-Assisted Physical Vapor Deposition

K. J. Lawson and J. R. Nicholls

19.1. Introduction 313
19.2. RF Magnetron Sputtering of Fluoropolymer Films 314
19.2.1. The Process 314
19.2.2. Sputter Deposition of PTFE, PVDF, and FEP 315
19.3. Results and Discussion 316
19.3.1. Deposition of PTFE. 316
19.3.2. Deposition of PVDF and FEP Films. 318
19.4. Conclusions. 318
19.5. References. 319

Index 321

I
Synthesis

1

Polyacrylates Containing the Hexafluoroisopropylidene Function in the Pendant Groups

V. SREENIVASULU REDDY, MARIA BRUMA, JOHN FITCH, and PATRICK CASSIDY

1.1. INTRODUCTION

Polymers prepared from acrylates and methacrylates have long been recognized for their optical clarity and stability upon aging under severe conditions.[1] Polyacrylates, whether linear or cross-linked, are excellent adhesives and are easily processed into many different components and shapes.[2] More and more attention is now being paid to the monomers and polymers containing hexafluoro-isopropylidene (6F) groups, which promise as film-formers, gas separation membranes, coatings, seals, and many other high-performance applications.[3] Frequently, the inclusion of 6F groups into the polymer structure will increase the thermal stability, flame retardancy, oxidation resistance, transparency, and environmental stability, while there is often a decrease in color, crystallinity, surface energy, and water absorption.[4] Fluoroacrylate polymers have also been suggested for medical applications, such as dental materials and artificial joints,[5]

V. SREENIVASULU REDDY · Polymer Research Group, Department of Chemistry, Southwest Texas University, San Marcos, Texas 78666. MARIA BRUMA · Institute of Macromolecular Chemistry, Iasi, Romania. JOHN FITCH and PATRICK CASSIDY · Polymer Research Group, Department of Chemistry, Southwest Texas State University, San Marcos, Texas 78666.

Fluoropolymers 1: Synthesis, edited by Hougham *et al.*, Plenum Press, New York, 1999.

3

owing to the ability to polymerize acrylate monomers containing a suspension of polytetrafluoroethylene that yields tough products with low water absorption and low-friction surfaces. Superior chemical resistance and useful surface properties often result when fluorine is incorporated into such polymers.

Herein we present the synthesis of two series of fluorinated acrylate polymers and copolymers derived from commercially available hexafluoro-2-hydroxy-2-(4-fluorophenyl)propane. The solubility, film-forming ability, thermal stability, and water absorption in these polymers have been studied.

1.2. EXPERIMENTAL

1.2.1. Materials

Hexafluoro-2-hydroxy-2-(4-fluorophenyl)propane (HFAF) was obtained from Central Glass Co., Japan, and was distilled under argon prior to use. Ethyl bromoacetate, methyl acrylate, acryloyl chloride, and methacryloyl chloride were obtained from Aldrich and were used as received. Hydroxyethyl methacrylate and hydroxypropyl methacrylate, also from Aldrich, were purified as described in the literature.[6]

1.2.2. Synthesis of the Monomers

Hexafluoro-2-(4-fluorophenyl)-2-propyl acrylate (**Ia**) and hexafluoro-2-(4-fluorophenyl)-2-propyl methacrylate (**Ib**)[7] were prepared by gradually adding a solution of 0.06 mol of acryloyl or methacryloyl chloride, respectively, in 20 ml of tetrahydrofuran (THF), to a cooled solution of 0.04 mol of HFAF in 50 ml of THF containing 0.06 mol of triethylamine. The mixture was stirred for 10–12 h at room temperature; it then was poured into 200 ml of water, and the new product was extracted with diethyl ether. After the diethyl ether was evaporated, the crude product was purified by column chromatography using a mixture of hexane and dichloromethane (1 : 1 by volume) as the elution solvent. Distillation afforded **Ia**, b.p. 72–74°C/10^{-3} mm Hg, and **Ib**, b.p. 48–52°C/10^{-3} mm Hg.

2-Hydroxyethylmethacrylate (**Va**) and 3-hydroxypropylmethacrylate (**Vb**) esters of 2-[hexafluoro-2-(4-fluorophenyl)-2-propoxy]acetic acid (**III**) were prepared through the following steps[8]: An aqueous solution of 2.81 g of KOH was added to a solution of 0.05 mol of HFAF in 50 ml of toluene. The reaction mixture was heated with stirring for 4 h, and water was removed as an azeotrope with toluene. Then 0.05 mol of ethylbromoacetate was added and heating was continued for 2 h at reflux. After cooling and neutralization with diluted HCl the organic layer was separated, and the solvent was evaporated. The residue was then heated with 25 ml of 10% aqueous NaOH for 4 h at 80°C. The homogeneous solution was cooled and neutralized with diluted HCl, and the resulting product was extracted with methylene chloride. Evaporation of methylene chloride and recrystallization from hexane afforded 2-[hexafluoro-2-(4-fluorophenyl)-2-propoxy]acetic acid (III) m.p. 99–101°C.

Reaction of III with excess thionyl chloride at reflux gave the corresponding acid chloride (IV). A solution of 0.03 mol of acid chloride IV in 25 ml of THF was added to a solution of 0.03 mol of hydroxyethyl methacrylate or hydroxypropyl methacrylate in 25 ml of THF containing 5 ml of triethylamine. The reaction mixture was stirred for 6 h at room temperature; then the solvent was evaporated and the residue was dissolved in diethylether and washed with water to remove the triethylamine hydrochloride salt. After the diethylether was evaporated, the resulting product was purified by column chromatography using methylene chloride as eluent to yield, respectively, the fluorinated ethyl methacrylic monomer Va, b.p. 60–62°C/10^{-3} mmHg, and the fluorinated propyl methacrylic monomer Vb, b.p. 75–77°C/10^{-3} mm Hg.

1.2.3. Synthesis of the Polymers

Homopolymerization of the new, fluorinated acrylate or methacrylate compounds or their copolymerization with other acrylate or methacrylate monomers as performed by bulk, free-radical polymerization at 80°C, using AIBN as initiator in the case of monomers I, and benzoyl peroxide in the case of monomers V. Reactions were conducted in an argon atmosphere. Monomers I were polymerized under a continuous flow of argon that was bubbled through the monomer and then passed through a reflux condenser. Highly viscous homopolymers II were obtained within about 20 min, while for the copolymers II' the maximum viscosity was attained after about 5 h. Homopolymers VI and copolymers VI' were prepared in sealed ampoules, which were immersed in a heating bath for 2 h at 80°C. Finally, all the polymers and copolymers were dissolved in THF and then precipitated in methanol, filtered, and dried *in vacuo* at 50°C.

1.2.4. Measurements

FTIR spectra were recorded with a Perkin-Elmer 1600-Series FTIR spectrometer, using KBr pellets or films. NMR spectra were obtained on an IBM 80-MHz spectrometer, using deuterated dimethylsulfoxide or deuterated chloroform as solvent. Elemental analyses were performed by Desert Analytics, Tucson, Arizona. Inherent viscosities were measured for polymer solutions in THF or DMF at a concentration of 0.25 dl/g, at 25°C, using a Cannon-Fenske viscometer. Thermogravimetric analysis was run with a DuPont 9900 thermal analyzer in nitrogen. The water absorption of the polymers was studied by drying a thin film to constant weight in a desiccator and then placing it in deionized water at room temperature. At 1-day intervals the film was removed, wiped dry, and weighed until a constant weight was again attained—usually 2 to 4 days.

1.3. RESULTS AND DISCUSSION

Four new fluorinated acrylic or methacrylic monomers were readily accessible using the commercially available HFAF as a means of introducing 6F groups into the monomers. Its reaction with acryloyl or metacryloyl chloride, as shown in Figure 1.1, gave the new monomers **I**.

By reacting first with ethyl bromoacetate, a fluorinated carboxylic acid (**III**) resulted, which was then converted into the acid chloride (**IV**) and subsequently reacted with hydroxyethyl methacrylate or hydroxypropyl methacrylate giving rise to the monomers **V**, as shown in Figure 1.2. All these monomers are clear, colorless liquids, and were characterized by NMR and FTIR spectra and elemental analyses.

Homopolymerization of monomer **I** or its copolymerization with 2-hydroxyethyl methacrylate or 3-hydroxypropyl methacrylate gave the first series of

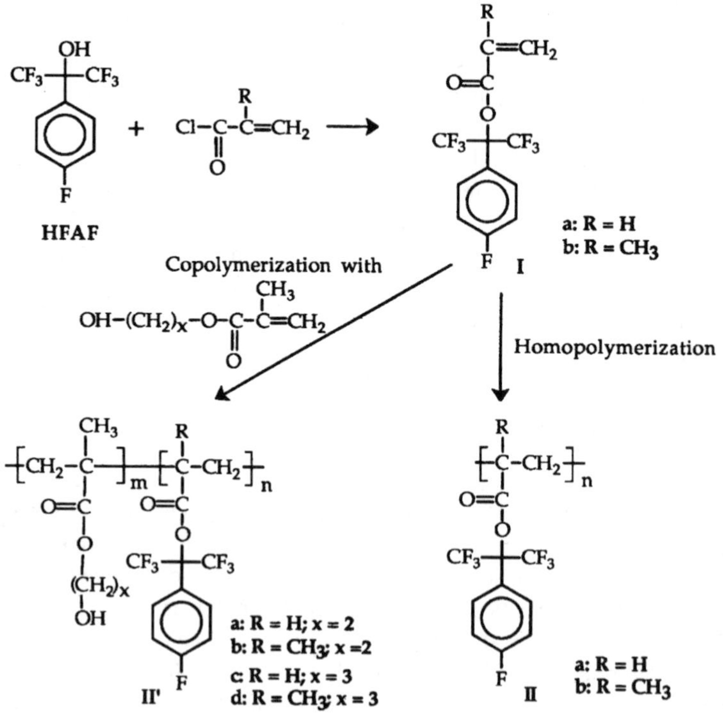

Figure 1.1. Synthesis of 6F-containing polyacrylates.

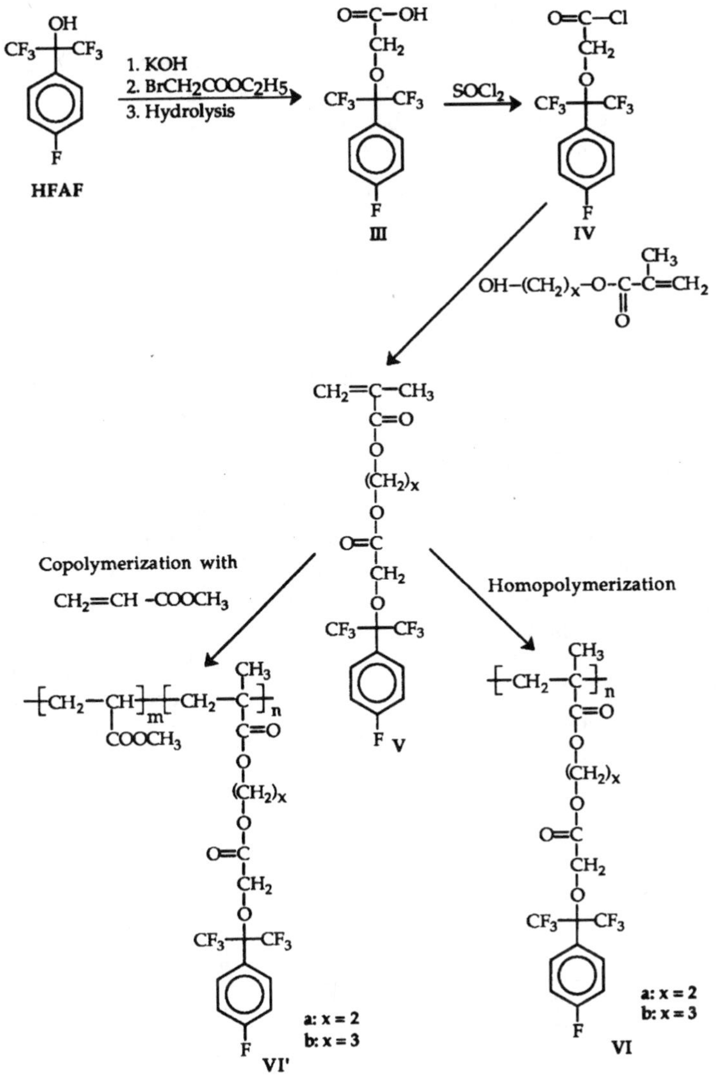

Figure 1.2. Synthesis of 6F- and ether linkage-containing polyacrylates.

polymers **II** and **II'**, respectively, containing 6F groups in the pendant chain (Figure 1.1). Homopolymerization of monomer **V** or its copolymerization with methyl acrylate gave the second series of polymers, **VI** and **VI'**, also containing 6F groups in the pendant chain (Figure 1.2). Homopolymerization and copoymerization were performed by bulk free-radical polymerization using azoisobutyronitrile or benzoyl peroxide as initiator. The copolymerizations were run using a 1:1 molar ratio of the two comonomers. All polymers and copolymers are soluble in chloroform and THF. Inherent viscosities span a wide range: 0.17–1.5 dl/g (Table 1). Transparent, colorless films have been obtained by casting polymer solutions in chloroform onto glass plates. Infrared spectra of polymer films show strong absorption bands at 1762–1755 cm^{-1} (carbonyl) and at 1276–1216 cm^{-1} (C–F stretching).

Thermogravimetric analysis in nitrogen shows that homopolymers **II** begin to decompose at 370°C, while copolymers **II′** decompose at about 320°C, which suggests that increasing the amount of fluorinated segment in the polymer structure leads to higher thermal stability. However, for homopolymers **VI**, TGA shows that decomposition begins at 300°C, while for related copolymers **VI′** decomposition begins at about 330°C. The lower decomposition temperature of all homopolymers **VI** may be due to the greater amount of long aliphatic side chain, which masks any fluorine effect. All these polymers and copolymers exhibited weight loss in a single step and decomposed completely with a negligible quantity of residue. The glass transition (T_g) of homopolymers **II** is 85–119°C, while that of copolymers **II′** is in the range 65–101°C. The higher T_g of homopolymers **II** is attributed to the larger number of 6F groups, providing a bulkier, less mobile polymer backbone.

Water absorption studies show that homopolymers **II** do not absorb moisture (to the nearest 0.1% by weight), which is consistent with reported data on related polymers.[3] Copolymers **II′** absorb water in the range 0.6–5.3 wt%.

Table 1.1. Polymer Properties[a]

Polymer	η_{inh} (dl/g)	T_g (°C)	IDT[b] (°C)	Water absorption (%)
IIa	0.22	85	370	0.0
IIb	0.17	119	372	0.0
II′a	0.76	80	325	0.6
II′b	0.52	74	328	5.3
II′c	0.46	65	320	1.3
II′d	0.44	83	320	2.7
VIa	0.90	NO[c]	300	NO
VIb	1.21	NO	303	NO
VI′a	1.50	NO	334	NO
VI′b	0.84	NO	332	NO

[a] All data given for copolymers **II′** and **VI′** refer to 1:1 ratio of comonomers
[b] IDT—initial decomposition temperature
[c] Not observed

1.4. CONCLUSIONS

New acrylic monomers containing the hexafluoroisopropylidene function in the pendant group have been synthesized, and their homopolymerization or copolymerization with nonfluorinated acryclic comonomers was performed by free-radical bulk techniques. Colorless, transparent films were cast from polymer solutions. The inclusion of a hexafluoroisopropylidene group in these polymers led to increased thermal stability and increased T_g. When the pendant group containing 6F also had some methylene units the thermal stability of the corresponding polymers and copolymers was influenced predominantly by these aliphatic groups rather than by the 6F groups. Fluorinated homopolymers showed no water absorption as measured to the nearest 0.1 wt%. Additional properties of these polymers are under investigation.

ACKNOWLEDGMENTS: We hereby acknowledge the financial support provided to Maria Bruma by the Robert A. Welch Foundation, Houston, Texas, USA (Grant #AI-0524). Our thanks also go to Central Glass Company, Japan, for the generous supply of HFAF, and to Mr. Rock Rushing from Texas Research Institute, Austin, Texas, USA, for conducting the thermal analyses.

1.5. REFERENCES

1. E. E. Gilbert and B. S. Farah, U.S. Patent 3,544,535 (1970).
2. J. R. Griffith, R. Jacques, and G. O'Rear, U.S. Patent 4,356,296 (1982).
3. P. E. Cassidy, T. M. Aminabhavi, and J. M. Farley, *J. Macromol. Sci.—Rev. Macromol. Chem. Phys.* C29, 365 (1989).
4. P. E. Cassidy, *J. Macromol Sci.—Rev. Macromol. Chem. Phys.* C34, 1 (1994).
5. J. R. Griffith, *Am. Chem. Soc., Div., Polym. Mat. Sci. and Eng. Preprints 50*, 304 (1984).
6. J. Montheard, M. Chatzopoulos, and D. Chapara, *J. Macromol. Sci.—Rev. Macromol. Chem. Phys.* C32, 1 (1992).
7. D. S. Gupta, V. Sreenivasulu Reddy, P. Cassidy, and J. Fitch, *Polym. Preprints 34*, 4173 (1993).
8. V. Sreenivasulu Reddy, P. Cassidy, and J. Fitch, *Polym. Preprints 34*, 435 (1993).

2

Fluoromethylene Cyanate Ester Resins

ARTHUR W. SNOW and LEONARD J. BUCKLEY

2.1. INTRODUCTION

Fluoromethylene cyanate ester resins are new and unique in the class of cyanate ester resins in that they are the only members of the class based on aliphatic monomer structures. This fluoraliphatic character gives them a correspondingly unique set of physical properties, as might be expected from a consideration of the variable lengths of fluoromethylene sequences that are incorporated into the resin structure (see Figure 2.1). Pushed to the extreme, one might envision a thermoset form of polytetrafluoroethylene, i.e., a resin with physical properties comparable to PTFE and processing characteristics of a good thermoset. Our interest in this resin system is focused on its merit as a very low-dielectric, easily processible thermoset system for microelectronic and electromagnetic transmission applications. We began work on this system in 1994, and what follows is a summary of its syntheses and properties.

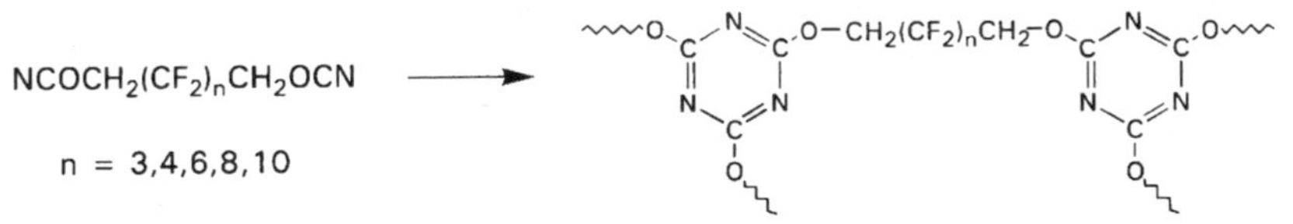

$$NCOCH_2(CF_2)_nCH_2OCN \longrightarrow$$

$$n = 3,4,6,8,10$$

Figure 2.1. Fluoromethylene cyanate ester structures and curing reaction.

ARTHUR W. SNOW and LEONARD J. BUCKLEY · Naval Research Laboratory, Washington, D.C. 20375

Fluoropolymers 1: Synthesis, edited by Hougham *et al.*, Plenum Press, New York, 1999.

2.2. BACKGROUND

Cyanate ester resins are characterized by a curing reaction wherein three cyanate functional groups form a symmetrical cyanurate heterocycle, which becomes a junction or cross-link in a macromolecular network as depicted in Figure 2.1. The chemistry of the cyanate ester functional group is such that only aromatic and fluoroaliphatic cyanate esters are stable enough to be handled as polymerizable monomers. Aromatic cyanate esters have been developed and commercialized as thermosetting resins over the last 20 years.[1,2] They are high T_g, low-dielectric, easily processible thermosetting systems with a market niche for printed circuit board and radome applications. In contrast, very little has been done with fluoroaliphatic cyanate esters as thermoset resin monomers. Two synthetic methods for monofunctional fluorocyanate esters were published in the 1960s,[3,4] and two patents for members of the resin series with $n = 3$ and 4 were issued in 1972 and 1973.[5,6] These syntheses were performed on a small scale, and the resins were thermally cured with very little characterization or property measurement provided.

A critical issue for preparation of fluoromethylene cyanate ester monomers is the availability of fluorinated diol precursors. Recent developments in commercial elemental fluorination of hydrocarbons now make it possible to obtain fluorinated diols in kilograms and larger quantities with flouromethylene sequences ranging from $n = 3$ to $n = 10$. The prospect now exists of generating cyanate resins from the higher members of this series. From a consideration of the current interest in very low-dielectric materials for microelectronic and communication applications,[7] the fluoromethylene cyanate resin system offers intriguing issues from both the scientific and technological perspectives. Such issues include: the degree to which the dielectric constant can be depressed by increasing the fluoromethylene sequence length; the effect of the fluoromethylene sequence length on other physical properties; determination of a fluoromethylene sequence length where an optimum trade-off between properties and processing is reached; development of a large-scale synthesis; and development of workable processing methods.

2.3. MONOMER SYNTHESIS

The series of monomers $NCO—CH_2(CF_2)_nCH_2—OCN$ ($n = 3, 4, 6, 8, 10$) can be synthesized by the original method of Grigart and Pütter,[3] in which a stoichiometric quantity of triethylamine is added to a cold solution of the alcohol and cyanogen halide (Reaction 1).

$$HOCH_2(CF_2)_nCH_2OH \quad + \quad BrCN \quad \xrightarrow{(C_2H_5)_3N} \quad NCOCH_2(CF_2)_nCH_2OCN$$

Reaction 1

We have found that the fluoromethylene dicyanates are more difficult to isolate in pure form than the aromatic cyanate ester monomers. The purification is particularly critical because the impurities from the synthesis reaction are strong catalysts for the trimerization reaction. Ambient storage of the monomers without sufficient purification results in gel particles being observed when the product is later formed into a melt or dissolution is attempted. We have found that a modification in the Grigat–Pütter procedure and some precautions in the synthesis and workup conditions are very helpful.[8,9] These include utilizing dry solvents and reagents, prereacting the alcohol with the triethyl amine, adding this adduct slowly to a cold solution of the cyanogen halide, and avoiding heating the crude product during the initial workup. The $n = 3$ and 4 members of the series are liquids with low vapor pressures. They are the most difficult to purify, with considerable losses occurring during vacuum distillation. The $n = 6$, 8, and 10 members are crystalline solids and become higher-melting and less soluble as the fluoromethylene sequence length increases. It is possible to recrystallize these monomers, but we find that better purity is obtained by a thorough washing with 2-propanol. This last observation has greatly expedited the synthesis, making possible the lab bench preparation of a kilogram quantity of the $n = 6$ monomer.

To assess the monomer purity, the melting point and DSC thermogram are particularly useful. Differential scanning calorimetry (DSC) thermograms of the $n = 6$ monomer at various stages of purification are presented in Figure 2.2. The melting point and the width of the melting endothermic peak correlate with the level of purity, but most important is the processing window between the melting endotherm and the onset of the exotherm of the curing reaction. This window should be at least 50 to 100°C for storage and melt-processing. The impurity that catalyzes the curing reaction has not been definitively identified. The IR and NMR spectra of impure batches show bands indicative of unreacted fluoroalcohol, carbamate, N,N-diethylcyanamid, triethylammonium hydrogen bromide, and traces of cyanurate formation.[9] These are reagents or by-products of the synthesis. Selective addition of these individual components to purified monomer results in a narrowing of the processing window, but not to the extent observed for the crude unpurified product.

The NCO—CH$_2$(CF$_2$)$_n$CH$_2$—OCN monomer series have been characterized by DSC, IR, ^{1}H-NMR, ^{19}F-NMR, ^{13}C-NMR, and elemental analysis.[8] Table 2.1 summarizes the characterization most pertinent to these cyanate ester monomers. The $n = 5$, 7, and 9 members are missing. This is a reflection of the difficulty in obtaining the odd hydrocarbon diol precursors. The trend of a rapid melting point increase with increasing fluoromethylene sequence length is an indication that monomers with $n > 10$ will probably not be melt-processible since the onset of the cure exotherm in most purified monomers occurs at 200°C.

The spectra of the cyanate monomers are remarkably simple and particularly useful for purity and conversion analyses. In the IR spectrum the cyanate band

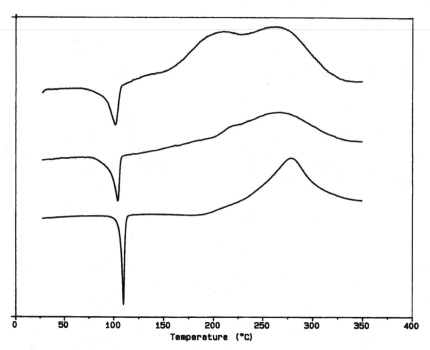

Figure 2.2. DSC thermograms of $n = 6$ monomer illustrating melting point processing window at three stages of purification (top, crude monomer product; middle, insufficiently purified monomer; bottom, sufficiently purified monomer).

Table 2.1. $NCOCH_2(CF_2)_nCH_2OCN$ Monomer Melting Point, IR, and NMR Spectroscopic Characterization[a]

Monomer	Melting point (°C)	ν_{OCN}	$\delta_{\underline{C}H2}$	$\delta_{C\underline{H}2}$	$\delta_{O\underline{C}N}$	δ_{CF2}
$n = 3$	-8	2266	5.37	73.4	111.7	-120.45
$n = 4$	15	2266	5.40	73.4	111.8	-120.28
$n = 6$	109	2265	5.46	73.6	111.9	-120.10
$n = 8$	152	2264	5.47	73.5	111.9	-120.08
$n = 10$	181	2264	5.47	73.5	111.9	-120.02

[a] ν: IR band (cm^{-1}); δ: NMR chemical shift (ppm); δ_{CF2}: resonance of terminal CF_2 unit.

(O—C≡N stretch, 2265 cm^{-1}) is a useful diagnostic for conversion measurements and can be referenced against the methylene bands (CH$_2$, 3040, 2980, 1450 cm^{-1}) as internal standards. In the NMR spectra the nuclei incorporated in the cyanate (^{13}C, 111.9 ppm), methylene (^{1}H, t, 5.4 ppm; ^{13}C, t, 73.5 ppm) and terminal fluoromethylene (^{19}F, −120.1 ppm) groups are easily resolved with little interference. The ^{19}F-NMR is illustrated in Figure 2.3. The terminal resonance at −120.0 ppm is particularly useful for conversion and impurity assessments in that it is shifted with total resolution when formed from the precursor diol (−121.5 ppm) and is further shifted when converted to a cyanurate (−118.9 ppm) or carbamate (−119.5 ppm).[9]

The hydrolytic susceptibility of the fluoromethylene cyanate ester is greater than that of the aromatic cyanate ester. Direct contact with water results in measurable hydrolysis to the carbamate in a 24-h period for the former while the latter is unaffected.[9]

2.4. RESIN CURE

It has been established by a variety of techniques that aromatic cyanate esters cyclotrimerize to form cross-linked cyanurate networks.[1] Analogously, the fluoromethylene cyanate monomers cure to cyanurate networks. In addition to the ^{19}F-NMR spectra shown in Figure 2.3, evidence includes an up-field shift of the methylene triplet (^{1}H-NMR, 0.21 ppm; ^{13}C-NMR, 9.4 ppm), the disappearance of the cyanate functional group (IR, 2165 cm^{-1}; ^{13}C-NMR, 111.9 ppm) and the appearance of the cyanurate functional group (IR, 1580 and 1370 cm^{-1}; ^{13}C-NMR, 173.6 ppm).[9] Typically, monomers are advanced to prepolymers by thermal treatment at 120°C or just above the melting point. The prepolymers are then cured at 175°C and are postcured at 225°C.

The curing reaction can be carried out thermally or with the addition of a catalyst. The thermal cure is strongly influenced by impurities associated with the synthesis. The greater the degree of monomer purity, the more slowly the thermal cure proceeds. If the monomer is sufficiently purified, the cure rate can be predictably controlled by the addition of catalysts. As with the aromatic cyanate esters, the fluoromethylene cyanate esters can be cured by the addition of active hydrogen compounds and transition metal complexes. Addition of 1.5 wt% of the fluorinated diol precursor serves as a suitable catalyst.[9] The acetylacetonate transition metal salts, which work well for the aromatic cyanate esters,[1] are also good catalysts.

Castings of the neat resin, prepared as described above, are clear amber in appearance. Significant shrinkage occurs during cure and, in particular, during postcure cooling. The latter is the result of a large thermal expansion coefficient. The castings are tough plastics and are machinable into desired forms.

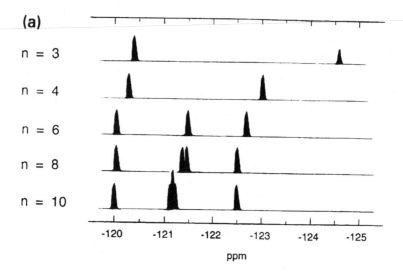

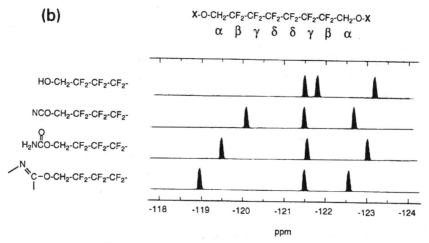

Figure 2.3. Schematic ^{19}F-NMR spectra illustrating effects of (a) fluoromethylene chain length on monomer resonances and (b) chemical transformation of cyanate functional group on fluoromethylene resonances of $n = 6$ monomer.

2.5. RESIN PROPERTIES

The fluorine content, density, critical surface energy, glass transitions, thermal expansion coefficient above and below the glass transition, and 300°C isothermal thermogravimetric stabilities of the fluoromethylene cyanate ester resin system with $n = 3, 4, 6, 8, 10$ are summarized Table 2.2. Also included for the purpose of comparison are the corresponding data for the aromatic cyanate ester resin based on the dicyanate of 6F bisphenol A (AroCy®F, Ciba Geigy).

The fluorine content for the fluoromethylene cyanate ester resin system is clearly significantly greater than that for AroCy®F and also significantly less than that for PTFE (76.0%). Correspondingly, the density reflects the fluorine content. As a homologous series, the contribution of incremental CF_2 units can be quantified. From the density data, the volume equivalent of $41.0 \text{ Å}^3/CF_2$ structural unit in an amorphous thermoset matrix has been determined.[8]

The glass transition displays a small dependence on the fluoromethylene chain length ranging from 84 to 101°C. Measured by DSC, it is relatively weak and broad, reflecting a small difference in heat capacities above and below the glass transition. This measurement is more precise with thermomechanical analysis (TMA), and the thermal expansion coefficients of the corresponding glassy and rubbery states display a dependence on the fluoromethylene chain length that correlates with cross-link density. The thermal expansion coefficients of the fluoromethylene cyanate ester resin series, particularly for the glassy state, are significantly greater than those for the aromatic cyanate ester resin and large for thermosets in general. This is attributed to the fluoroaliphatic character of this thermoset, where very weak intermolecular forces are combined with a moderately flexible chain structure. The mobility in the glassy state is enhanced, as reflected in its thermal expansion coefficient, and smaller differences between the glassy and rubbery heat capacities make the glass transition difficult to detect by DSC.

Table 2.2. NCOCH$_2$(CF$_2$)$_n$CH$_2$OCN and AroCy®F Resin Physical Property Characterization[a]

Resin	%F	ρ (g/cm³)	γ_c (dyn/cm)	T_g (°C)	α_G (ppm/°C)	α_R (ppm/°C)	%H$_2$O absorption	$(\Delta m/\Delta t)_{300°C}$ (%/min)
$n = 3$	43.5	1.771	40	86	109	238	1.67	−0.0196
$n = 4$	48.7	1.814	36	84	124	244	0.93	−0.0191
$n = 6$	55.3	1.856	31	92	144	248	0.75	−0.0156
$n = 8$	59.4	1.889	27	92	145	255	0.74	−0.0133
$n = 10$	62.1	1.908	23	101	152	275	0.68	−0.0064
AroCy®F	29.4	1.471	40	256	65.9	223	1.61	−0.0024

[a] ρ: density at 25°C; γ_c: critical surface tension; α_G: glassy state thermal expansion coefficient; α_R: rubbery state thermal expansion coefficient; %H$_2$O absorption: 100°C immersion/96 h; $(\Delta m/\Delta t)_{300°C}$: percent weight loss per minute at 300°C.

The surface properties are of particular interest for composites and coatings. The $n = 6$ monomer will wet Teflon, and PTFE filled composites can be prepared. The critical surface tension of wetting for the fluoromethylene cyanate ester resin series has been determined from contact-angle measurements on cured resin surfaces. As indicated in Table 2.2, it parallels the fluorine composition and begins to approach the PTFE value of 18 dyn/cm.

The absorption of moisture critically affects other important resin properties, particularly those associated with low-dielectric and thermomechanical applications. Results of a 96-h boiling water immersion test are presented in Table 2.2. The moisture absorbed decreased substantially with fluoromethylene chain length from $n = 3$ to $n = 6$, followed by only modest decreases for $n = 8$ and 10. This latter behavior was somewhat unexpected and may be the effect of decreased cross-link density counteracting the increased fluorine content. These 100°C measurements are just above the glass transition and the situation may be different at room temperature. These measurements are in progress.

The thermal stability is dependent on the property to which it is being indexed and the conditions of the measurement. The common scanning thermogravimetric stability provides an indication of the onset of catastrophic decomposition. By this measurement, the fluoromethylene cyanate resins progressively increase in stability with increasing n, and at $n = 6$ are equivalent to the aromatic cyanate ester resins (decomposition onset at 430°C). A more meaningful measurement is the isothermal thermogravimetric analysis (TGA) at a temperature where decomposition is very slow (e.g., <5 wt% over 2 h). This process is more linear with time, and the slope provides a single parameter to discriminate among similar but different resins. Isothermal TGA results (300°C) for the fluoromethylene cyanate ester resins and AroCy®F are entered in Table 2.2. AroCy®F is the more thermogravimetrically stable at 300°C. However, at 360°C, the isothermal thermogravimetric stabilities of the $n = 6$ fluoromethylene cyanate ester and AroCy®F are reversed with the latter losing weight at a faster rate. Also, at this temperature AroCy®F loses its dimensional stability while the fluoromethylene cyanate ester casting retains its machined dimensions.

The tensile mechanical properties of the $n = 6$ fluoromethylene cyanate ester resin are presented in Table 2.3 along with those for a T_g comparable epoxy and AroCy®F. The measurements for AroCy®F reflect its nature as a high-T_g, aromatic, and somewhat brittle resin. The T_g-comparable epoxy is based on the diglycidyl ether of bisphenol A (Epon 828, Shell) cured with the amine terminated 230-molecular-weight propylene oxide oligomer (Jeffamine D230, Huntsman), and it was selected as a more relevant comparison resin to illustrate the effect of the fluoroaliphatic structure. This fluoroaliphatic character results in a significantly lower modulus and strength. The fluoromethylene resin also has a slightly greater deformation and recovery. These properties correlate with the weaker intermolecular forces of the fluorocarbon system.

Table 2.3. Tensile Mechanical Properties

Resin	Modulus (Mpa)	Strength (Mpa)	Break strain (%)	Permanent set (%)	T_g (°C)
Cyanate ($n = 6$)	1100	35	11.5	0.8	92
Epoxy	3500	55	9	1.2	83
AroCy®F	3100	74	2.8	—	256

2.6. DIELECTRIC PROPERTIES

The dielectric constant and tan delta of the fluoromethylene cyanate ester resin series are shown in Figure 2.4, which presents the total complex permittivity (dielectric storage and dielectric loss) of the systems. The data represent an average value taken over a frequency range from 500 MHz to 1.5 GHz. It shows a decrease in the dielectric constant for the $n = 3$ to $n = 6$ materials beyond which the values appear to increase slightly with additional fluoromethylene units. The complex permittivity has been shown to be sensitive to several structural factors in a macromolecule.[10] These factors include the total polarizability owing to electronic, ionic, and dipolar contributions within the compound in addition to the morphological features such as free volume and molecular order. In general, the greater the free volume along with the absence of any molecular order, the lower the permittivity of the material. The polarizability contributions can be influenced in conflicting ways with fluorine incorporation. For example, Hougham *et al.* has shown that the substitution of a trifluoromethyl group for a methyl group in a polyimide system decreases the electronic contribution to the dielectric constant but increases the dipolar contribution and so has no net effect.[11,12]

The lengthening of the fluoromethylene chain segment in the cyanurate series shows how the polarizability contributions can change the permittivity in different ways. At optical frequencies, the electronic polarizability contribution can be assessed from the square of the refractive index by way of the Maxwell relation. The n^2 values, given in Table 2.4, show a steady decrease from $n = 3$ to $n = 10$ owing to a decreasing electronic contribution to the dielectric constant. This corresponds to an increasing fluorine content in the molecule. The refractive index values were determined from measurements at 589 nm on polished surfaces of resin castings. The 1.0-GHz data show an initial decrease for the $n = 3$ to $n = 6$ compounds, but then show a very slight increase for the $n = 8$ and $n = 10$ compounds. These measurements were made on resin castings using a contact method where the complex admittance was determined from the capacitive and inductive currents within the sample at a specific frequency.[13]

This apparent anomaly may be due to increased dipolar contributions as well as to possible decrease in the free volume of polymer. Free volumes of the resin

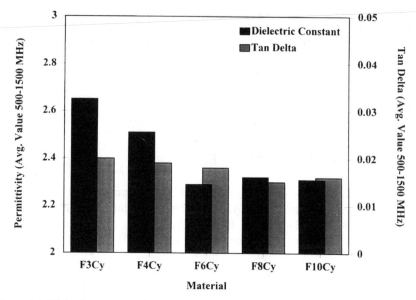

Figure 2.4. Dielectric constant and tan delta (average values 0.5 to 1.5 GHz) of fluoromethylene cyanate ester resin series.

series were calculated from the ratio of the Bondi occupied volume to the experimental density-derived volume of the molecular repeat unit.[14] By this method, the changes in the free volume are 0.5% or less, which indicates insignificant changes in the packing efficiency for more than six fluoromethylene units in the monomer.

Packing efficiency can also be described by the extent of short-range order in the amorphous state. Mitchell has shown through X-ray scattering studies that, while the local molecular organization of noncrystalline polymers is random, in many cases, there are additional correlations that do not perturb the chain trajectory but will impact polymer properties.[15] These correlations have a limited spatial range (<50 Å) but will have a particular impact on bulk properties

Table 2.4. Dielectric and Tan Delta Characterization of $NCOCH_2(CF_2)_nOCN$ Resins

Resin	%F	ϵ'	n^2	tan delta
$n = 3$	43	2.66	2.09	0.02
$n = 4$	49	2.51	2.05	0.019
$n = 6$	55	2.29	1.98	0.018
$n = 8$	59	2.32	1.95	0.015
$n = 10$	62	2.31	1.91	0.016

involving mutual movement of polymer chain segments that, in the present case, can form dipoles. Hence, if dipole alignment is enhanced with this short-range order, then the dipolar or orientational contribution to the polarizability will increase. Free volume will remain constant but the polarizability will increase owing to a larger dipole effect from the short-range molecular order.

Water absorption can also cause significant changes in the permittivity and must be considered when describing dielectric behavior. Water, with a dielectric constant of 78 at 25°C, can easily impact the dielectric properties at relatively low absorptions owing to the dipolar polarizability contribution. However, the electronic polarizability is actually lower than solid state polymers. The index of refraction at 25°C for pure water is 1.33, which, applying Maxwell's relationship, yields a dielectric constant of 1.76. Therefore, water absorption may actually act to decrease the dielectric constant at optical frequencies. This is an area that will be explored with future experiments involving water absorption and index measurements.

2.7. PROCESSING

As indicated above, the dielectric properties are the chief consideration for applications of these fluoromethylene cyanate ester resins. As such, processing as coatings for microelectronic applications and as nonstructural castings and composites for radome-related applications is being investigated. In this respect the $n = 6$ system has been identified as the best compromise for synthesis, properties, and processing. Results in this section pertain only to this member of the series, which is currently under evaluation for the above applications.[13]

For microelectronics applications, the processing objective is the preparation of uniform submicron-thick coatings to serve as interlevel dielectrics in planar silicon device fabrication processes. The monomer is not suitable if the spin-coating method of processing is employed as its melt or concentrated solution viscosity is too low to use the material efficiently. Monomer crystallization and evaporation under curing conditions are also detracting effects. These difficulties are avoided by advancing cure of the monomer to a soluble prepolymer conversion. A conversion of 30 to 40% will produce a soluble prepolymer with a suitable solution viscosity for spin-coating. The prepolymer thin film can be rapidly cured by simple heating. The cured coatings exhibit good gap fill and planarization. An SEM photograph in Figure 2.5 illustrates the gap fill of the fluoromethylene cyanate ester coating between closely spaced aluminum features on a silicon wafer. Adhesion of this cyanate resin to silicon, as assessed by a tape pull-off test, is less than but comparable to epoxies and surpasses polyimides and other fluoromethylene-containing polymers such as Teflon AF (copolymer of tetrafluoromethylene and 2,2-bistrifluoromethyl-4,5-difluoro-1,3-dioxole).[13]

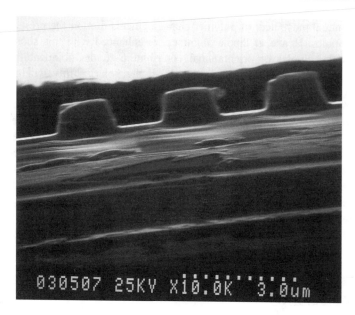

Figure 2.5. SEM photograph of a silicon wafer cross section illustrating gap fill quality of a cured, spin-coated fluoromethylene resin ($n = 6$) over aluminum features having width and spacing.

In the preparation of molded objects and composites, a monomer melt has been used successfully for mold-filling and prepegging. However, there is significant cure shrinkage and resin bleeding, which can be lessened by utilizing a 30% conversion prepolymer. Special consideration should also be given to the resin's adhesion to surfaces and to the relatively large coefficient of thermal expansion. The performance of release agents seems to vary with the nature of the metal surface and to some extent with the geometry of the mold. For aluminum or steel surfaces, silicone release agents appear to be slightly more effective than the PTFE agents. Mold geometries that have large surface-to-volume ratios, particularly thin-panel molds ($6 \times 8 \times 0.1$ in), have caused some difficulty, with crack formation occurring in the molded object. At present it is believed that a more efficient surface release is needed. This cracking behavior has only been observed in thin-panel casting preparations and not in the casting of rods or thick blocks or in the curing of thin films described in the paragraph above.

2.8. SUMMARY

The synthesis and properties of the fluoromethylene cyanate ester monomer series, $NCOCH_2(CF_2)_nCH_2OCN$, $n = 3, 4, 6, 8, 10$, and corresponding resins have been described. The resin property of particular importance is the dielectric constant, which decreases from 2.6 at $n = 3$ to 2.3 (1 GHz) at $n \geq 6$. Based on yield, purification, melting point, cured resin dielectric, and moisture absorption properties, the $n = 6$ system has been identified as having the optimum trade-off between properties and processing. The synthesis has been scaled up to a kilogram, and mechanical properties are reported. Processing as thin films, molded objects, and composites is also described.

ACKNOWLEDGMENT: The authors gratefully acknowledge the Office of Naval Research for financial support.

2.9. REFERENCES

1. T. Fang and D. A. Shimp, in *Progress in Polymer Science, Vol. 20* (O. Vogl. Ed.), Elsevier Science, Oxford, (1995), PP. 61–118.
2. I. Hamerton (ed.), *Chemistry and Technology of Cyanate Ester Resins*, Chapman and Hall, Glasgow (1994).
3. E. Grigat and R. Pütter, *Chem. Ber. 97*, 3012–3017 (1964).
4. C. Krespan, *J. Org. Chem. 34*, 1278–1281 (1969).
5. B. L. Loudas and H. A. Vogel, U.S. Patent 3,681,292 [*CA 77*, 127, 425n (1972)].
6. B. L. Loudas and H. A. Vogel, U.S. Patent 3,733,349 [*CA 79*, 92, 790q (1973)].
7. J. H. Lai (ed.), *Polymers for Electronic Applications*, CRC Press, Boca Raton (1989).
8. A. W. Snow and L. J. Buckley, *Macromolecules 30*, 394–405 (1997).
9. A. W. Snow, L. J. Buckley, and J. P. Armistead, *J. Poly. Sci. Pt. A 37*, 135–150 (1999).
10. A. von Hippel, *Dielectric Materials and Applications*, Cambridge, Boston (1954).
11. G. Hougham, G. Tesoro, A. Viehbeck, and J. Chapple-Sokol, *Macromolecules 27*, 5964–5971 (1994).
12. G. Hougham, G. Tesoro, and A. Viehbeck, *Macromolecules 29*, 3453–3456 (1996).
13. L. J. Buckley and A. W. Snow, *J. Vac. Sci. Technol. 15* (2), 259 (1997).
14. A. Bondi, *Physical Properties of Molecular Crystals, Liquids and Glasses*, John Wiley and Sons, New York (1968), Ch. 14.
15. G. Mitchell, *Order in the Amorphous State of Polymers*, Plenum Press, New York (1986).

3

Polymers from the Thermal (2π+2π) Cyclodimerization of Fluorinated Olefins

DAVID A. BABB

3.1. INTRODUCTION

The use of fluorocarbon functionality in polymers designed for high-performance applications has taken on greater significance in recent years, as the benefits of working with fluorinated polymers have become more widely known. Some of the benefits gained by the incorporation of fluorocarbon functionality into polymers include increased thermal/oxidative stability, optical transparency, greater solvent compatibility, and increased environmental stability.[1,2] One recent advance in the incorporatin of fluorine into high-performance polymers has involved an unlikely chemical reaction, the thermal [2π + 2π] cyclodimerization of fluorinated olefins. This reaction proceeds at relatively mild temperatures and produces a covalent linkage that is chemically, thermally, and oxidatively stable. This paper discusses the evolution of the understanding of this unusual chemical reaction and its use in the synthesis of polymers.

3.1.1. Cyclodimerization of Fluorinated Olefins

The first report of the cyclodimerization of fluorinated olefins was provided by Lewis and Naylor,[3] working at E.I. DuPont de Nemours & Co., in 1947. While studying the pyrolysis of polytetrafluoroethylene (PTFE), the compound octafluorocyclobutane was isolated from the pyrolysis off-gas stream. The researchers identified the product and speculated that it was formed by the cyclodimerization

DAVID A. BABB · Central Research and Development, Organic Product Research, Dow Chemical Company, Freeport. Texas 77541.
Fluoropolymers 1: Synthesis, edited by Hougham *et al.*, Plenum Press, New York, 1999.

of tetrafluoroethylene, the primary pyrolytic decomposition product, at high temperatures. They also proposed that this dimerization proceeds by a two-step radical mechanism, a hypothesis that was subsequently borne out by further investigation.

In the years following more details of this reaction began to appear in the literature. In 1952 Lacher *et al.* published the kinetics of the dimerization of tetrafluoroethylene and assigned an activation energy of 26.3 kcal/mol for the reaction.[4] Shortly afterward Atkinson and Trentwith published another paper on the pyrolysis PTFE.[5] They recognized the reaction $2[C_2F_4] \rightleftharpoons [C_4F_8]$ as an equilibrium at high temperatures. They quantified the rate constant for the thermal dimerization, assigning an activation energy for the forward and reverse reactions of 25.4 kcal/mol, and 74.1 kcal/mol, respectively, with an enthalpy of reaction of −50 kcal/mol. These values reflect the thermal stability of the perfluorocyclobutane ring.

Following the initial report of Lewis and Naylor on the cyclodimerization of tetrafluoroethylene, a number of studies were published on the dimerization of other fluorinated olefins. Structurally, the olefins appear to require a terminal difluoromethylene group ($=CF_2$) in order to undergo the dimerization reaction; 1,2-difluoro olefins do not participate in this type of cyclodimerization.[6] In general, perfluorovinyl groups that are substituted with electron-donating substituents tend to dimerize more readily than those with electron-withdrawing groups. A more comprehensive coverage of the dimerization of fluorinated olefins can be found in a number of excellent review articles.[7,8]

Early experimental results suggested a stepwise radical mechanism for this $[2\pi + 2\pi]$ cycloaddition, with initial bond formation between the terminal difluoromethylene carbons. This bond formation then produces a biradical intermediate, which is stabilized by electron-donating substituents on the radical-bearing carbons. Fluorinated olefins that possess such electron-donating substitutents therefore show a great preference for the formation of 1,2-substituted hexafluorocyclobutane rings. Fluorinated olefins with electron-withdrawing substituents, such as chlorotrifluoroethylene, constitute a much greater proportion of the 1,3-substituted hexafluorocyclobutane. These cyclodimers are also a mixture of *cis* and *trans* isomers, often approaching a statistically random distribution of the two. The speculation about this mechanism was eventually settled by the very elegant work of Paul Bartlett *et al.*,[9] which was published in 1963. By selectively cross-reacting the *cis-cis*, *cis-trans*, or *trans-trans* isomers of 2,4-hexadiene with 1,1-dichloro-2,2-difluoroethylene, and thereafter identifying and quantifying the reaction products, Bartlett was able to clearly demonstrate that the isomer configuration of the reacting hydrocarbon olefin moiety becomes scrambled during the cyclodimerization, thereby proving the existence of a biradical intermediate that is long-lived on a bond-rotational time scale. He also confirmed that the addition takes place in a head-to-head orientation, in a manner that creates the

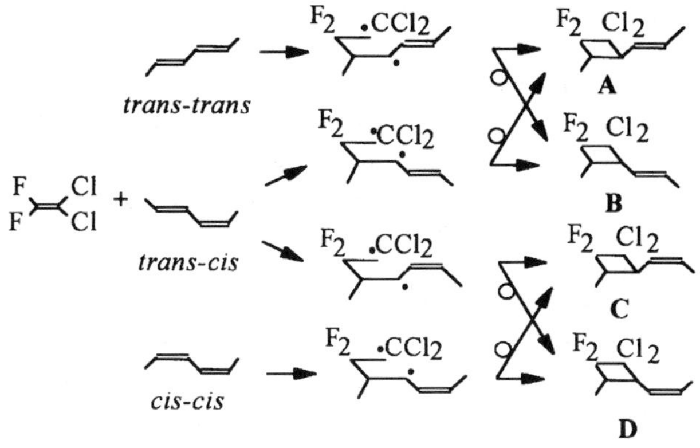

Figure 3.1. Bartlett's determination of the biradical intermediate in the $[2\pi + 2\pi]$ cyclodimerization reaction. Reprinted with permission from L. K. Montgomery, K. Schueller, and P. D. Bartlett, *J. Am. Chem. Soc.* *86*, 622–628 (1964). Copyright 1964 American Chemical Society.

most stable biradical intermediate. Figure 3.1 illustrates the design of the experiment.

The perfluorocyclobutane ring has proven to be unusually thermally stable with respect to its hydrocarbon analogue. In their paper on the pyrolysis of PTFE,[5] Atkinson and Trentwith indicated that the ring-opening reversion of the octa-fluorocyclobutane structure does not proceed at an appreciable rate until temperature in excess of 500°C are attained. The stability of the perfluorocyclo-butane structure with respect to its hydrocarbon analogue was the subject of a paper by W. A. Bernett in 1969, in which the stabilization effect was attributed to the relative differences in bond strain energies between the two cyclobutane rings and their olefin precursors.[10] By comparing the bond strain energies of the olefin–cyclodimer pairs, Bernett demonstrated that tetrafluoroethylene undergoes a decrease in bond strain of -9.2 kcal/mol by cyclizing to octafluorocyclobutane, while the equivalent or cyclodimerization of ethylene results in an increased bond strain energy of 3.8 kcal/mol. Bernett used bond angle determinations to argue that the unsaturated carbon bonds in TFE are actually sp^3-hybridized, and can be best described by a bend-bond model.[11] The stabilization effect then takes the form of relief of angular orbital strain in going from an unsaturated bent-bond olefin to a saturated four-membered cyclic bonding system.

3.1.2. Polymers Containing the Hexafluorocyclobutane Ring

While several fluorocarbon polymers that contain the hexafluorocyclobutane ring have been prepared, there is little precedent in the chemical literature for the use of the cyclodimerization of perfluorovinyl groups in their preparation. These perfluorocyclobutane ring-containing polymers have been prepared by the coupling of 1,2-dihalohexafluorocyclobutane with other fluorocarbon dihalides in a metal-mediated cross-coupling reaction,[12] or by the nucleophilic ring-opening initiation of perfluorocyclobutyl epoxide to form a polyether.[13]

The first use of the dimerization of fluoroolefins for increasing molecular weight in polymeric systems was reported in the preparation of oligomeric precursors to perfluoroalkyl either thermoset polymers.[14] Perfluoroalkyl ethers that were end-capped with trifluorovinyl ether functionality were heated at 135–150°C for up to 6 h to form short-chain oligomers containing perfluorocyclobutyl ether groups. These oligomers were then linearly polymerized through the olefin end groups to form thermoset resins containing perfluorocyclobutane rings (Figure 3.2).

In 1988 Heinze and Burton reported a facile synthesis of various α, β, β-trifluorostyrenes.[15] These trifluorostyrene compounds were reported to be unstable to cyclodimerization at room temperature when stored neat, especially the compounds that were *para*-substituted with electron-donating substituents. They described the preparation of one compound, 1,4-bis(trifluorovinyl)benzene with the observation that the material gelled when allowed to stand neat overnight. They offered the explanation that the gel was a "polymer network connected with fluorinated cyclobutanes." Burton later went on to utilize this dimerization reaction for the cross-linking of polyimide thermoplastics.[16]

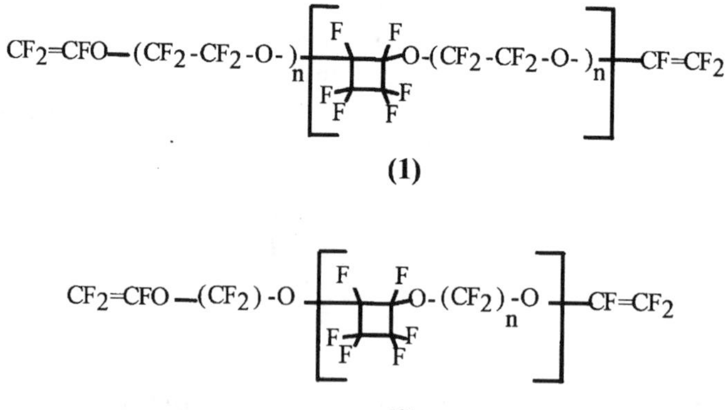

$$CF_2=CFO-(CF_2-CF_2-O-)_n \left[\begin{array}{c} F \quad F \\ \\ O-(CF_2-CF_2-O-)_n \\ F \quad F \\ F \quad F \end{array} \right] CF=CF_2$$

(1)

$$CF_2=CFO-(CF_2)-O \left[\begin{array}{c} F \quad F \\ \\ O-(CF_2)-O \\ F \quad F \\ F \quad F \end{array} \right]_n CF=CF_2$$

(2)

Figure 3.2. Perfluorocyclobutane ring containing perfluoroalkyl ether oligomers prepared by Beckerbauer.

3.1.3. Perfluorocyclobutane Aryl Ether Polymers

Our own work in the area of perfluorocyclobutane ring-containing polymers has involved the cyclodimerization of aryl trifluorovinyl esters.[17–21] Compounds consisting of aromatic core structures with multiple pendant trifluorovinyl ether groups serve as the monomers for these polymers. Upon heating, the trifluorovinyl units cyclodimerize to form polymers with alternating perfluorocyclobutane and aromatic ether groups along the backbone of the polymer. Monomers containing two dimerizable fluoroolefins react to form linear thermoplastic polymers, while monomers with three or more dimerizable fluoroolefins form network polymers. The work began by functionalizing a series of bisphenols to form monomers for thermoplastic perfluorocyclobutane aryl ether polymers. Monomers were subsequently polymerized, either neat or in solution, to provide the perfluorocyclobutane ring-containing thermoplastic polymers. The synthetic route followed the general method outlined in Figure 3.3, which uses resorcinol as an exemplary starting phenolic.

This chemistry was also performed on one trisphenol, 1,1,1-tris(4-hydroxyphenyl)ethane (THPE). The resulting tris(trifluorovinyl ether) monomer forms a thermoset polymer upon curing.

This research was an attempt to develop new polymers with the mechanical properties of polyarylene ethers and the dielectric properties of fluoropolymers. After initially testing the viability of the $[2\pi + 2\pi]$ cyclodimerization reaction for preparing high-molecular-weight polymers and testing the dielectric properties of these polymers, two polymers (one thermoplastic and one thermoset) were prepared in larger quantities to evaluate the thermal and mechanical performance of these novel compositions. The high T_g thermoset was also quantitatively tested for thermal/oxidative stability.

3.2. EXPERIMENTAL

All gas chromotography/mass spectrometry (GC/MS) analyses of monomers and intermediates were performed on a Finnigan 1020 GC/MS using a 30-m RSL-150 fused silica capillary column. Liquid chromatography/mass spectrometry (LC/MS) was performed on a Finnigan 4500 mass spectrometer using acetonitrile–water eluent and a moving belt LC/MS interface.

Dynamic mechanical anlaysis (DMA) measurements were done on a Rheometrics RDS-7700 rheometer in torsional rectangular geometry mode using $60 \times 12 \times 3$ mm samples at 0.05% strain and 1 Hz. Differential scanning calorimetry (DSC), thermomechanical analysis (TMA), and thermogravimetric analysis (TGA) were performed on a Perkin-Elmer 7000 thermal analysis system.

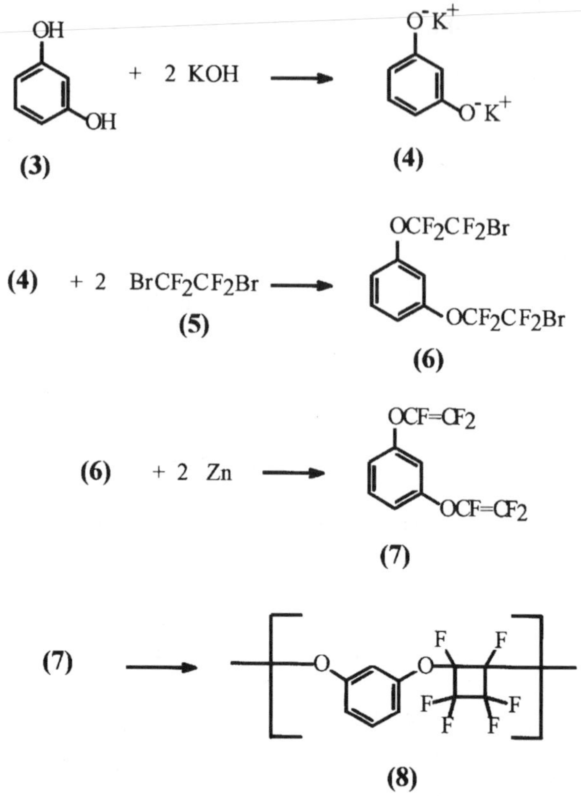

Figure 3.3. Synthesis of trifluorovinyl ether monomers and perfluorocyclobutane aromatic ether polymers.

Dielectric constant and dissipation factor measurements were conducted according to the procedures of ASTM D-150-87. Tensile strength and modulus and percent elongation were measured on an Instron model 1125 according to the procedures of ASTM-D-882-83.

Gel permeation chromatography (GPC) was carried out on a Waters 720 GPC instrument using a methylene chloride eluant and a series of Micro-styragel® columns of 10,000, 1000, 500, and 100 Å pore sizes. Reported values were standardized against polystyrene.

Granular zinc was activated by washing in 0.1 N hydrochloric acid (HCl) followed by drying in a vacuum oven at 0.5 mm Hg and 140°C for 10 h.

Infrared (IR) spectra were measured on a Beckmann Microlab 600 model spectrophotometer. Nuclear magnetic resonance (NMR) spectra were measured on a Varian EM360 spectrometer, with ^{19}F-spectra collected using trifluoroacetic acid as a standard, or with [1]H-spectra collected using tetramethylsilane as a standard.

3.2.1. Synthesis of 2-Bromotetrafluoroethyl Aryl Ethers

3.2.1.1. Method I

Dimethylsulfoxide (DMSO) (1800 ml) was placed in a 5-liter five-necked flask fitted with a mechanical stirrer, a Dean–Stark phase separating trap topped with a reflux condenser under nitrogen, and a thermocouple attached to a temperature controller that controlled a heating mantle on the flask. The solvent was stirred and purged of oxygen by blowing nitrogen in through a gas dispersion tube placed below the surface of the liquid while the phenolic starting material (2.44 mol) was added to the flask. The system was stirred and purged for 20 min, then the correct stoichiometric amount of potassium hydroxide (85% pellets) was added slowly. The stirred mixture was then heated to 120°C. The temperature was held at 120°C for 30 min; then the heat was turned off and the mixture was allowed to cool to room temperature. Toluene (600 ml) that had been thoroughly purged with nitrogen was added to the solution and the resulting mixture was heated to reflux (135°C).

Water was azeotropically removed from the reactor through the Dean–Stark trap for a total of 36 h, cooling the reactor once after 24 h to allow for salt formation to be broken up by opening the flask under a nitrogen sweep and scraping the sides with a spatula. After 36 h the Dean–Stark trap was removed and replaced with a Soxhlet extractor containing anhydrous sodium sulfate. The toluene was then refluxed through the Soxhlet extractor for 7 h to dry the toluene. After 7 h, the Soxhlet was replaced with a Dean–Stark trap, and toluene (300 ml) was removed from the reactor by simple distillation. The reaction mixture was then cooled to 30°C in an ice-water bath and 1,2-dibromotetrafluoroethane (1300 g, 5.00 mol) was added slowly dropwise over 3 h at a rate that maintained a reactor temperature of 30° ± 2°C. When the addition was complete the reaction temperature was allowed to stabilize and then a heating mantle was applied to the flask. The reactor was heated to 50°C for 8 h, then allowed to cool to room temperature with constant stirring. The crude reaction mixture was filtered to remove the potassium bromide salts and the precipitate was washed with acetone. The filtrates were combined and thoroughly evaporated to remove acetone, DMSO, and residual toluene. The residue was further purified, either by fractional distillation under reduced pressure or by column chromatography on neutral alumina using hexane as an eluent, to provide the 2-bromotetrafluoroethyl ether product.

(a) *1,3-Bis(2-Bromotetrafluoroethoxy)benzene.* This product was prepared according to Method I, using resorcinol (27.5 g, 0.25 mol) in DMSO (500 ml) and toluene (200 ml). The crude product was distilled on a 6-in. Vigreaux column (105–110°C), 3.5 mm Hg) to provide the pure product (76.0 g, 65.0% yield) as a water white oil. GC/MS, m/e (%): 129 (28.0), 131 (28.3), 179 (10.2), 181 (10.2), 271 (38.2), 272 (10.1), 273 (36.7), 287 (10.6), 466 (48.1), 468 (100), 470 (46.4).

(b) *4,4'-Bis(2-Bromotetrafluoroethoxy)Biphenyl.* This product was prepared according to Method I, using 4,4'-biphenol (454 g, 2.44 mol) in DMSO (1800 ml) and toluene (600 ml). The crude product was fractionally distilled (138–148°C, 0.35 mm Hg) to provide 1031.1 g (1.90 mol, 77.9% yield) of 4,4'-bis(2-bromotetrafluoroethoxy)biphenyl, melting point 71–73°C. GC/MS, m/e (%): 63 (34.3), 76 (41.1), 102 (32.9), 127 (33.2), 128 (100), 129 (37.4), 139 (90.1), 140 (36.7), 156 (78.3), 168 (33.7), 335 (34.7%), 337 (30.3), 363 (50.9), 365 (48.7), 541 (23.8), 543 (48.9), 545 (29.8).

(c) *4,4'-Bis(2-Bromotetrafluoroethoxy)Diphenyl Sulfide.* This product was prepared according to Method I, using 4,4'-dihydroxydiphenyl sulfide (200.0 g, 0.916 mol) in DMSO (750 ml) and toluene (250 ml). The crude product was distilled (141–143°C, 0.40 mm Hg) to provide the purified product (376.7 g, 71.4% yield) as a water white oil. GC/MS, m/e (%): 115 (16.0), 129 (21.9), 131 (19.8), 171 (46.5), 179 (21.4), 181 (21.6), 303 (31.7), 305 (38.9), 395 (48.6), 397 (62.3), 574 (54.6), 576 (100), 578 (64.6).

(d) *2,2-Bis[4-(2-Bromotetrafluoroethoxy)phenyl]Propane.* This product was prepared according to Method I using 2,2-bis(4-hydroxyphenyl)propane (*para*-bisphenol A) (100.0 g, 0.438 mol) in DMSO (300 ml) and toluene (100 ml). The crude reaction mixture was distilled (144–148°C, 0.14 mm Hg) to provide the product (113.89 g), 44.4% yield) as a water white oil in >99% purity by GC analysis. GC/MS, m/e (%): 115 (32.4), 129 (36.9), 131 (33.6), 152 (30.9), 165 (37.8), 167 (29.1), 179 (37.6), 181 (31.1), 297 (30.4), 299 (28.0), 313 (28.1), 315 (27.0), 569 (52.1), 571 (100), 573 (54.5), 584 (4.5), 586 (9.2), 588 (4.5).

(e) *2,2-Bis[4-(2-Bromotetrafluoroethoxy)phenyl]Hexafluoropropane.* According to the procedure of Method I this compound was prepared using 2,2-bis(4-hydroxyphenyl)hexafluoropropane (6F diphenol) (500.0 g, 1.5 mol) in DMSO (750 ml) and toluene (250 ml). The crude product was purified by column chromatography on neutral alumina using hexane as eluent to provide the purified product (771 g, 73.3% yield) as a light yellow oil in 98% purity by GC analysis. GC/MS, m/e (%): 129 (30.0), 131 (31.2), 26.8 (151), 152 (33.5), 179 (28.7), 181 (34.3), 201 (37.6), 347 (10.2), 349 (10.0), 377 (14.7), 379 (10.1), 421 (26.3), 423 (26.7), 426 (11.7), 428 (12.3), 497 (11.9), 499 (10.8), 622 (54.2), 624 (100), 626 (57.3), 692 (19.5), 694 (37.4), 696 (20.1).

(f) *9,9-Bis[4-(2-Bromotetrafluoroethoxy)phenyl]Fluorene.* This product was prepared according to Method I using 9,9-bis(4-hydroxyphenyl)fluorene

(200.0 g, 0.57 mol) in DMSO (650 ml) and toluene (200 ml). The product was purified by column chromatography on neutral alumina using hexane as an eluent to provide the product 9,9-bis[4-(2-bromotetrafluoroethoxy)phenyl]fluorene (331.4 g, 0.468 mol, 82% yield), melting point 157–158°C. LC/MS, m/e (%): 129 (23.1), 131 (27.6), 157 (47.2), 202 (27.7), 226 (47.3), 227 (38.9), 228 (36.2), 239 (35.9), 289 (19.5), 290 (33.9), 355 (15.7), 435 (55.8), 436 (14.7), 437 (52.4), 438 (12.8), 511 (28.5), 513 (28.4), 706 (49.8), 707 (23.3), 708 (100.0), 709 (34.0), 710 (53.0).

3.2.1.2. Method II

Potassium hydroxide solution (45.3% w/w, 1800 ml, 21.04 mol) was added to a 5-liter three-necked flask fitted with a mechanical stirrer, a thermometer, a nitrogen sweep, and a heating mantle. The solution was stirred under a sweep of nitrogen as the starting phenolic compound (amount corresponding to 21 equivalents of phenolic functionality) was added over 30 min. After an initial heat of reaction, water (800 ml) was added and the mixture was heated to maintain a temperature of 50°C overnight. This solution was then fed into a double-drum dryer operating at 120°C and the dried powder was collected and dried under vacuum in a vacuum oven at 100°C and 0.50 mm Hg overnight to provide the potassium salt of THPE at approximately 3.5% water by weight.

A 12-liter five-necked round bottom flask was fitted with a mechanical stirrer, a Dean–Stark trap topped with a reflux condenser under nitrogen, and a thermocouple attached to a temperature controller that controlled a heating mantle on the flask. A mixture of DMSO (5.0 liters), toluene (1.5 liters), and the potassium salt of the phenolic starting compound (3.5% water w/w, 2.97 mol) was added to the flask under nitrogen purge. The mixture was then stirred at reflux for 48 h with azeotropic removal of water. After 48 h the Dean–Stark trap was removed and replaced with a Soxhlet extractor containing dry 4A molecular sieves. The toluene was then refluxed through a Soxhlet extractor for 7 h to dry the toluene. After 7 h, the Soxhlet was replaced with a Dean–Stark trap, and toluene (700 ml) was removed from the reactor by simple distillation. The resulting suspension was colled to 25°C in an ice bath and 1,2-dibromotetrafluoroethane (850 g, 3.27 mol) was added at a rate that maintained a temperature of 30° ± 2°C. When the addition was complete, the mixture was heated to 50°C with continuous stirring for 3 h. The crude reaction mixture was then transferred to a 2.5-liter carboy, acidified with 100 ml of 6 N HCl, and extracted six times with 2.0-liter portions of hexane. The hexane extracts were combined to provide the 2-bromotetrafluoroethoxy aryl ether product.

1,1,1-Tris[4-(2-Bromotetrafluoroethoxy)phenyl]Ethane. According to the procedure of Method II this compound was prepared from 1,1,1-tris(4-hydroxyphenyl)ethane (2.97 mol) in DMSO (5.0 liters) and toluene (1.5 liters) to provide the product of 1,1,1-tris[4'-(2''-bromotetrafluoroethoxy)phenyl]ethane (2400 g, 2.85 mol, 95.8% yield) as a white crystalline solid, m.p.= 111–112°C. LC/MS, m/e (%): 43 (64), 59 (100), 101 (59), 152 (36), 165 (36), 167 (27), 177 (18), 178 (21), 179 (28), 215 (33), 297 (30), 299 (30), 569 (18), 571 (33), 583 (17), 825 (34), 826 (10), 827 (94), 828 (26), 829 (93), 830 (25), 831 (34), 832 (9), 840 (2), 841 (1), 842 (6), 843 (2), 844 (6), 845 (2), 846 (2), 847 (1).

3.2.2. Synthesis of Trifluorovinyloxy Aryl Ether Monomers

3.2.2.1. Method III

Into a 1-liter five-necked flask equipped with a mechanical stirrer, a thermocouple attached to a temperature controller that controlled a heating mantle on the flask, an addition funnel, and a reflux condenser was placed freshly distilled diglyme (200 ml) and fresh zinc powder (36.0 g, 0.55 mol). The mixture is stirred and heated to 130°C. The 2-bromotetrafluoroethoxy aryl ether product (0.184 mol) was added very slowly via the addition funnel over 3.5 h. The mixture was then stirred mechanically at 115°C for 1 h, after which the heat was turned off and the mixture allowed to cool to room temperature. The solution was centrifuged to remove the zinc salts. The liquid was decanted and the zinc salts were washed with acetone and centrifuged again. The liquid portions were combined and evaporated thoroughly, and the residue was purified further, either by recrystallization in an ethanol/water (10 : 1) mixture or by column chromatography on neutral alumina using hexane as an eluent to provide the trifluorovinyloxy aryl monomer product in ≥ 99.8% purity.

(a) 1,3-Bis(Trifluorovinyloxy)Benzene. This product was prepared from 1,3-bis(2-bromotetrafluoroethoxy)benzene (200.0 g, 0.427 mol) and zinc (69.6 g, 1.1 mol) in tetraglyme (250 ml) according to Method III. The crude product was purified first by vacuum distillation (105–112°C, 110 mm Hg), then by column chromatography on neutral alumina using hexane as eluent to provide the pure product (85.7 g, 74.3% yield) as a water white oil. GC/MS, m/e (%): 50 (44.7), 76 (100), 145 (9.6), 173 (17.7), 270 (24.2).

(b) 4,4'-Bis(Trifluorovinyloxy)Biphenyl. This product was prepared from 4,4'-bis(2-bromotetrafluoroethoxy)biphenyl (100 g, 0.184 mol) according to Method III to provide 62.45 g (0.180 mol) of 4,4'-bis(trifluorovinyloxy)biphenyl of 94.5% purity by GC analysis in 98% yield. This product was recrystallized from an ethanol/water mixture in greater than 70% recovery to provide the product in 99.8% purity by GC analysis, melting point 44.5–46°C. IR (cm^{-1}): 1833 (w, CF=CF$_2$); 1601, 1491 (C=C); 1231, 1196–1132 (C—O, C—F); 818 (aromatic). GC/MS, m/e (%): 63 (14.9), 76 (14.9), 150 (11.7), 151 (27.0), 152

(100.0), 153 (13.8) 346 (31.3). DSC analysis of the 4,4'-bis(trifluorovinyloxy)biphenyl monomer 20 to 360°C/min) indicated a sharp endotherm of melting beginning at 45°C, followed by a broad exotherm beginning at about 135°C, corresponding to the heat of cyclization of the trifluorovinyl groups to form hexafluorocyclobutane rings.

(c) *4,4'-Bis(Trifluorovinyloxy)Diphenyl Sulfide.* This product was prepared from 4,4'-bis(2-bromotetrafluoroethoxy)diphenyl sulfide (150.0 g, 0.26 mol) and zinc (35.0 g, 0.53 mol) in tetraglyme (150 ml) according to Method III. The crude product was purified by column chromatography to provide the purified product (63.0 g, 64.1% yield) as a light orange colored oil. GC/MS, m/e (%): 108 (31.8), 139 (37.9), 152 (26.0), 158 (11.3), 184 (100), 378 (22.4).

(d) *2,2-Bis(4-Trifluorovinyloxyphenyl)Propane.* This product was prepared from 2,2,-bis[4-(2-bromotetrafluoroethoxy)phenyl] propane (103.4 g, 0.176 mol) and zinc (25.0 g, 0.38 mol) in diglyme (150 ml) according to Method III. The crude reaction mixture was evaporated and the residue was purified by column chromatography on neutral alumina using hexane as an eluent to provide the product (42.8 g, 63% yield) as a viscous water white oil in > 99.6% purity by GC analysis. GC/MS, m/e (%): 115 (17.1), 117 (18.3), 118 (24.0), 152 (15.6), 178 (50.0), 179 (24.8), 199 (12.8), 215 (46.6), 276 (30.4), 373 (100), 388 (17.5).

(e) *2,2-Bis(4-Trifluorovinyloxyphenyl)Hexafluoropropane.* According to the procedure of Method III this product was prepared from 2,2,-bis[4-(2-bromotetrafluoroethoxy)phenyl]hexafluoropropane (500.0 g, 0.72 mol) and zinc (96.0 g, 1.47 mol) in diglyme (250 ml). The crude product was purified by column chromatography on neutral alumina using hexane as eluent to provide the purified product (68.0 g, 19% yield) as a water white oil in > 99.6% purity by GC analysis. GC/MS, m/e (%): 116 (33.6), 173 (55.1), 183 (88.7), 233 (100), 302 (19.1), 323 (13.3), 399 (16.2), 427 (38.9), 496 (84.1).

3.2.2.2. Method IV

Into a 500-ml five-necked flask fitted with a mechanical stirrer, a reflux condenser, and a thermocouple attached to a temperature controller is placed freshly activated granular zinc (4.3 g, 0.066 mol) and 25 ml dry acetonitrile. This mixture is stirred and heated to reflux (83°C) under nitrogen while the 2-bromotetrafluoroethoxy aryl product (amount corresponding to 0.021 mol equivalents of 2-bromotetrafluoroethoxy functionality) is dissolved in 21 ml of acetonitrile and added dropwise. The resulting mixture is stirred at 85°C for 6 h, then cooled and filtered. The filtrate is evaporated at 60°C under vacuum to remove the acetonitrile and the residue is purified by distributing the desired monomer product and the impurities between hexane and acetonitrile in a liquid/ liquid countercurrent extraction column with agitated plates, using hexane as the lighter, continuous phase and acetonitrile as the heavier, dispersed phase. The distribution coefficients of the components and the flow rates of the two solvents are calculated according to the method of Bush and Densen.[22] The desired monomer product elutes with the hexane phase and is recovered by evaporation to provide the final trifluorovinyloxy aryl monomer product in greater than 99.6% purity.

(a) *9,9-Bis(4-Trifluorovinyloxyphenyl)Fluorene.* This product was prepared from 9,9-bis[4-(2-bromotetrafluoroethoxy)phenyl]fluorene (736.2 g, 1.04 mol) and zinc (230.75 g, 3.53 mol) in acetonitrile (4.0 liters) according to Method IV to provide the tribfluorovinyl aryl ether monomer product 9,9-bis(4'-trifluorovinyloxyphenyl)fluorene (316.6 g, 0.621 mol, 59.7% yield), melting point 112–112.5°C. LC/MS, m/e; 120 (15.1), 144 (16.5), 145 (18.3), 150 (28.8), 155 (14.4), 157 (53.1), 158 (28.7), 207 (14.1), 237 (15.6), 239 (100.0), 240 (52.8), 241 (15.5), 313 (12.8), 315 (19.7), 316 (16.1), 337 (27.2), 510 (91.9), 511 (29.3).

(b) *1,1,1-Tris(4-Trifluorovinyloxyphenyl)Ethane.* According to the procedure of Method IV this compound was prepared from 1,1,1-tris[4-(2-bromotetrafluoroethoxy)phenyl]ethane (1875 g, 2.22 mol) and zinc (465 g, 7.1 mol) in acetonitrile (8.8 liters) to provide the product 1,1,1-tris(4'-trifluorovinyloxyphenyl)ethane (1028.7 g, 84.9% yield) as a white crystalline solid, m.p. = 29–30°C. GC/MS, m/e (%): 76 (29.5), 77 (15.9), 102 (31.7), 107 (18.8), 113 (27.3), 118 (25.6), 119 (70.3), 120 (39.1), 126 (28.7), 127 (20.3), 151 (17.8), 152 (31.9), 163 (17.3), 176 (25.4), 177 (17.8), 178 (100.0), 199 (19.3), 239 (73.9), 240 (28.1), 276 (16.9), 373 (24.4), 434 (17.9), 531 (44.0), 546 (3.2).

3.2.3. Polymerization of Bis(Trifluorovinyloxy)Aryl Ether Monomers

3.2.3.1. Method V

The bis(trifluorovinyloxy)aryl ether monomer (0.173 mol) was placed in a 250-ml three-necked round bottom flask with 75 ml of solvent. The flask was fitted with a mechanical stirrer, a heating mantle, and reflux condenser under nitrogen. After purging the flask thoroughly with nitrogen, the mixture was stirred and heated to reflux (215°C in perfluorotetradecahydrophenanthrene or 255°C in diphenyl ether). After stirring at reflux for 45 min to 2 h, a polymer phase separated; after stirring at reflux for a total of 3 to 6 h, the phase-separated polymer beame viscous enough to seize the stirring shaft. The cooled polymer was removed from the flask and evaporated under high vacuum (approximately 0.50 mm Hg) at about 220°C for 3 h to remove residual solvent. A small portion of the polymer was removed from the flask and analyzed by DSC to determine the approximate glass transition temperature of the polymer, and by GPC in dichloromethane to determine weight average molecular weight (M_w) and polydispersivity (M_w/M_n). The remaining polymer residue was removed from the flask and pressed into a bar $0.50 \times 3.0 \times 0.125$ in. by compression-molding at approximately $35–40°C$ above the glass transition temperature as indicated by DSC. This bar was used to analyze the polymer by DMA.

(a) *1,3-Phenylene Perfluorocyclobutyl Ether Polymer.* According to the procedure of Method V this polymer was prepared from 1,3-bis(trifluorovinyloxy)benzene (50 g) in perfluorotetradecahydrophenanthrene (50 g). The glass transition temperature by DMA was 32°C. GPC: $M_w = 41,400$ ($M_w/M_n) = 3.35$.

(b) *4,4′-Biphenylene Perfluorocyclobutyl Ether Polymer.* According to the procedure of Method V the monomer 4,4′-bis(trifluorovinyloxy)biphenyl (60.0 g, 0.173 mol) was placed in a 250-ml three-necked round bottom flask with 75 ml of perfluorotetradecahydrophenanthrene. After stirring at reflux for approximately 45 min a polymer phase separated. After stirring at reflux for a total of 3 h, the phase-separated polymer became viscous enough to seize the stirring shaft. The cooled polymer was removed from the flask and evaporated under high vacuum (approximately 0.50 mm Hg) at about 220°C for 3 h to remove residual solvent. A portion of this polymer was compression-molded at 250°C to provide a light yellow, transparent flexible plastic film. Another portion was dissolved in tetrahydrofuran and placed in an evaporating dish to make a solvent-cast film. After the solvent was evaporated overnight, a light yellow thin film was peeled from the dish. This sample exhibited excellent flexibility and transparency. The glass transition temperature by DMA was 165°C. GPC: $M_w = 103,300$ ($M_w/M_n) = 2.55$. IR (cm^{-1}): 1601, 1490 (C=C); 1302, 1194–1115 (C—O, C—F), 985 (hexafluorocyclobutane), 818 (aromatic).

(c) *4,4′-Diphenylene Sulfide Perfluorocyclobutyl Ether Polymer.* This polymer was prepared according to Method V from 4,4′-bis(trifluorovinyloxy)diphenyl sulfide (5.0 g) in perfluorotetradecahydrophenanthrene (12.0 g). The glass transition temperature by DMA was 78°C. GPC: $M_w = 42,500$ ($M_w/M_n) = 1.74$.

(d) *2,2-Bis(4-Phenylene)Propane Perfluorocyclobutyl Ether Polymer.* This polymer was prepared according to Method V from 2,2-bis(4-trifluorovinyloxyphenyl)propane (13.6 g) in perfluorotetradecahydrophenanthrene (14.0 g). The glass transition temperature by DMA was 98°C. GPC: $M_w = 102,700$ ($M_w/M_n) = 2.39$.

(e) *2,2-Bis(4-Phenylene)Hexafluoropropane Perfluorocyclobutyl Ether Polymer.* This polymer was prepared according to Method V from 2,2-bis(4-

trifluorovinyloxyphenyl)propane (13.6 g) in perfluorotetradecahydrophenanthrene (14.0 g). The glass transition temperature by DMA was 125°C. GPC: $M_w = 23,500$ $(M_w/M_n) = 6.0$. The GPC analysis of this material indicated a large quantity of polymer with mass between 96,400M_w and 12,800M_w, with a second peak between 2000M_w and 800M_w, which skewed the distribution and broadened the polydispersity.

(f) *9,9-Bis(4′-Phenylene)fluorene Perfluorocyclobutyl Ether Polymer.* According to the procedure of Method V this polymer was prepared from 9,9-bis(4′-trifluorovinyloxyphenyl)fluorene (3.0 g, 0.0059 mol) and diphenyl ether (5.0 ml). The mixture was stirred and heated to reflux (225°C) for 22 h. The diphenyl ether (DPO) solvent was evaporated under high vacuum on a 100-ml Kugelrohr bulb-to-bulb apparatus (0.03 mm, 165°C) to provide the polymer product, which was dissolved in methylene chloride and cast into a thin film. The glass transition temperature by DMA was 228°C. GPC: $M_w = 110,700$ $(M_w/M_n) = 3.0$.

3.2.3.2. Method VI

4,4′-Biphenylene Perfluorocyclobutyl Ether Polymer. The monomer 4,4′-bis(trifluorovinyloxy)biphenyl (15.0 g, 0.043 mol) was placed in a nitrogen-purged 100-ml round bottom flask and polymerized by heating at 210°C for 2 h without stirriing. After cooling, a small sample was removed for analysis by DSC. The sample showed a small crystalline melt with a peak at 60°C, followed by a broad exotherm beginning at about 200°C. The bulk sample was heated again at 235°C for an additional 3 h. Again a sample was removed and analyzed by DSC. The analysis indicated a very small crystalline melt with a peak at 60°C, followed by a low-intensity exotherm beginning at about 230°C. The bulk sample was heated again to 265°C for 45 min. Analysis of this sample indicated no crystalline melt and no exothermic activity up to and including 325°C, with the emergence of an endothermic glass transition T_g at 148°C.

3.2.4. Fabrication of Polymer Plaques for Mechanical Testing

A sample of monomer ($\sim$50 g) was placed in a 100-ml round bottomed flask, which was attached to a rotary evaporator. The flask was heated to 150°C with the aid of an oil bath while stirring under vacuum (0.20 mm Hg). The monomer was stirred and heated under vacuum for a period of 45 min to 1 h in order to simultaneously degas and B-stage the monomer. The B-staged monomer was then held under nitrogen and poured into a preheated mold, which consisted of two polished aluminum plates (6 × 6 × 1/2 in.) with a three-sided 1/8-in. thick Teflon® spacer between the two plates. The spacer was cut from a 6 × 6 × 1/8 in.

piece of Teflon such that a 1-in. wall was left on three sides, with the fourth side open. Thus the maximum size of a cast plaque was 4 × 5 × 1/8 in. The mold was held together with large ACCO binder clips around the edge and then was placed in a vacuum oven at 205°C to be preheated. After the monomer was poured into the mold it was placed in the oven, which was evacuated to 200 mm Hg absolute pressure, then brought to ambient pressure with nitrogen. This evacuation/refilling process was repeated twice and the oven was then left under nitrogen at 205°C for 18 h. It was then turned off and allowed to cool to about 50°C over a period of 3 h, after which the mold was removed from the oven and the cast plaque removed from the mold. Test specimens were then cut using a diamond-edged wet saw and were sanded to remove imperfections.

3.2.5. Fabrication of Polymer Film Samples for Thermal Stability Testing

Samples of the thermoset polymer were prepared for quantitative thermal analysis by dissolving the monomer 1,1,1-tris(4-trifluorovinyloxyphenyl)ethane (58% w/w) in mesitylene and heating the solution to a temperature of 145°C for up to 16 h. Samples were removed from the solution periodically and analyzed to determine weight average molecular weight by GPC. When the sample reached a weight average molecular weight of 15,000, the solution was cooled and filtered. A portion (approximately 4 ml) was spin-coated (spread speed 500 rpm for 3 s, spin speed 1300 rpm for 30 s) onto silicon wafers that had previously been sputter-coated with 2 μm of aluminum. The polymer films were cured under nitrogen in a temperature-programmable oven according to the following cure schedule: ramp to 100°C over 30 min, hold at 100°C for 1 h, ramp to 220°C over 1 h, hold at 220°C for 1 h, cool to < 100°C. The resulting polymer films (10 μm thick) were lifted from the aluminized silicon wafers by soaking the wafers in an aluminum etch batch to dissolve the aluminum layer. The resulting free films were soaked in deionized water for 10 min to remove residual acid, then baked in a vacuum over with a slow nitrogen bleed at 80°C and 0.20 mm Hg overnight to dry the samples.

3.3. RESULTS AND DISCUSSION

3.3.1. Monomer Synthesis

The monomers for the perfluorocyclobutane polyarylates were prepared in two steps from phenolic starting materials.

In the first step, phenolic monomers were reacted with potassium hydroxide in DMSO solution to form the potassium arylate salts, which were then dried *in situ* by azeotropic distillation of the water with toluene. After removal of a portion of the toluene azeotroping agent by simple distillation, the mixture was cooled to 20°C and 1,2-dibromotetrafluoroethane was added in a 10% stoichiometric excess at a rate that maintained a reaction temperature of 25 to 35°C. This reaction scheme succeeded in providing the intermediate products in moderate to good yields.

This fluoroalkylation reaction is not a classical S_n2 displacement mechanism, but instead follows an unusual pathway involving nucleophilic attack of phenoxide on bromine to form an aryl hypobromite, according to the mechanism outlined in Figure 3.4.

This reaction mechanism has already been reported.[23,24] The mechanism is actually quite significant for the economics and efficiency of the preparative chemistry. The fact that an S_n2 mechanism is not in effect here means that only a slight stoichiometric excess of the fluorocarbon reagent is required for good yields of the intermediate products. Also, the action of an S_n2 mechanism would make the desired product itself susceptible to a second nucleophilic attack on the terminal —CF_2Br, resulting primarily in the production of polymer from this reaction instead of the observed product.

Two primary side reactions may be observed in this chemistry. Both result in the formation of the same by-product, making purification of the product mixture

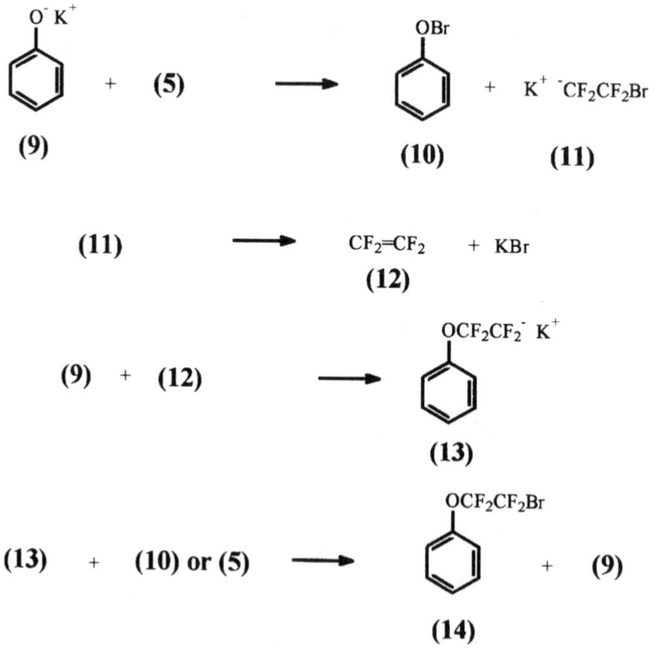

Figure 3.4. Mechanism for the fluoroalkylation of phenols.

a somewhat simpler process. The first side reaction occurs owing to the presence of water or other protic substances in the reaction mixture. The fluorocarbanion formed in Step 3 of Figure 3.4 is a sufficiently strong base to deprotonate water to form a tetrafluoroethyl ether by-product. The tetrafluoroethyl ether by-product represents a yield loss to the process. Attempts to chemically and thermally dehydrofluorinate this species to form the trifluorovinyl ether have been unsuccessful.[25,26]

The second side reaction of this chemistry is observed in aromatic sytems that are highly activated toward electrophilic aromatic substitution, such as the dipotassium salt of resorcinol. If temperatures are not carefully controlled below 30°C, the aryl hypobromite formed in Step 1 of Figure 3.4 can act as a brominating agent, resulting in an aromatic ring-brominated product. Figure 3.5 illustrates these by-product reactions using phenol as an exemplary phenolic.

As indicated, this reaction also generates one equivalent of phenol, which acts as a protic substance to produce one equivalent of tetrafluoroethyl ether by-product as well.

After isolation, the intermediate products were reacted with zinc in acetonitrile or a glyme solvent to provide the trifluorovinyl ether monomers. This reaction with zinc is also sensitive to the presence of protic substances in the reaction mixture, and forms the same tetrafluoroethyl ether by-product in an undesirable side reaction. The reaction of the 2-bromotetrafluoroethyl ether reactants with zinc involves the formation of zinc organometallic species as an intermediate, and this species is sensitive to hydrolysis by water or acidic substances (Figure 3.6).

In order to balance the chemistry of water hydrolysis, one should invoke the formation of 1 mol of HBr. The HBr thus formed would react either with the organometallic species to produce more of the tetrafluoroethyl ether by-product and zinc bromide, or with zinc metal to form zinc bromide and hydrogen gas.

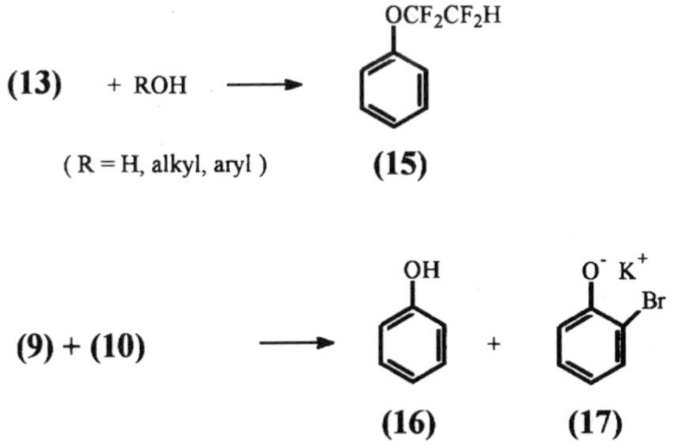

Figure 3.5. By-product formation during fluoroalkylation.

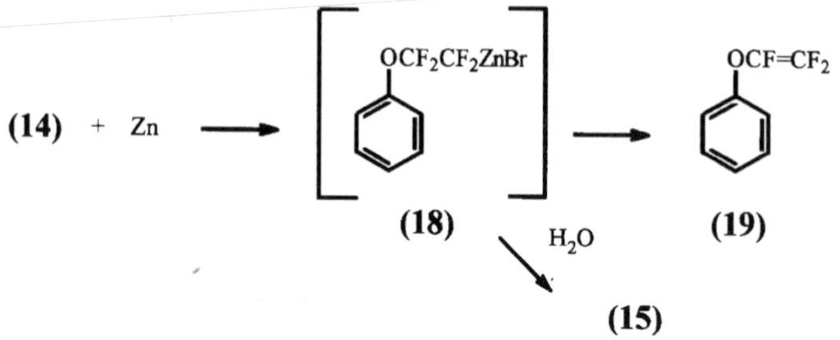

Figure 3.6. By-product formation during zinc dehalogenation.

The monomers were purified either by column chromatography or by a liquid/liquid countercurrent extraction process in which the monomer/by-product mixture was distributed between hexane and acetonitrile (see Method IV). Purification was performed to provide monomers that were $\geq 99.6\%$ pure by GC and/or LC anlaysis.

3.3.2. Polymerization

Polymerizations were carried out either neat or in solution. Solution polymerizations were performed in a variety of sovlents, with perfluorotetradecahydrophenanthrene giving the best results. This solvent could be heated to 215°C for hours with no detectable degradation. The advancing polymers generally remained in solution until moderately high molecular weights were achieved, at which point they precipitated from solution. Diphenyl ether was also used as a polymerization solvent. This solvent, while providing better solubility of the high-molecular-weight polymers, suffered from discoloration during polymerization, even when carefully deoxygenated. Diphenyl ether, however, had the additional advantage of allowing temperatures as high as 260°C during the polymerization reaction. Polymers were generally recovered from these solvents by adding the polymer solution to a nonsolvent such as hexane and collecting the powdered polymer by filtration. Precipitated polymers were heated under vacuum (< 0.20 mm Hg) at 140°C for approximately 2 h on a rotary evaporator. This had the dual effect of devolatilizing the polymers and continuing to advance the molecular weights of the polymers with lower glass transition temperatures.

These polymerization methods were carried out to produce small quantities of polymers for DSC and GPC analysis. These analyses were used to measure the efficiency of the polymerization and the glass transition temperature of the resulting polymers, as a screen for further development of selected materials. It was deemed at the time of the tests that a high glass transition temperature was desirable for further development of a selected polymer. Some of the thermoplastic polymers were also compression-molded into plaques for dielectric testing. The results of these screening tests are presented in Table 3.1.

The data in Table 3.1 clearly demonstrates that the cyclodimerization reaction is an efficient one for the preparation of high-molecular-weight polymers. It is noteworthy that high-molecular-weight polymers can be prepared from these bis(trifluorovinyl) aryl ether monomers, since this type of step-growth addition polymerization requires conversions in excess of 99% in order to obtain these types of molecular weights.[27] Therefore, high molecular weights can only be obtained from this chemistry by using highly purified bis(trifluorovinyl ether) monomers. This primarily involves the removal of the aryl ether by-product, which one has one 1,1,2,2-tetrafluoroethyl group and one trifluorovinyl ether group. This impurity acts as a polymer chain terminator. The relatively poor molecular weight obtained for the hexafluoroisopropylidene (6F diphenol) polymer may well have been a result of inadequate purification of the bis(trifluorovinyl ether) monomer.

The result of the dielectric testing on these polymers was quite encouraging. All of the polymers that were tested demonstrated low dielectric constant and dissipation factor values.

Samples of one polymer, prepared from 4,4′-biphenol, appeared to have good mechanical flexibility and toughness as well as good dielectric properties. This polymer was prepared in sufficient quantity to allow a more extensive evaluation of the polymer properties, including thermal, mechanical, and dielectric properties. Results of the tests are listed in Table 3.2.

Table 3.1. Properties of the PFCB Thermoplastic Polymers

Ar	T_g (DMA) (°C)	Dielectric constant 10 kHz	Dissipation factor 10 kHz	M_w	M_w/M_n
1,3-Phenylene	32	2.41	—	41,400	3.35
4,4′-Biphenyl	165	2.40	0.0003	116,400	1.89
4,4′-Diphenyl sulfide	78	2.62	0.0005	42,500	1.74
Isopropyl-2,2-diphenylene	98	—	—	102,700	2.39
Hexafluoroisopropyl-2,2-diphenylene	125	—	—	23,500	6.0
9,9-Bis(4′-phenylene) fluorene	228	—	—	110,700	3.0

Table 3.2. Properties of the Biphenyl PFCB Thermoplastic Polymer

Tensile strength (Mpa)	50.3 ± 1.4
Tensile modulus (MPa)	1700 ± 79.3
Percent elongation (break)	8.1 ± 0.6
Flexural strength (MPa)	92.4 ± 2.1
Flexural Modulus (MPa)	1780 ± 65.4
Limiting oxygen index	0.42
Water absorption (24 h)	0.040%

In addition, the thermoset polymer prepared from 1,1,1-tris(4-trifluoroviny-loxyphenyl)ethane was cast into plaques according to the procedure outlined in the Section 3.2, and was subjected to thermal, mechanical, and dielectric tests. The results of these tests are listed in Table 3.3.

3.3.3. Thermal Stability Testing

A number of methods have been developed for the determination of the thermal decomposition kinetics of organic and inorganic materials. These include both isothermal and dynamic TGA methods.[28-30] Our studies of the thermal stability of the thermoset polymer utilized isothermal evaluation of zeroth-order decomposition rate kinetics.[19]

Prepolymer samples were spin-coated onto silicon wafers and fully cured. The cured polymer films were removed from the wafer and tested by TGA. Analyses were carried out to measure the rate of weight loss at various temperatures in air and in nitrogen. From these results the time to a 1% weight loss was determined and these results are depicted graphically in Figure 3.7.

The rate of weight loss for the thermal decomposition in an inert atmosphere was found to follow the simple rate expression:

(1) $$\text{Rate} = k$$

where k is the rate constant in percent per minute. This type of rate expression is consistent with zeroth-order kinetics. In these calculations the change in sample weight is neglected (the concentration of the polymer does not change), a reasonable approximation when measurements are taken at 1% weight loss.

Table 3.3. Properties of the PFCB Thermoset Polymer

Tensile strength (MPa)	66 ± 1.4
Tensile modulus (MPa)	2.270 ± 79.3
Percent elongation (break)	4.1 ± 0.6
Flexural strength (MPa)	74 ± 12.4
Flexural modulus (MPa)	2320 ± 13.1
T_g (DMA, °C)	400°C
Limiting oxygen index	0.47
Water absorption (24 h)	0.021%

Frequency	Dielectric constant (dissipation factor)
1 kHz	2.50 (0.0007)
10 kHz	2.45 (0.0004)
1 MHz	2.41 (0.0004)
10 GHz	2.35 (0.0012)

By then utilizing the Arrhenius relationship:

$$k = Ae(-E_a/RT) \tag{2}$$

and rearranging to the familiar form $y = mx + b$:

$$\ln k = -E_a(1/RT) + \ln A \tag{3}$$

a plot of the natural log of the rate of weight loss, measured at 1% weight loss (with k in percent per minute) vs. the inverse of temperature ($1/T$ in K) yields a line with a slope equal to the activation energy divided by the ideal gas constant (E_a/R) and an ordinate intercept corresponding to the natural log of the Arrhenius preexponential factor A.

While the rate of decomposition in air was more rapid than in nitrogen, the kinetics of oxidative decomposition cannot be estimated reliably by isothermal weight loss because of the possibility of competing oxidative weight gain process. For this reason the kinetics were not estimated from the available data

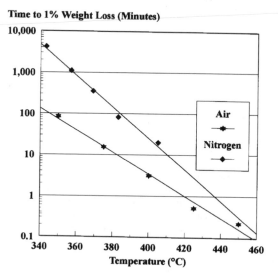

Figure 3.7. Thermal/oxidative stability of the PFCB thermoset polymer.

After performing a least-squares best fit calculation of the data points, the plot of ln k vs. $1/T$ for the decomposition in nitrogen yields an activation energy of 74.4 kcal/mol ($r^2 = 0.991$).

3.3.4. Thin-Film Processing

One of the advantages of working with the PFCB polymers emerges in the process of fabricating parts and films of the thermoset polymer. The step-growth polymerization that results from the thermal dimerization of the trifluorovinyl ether functionality provides a convenient method for controlling the molecular weight, and thereby the viscosity, of the thermoset prepolymer. The molecular weight of the prepolymer is a function of the thermal history. When the monomer is dissolved in a solvent and heated to advance the molecular weight, polymerization can be halted at any point by removing the heat from the sample. The viscosity of the prepolymer solution is then controlled by a combination of the molecular weight and the solution concentration. This control of molecular weight and viscosity has practical significance in the preparation of thin-film multilayer electronic and optical devices, where film thickness and dimensional integrity are important considerations in fabrication processes. In these processes polymer films are typically deposited from solution by spin-coating, and a degree of latitude in available film thickness is a desirable process variable. Coating thickness can be controlled by varying both solution viscosity and spin speed, as is demonstrated in Figure 3.8.

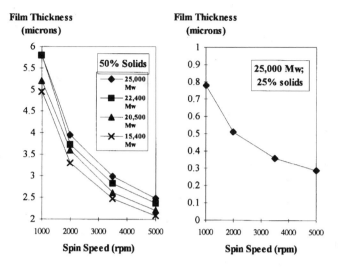

Figure 3.8. Spin curves for the thermoset prepolymer at various molecular weights and percent solids.

Prepolymer solutions of slightly lower molecular weight also exhibit excellent planarization of microelectronics features in multilevel devices. Figure 3.9 depicts the performance of a $10,000 M_w$ prepolymer solution (50% solids) in the planarization of a 2-μm step height over a range of feature sizes, and this is reported as degree of planarization (DOP) versus feature size. DOP is calculated according to the simple equation shown in Figure 3.9.

The ability to apply a planarizing, optically transparent, thermally stable polymer system that cures under relatively mild conditions has recently been demonstrated to have utility in the fabrication of multilayer devices such as advanced color liquid-crystal displays.[31,32] The ability to apply this material as a dielectric or optical coating for thin-film electronics devices has also recently been demonstrated with the fabrication of an optical wave guiding device.[33]

3.4. CONCLUSIONS

A new method for the syntheses of fluorocarbon polyarylate polymers has been demonstrated. The chemistry utilizes the $[2\pi + 2\pi]$ cyclodimerization of fluorinated olefins and generates polymers of novel composition. The first generation of polymers prepared by this method are polyarylate homopolymers. Theremoplastic polymers of high molecular weight can be achieved via neat or solution polymerization. One example of a thermoset polymer prepared by this method has a high T_g, low dielectric constant and dissipation factor, low moisture

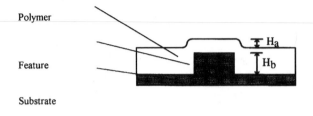

Polymer

Feature

Substrate

$$DOP = (H_b-H_a/H_b) * 100\%$$

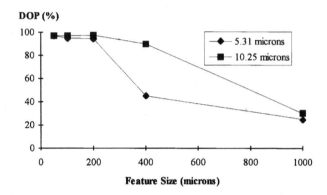

Figure 3.9. Degree of planarization of the thermoset prepolymer: Coated from 50% solids solution $M_w = 12,500$.

absorption, and excellent processability. This polymer also exhibits good thermal/oxidative stability.

A variety of polymer compositions that use this type of polymerization chemistry can be envisioned. In addition to the polyarylate homopolymers that have been described in this chapter, random or block copolymers can be prepared with reasonable ease by the combination of different monomers or oligomers. These compositions can be designed to optimize thermal, mechanical, dielectric, or optical properties of a polymer system. Also, the trifluorovinyl ether functionality can be incorporated into other high-performance polymer systems with relative ease.[34,35] The perfluorocyclobutane polyarylate chemistry is a versatile approach to the preparation of high-performance polymers, which is just beginning to demonstrate its utility.

ACKNOWLEDGMENTS: Many Dow researchers have contributed their time and talents to the chemistry and analytical data presented in this manuscript. A complete list of the contributors is not practical, but a few individual recognitions are appropriate. The contributions of the coinventors in this technology, Drs. Bob Ezzell, Katherine Clement, and Frank Richey have broadened the scope of this technology beyond that which appears in this manuscript and are documented in a long list of patients that are only now being issued. My greatest appreciation goes to Prof. Alvin Kennedy, now at North Carolina A&T University, for his work on defining the dielectric performance of these materials. Thanks to Vernon Snelgrove for making significant contributions to the synthesis of the monomers and in understanding the processability of the polymers. Special thanks go to Jeff Bremmer and Andrew Pasztor for the quantitative thermal analysis of the thermoset polyer.

3.5. REFERENCES

1. K. M. Kane, P. E. Cassidy, L. A. Hoelscher, and D. L. Meuer, *5th Intern. SAMPE Electronics Conf.* (1991) pp. 352–358.
2. P. E. Cassidy, T. M. Aminabhavi, and J. M. Farley, *J. Macromol. Sci.—Rev. Macromol. Chem. Phys.* C29 (2&3), 365–372 (1989).
3. E. E. Lewis and M. A. Naylor, *J. Am. Chem. Soc. 69*, 1968–1970 (1947).
4. J. R. Lacher, G. W. Thompkin, and J. D. Park, *J. Am. Chem. Soc. 74*, 1693–1696 (1952).
5. B. Atkinson and A. B. Trentwith, *J. Chem. Soc. 75*, 2082–2087 (1953).
6. R. D. Chambers, *Fluorine in Organic Chemistry*, John Wiley and Sons, New York (1973), pp. 179–186.
7. M. Hudlickey, *Chemistry of Organic Fluorine Compounds, 2nd Ed.*, Simon and Schuster, New York (1992), pp. 450–462.
8. W. H. Sharkey, *Fluorine Chem. Rev. 2*, 1–53 (1968).
9. L. K. Montgomery, K. Schueller, and P. D. Bartlett, *J. Am. Chem. Soc. 86*, 622–631 (1964).
10. W. A. Bernett, *J. Org. Chem. 34*(6), pp. 1772–1776 (1969).
11. R. S. Mulliken, *Tetrahedron 6*, 68–87 (1959).
12. R. W. Anderson and H. R. Frick, U.S. Patent 3,840,603 (Oct. 8, 1974).
13. D. P. Carlson, U.S. Patent 3,303,145 (Feb. 7, 1967).
14. R. Beckerbauer, U.S. Patent 3,397,191 (Aug. 13, 1968).
15. P. L. Heinze and D. J. Burton, *J. Org. Chem. 53*, 2714 (1988).
16. M. Yamomoto, D. C. Swenson, and D. J. Burton, *Macromol. Symp. 82*, 125–141.
17. D. A. Babb, B. R. Ezzell, K. S. Clements, W. F. Richey, and A. P. Kennedy, *J. Polym. Sci. Pt. A: Polym. Chem. 31*, 3465–3477 (1993).
18. D. A. Babb, K. S. Clement, W. F. Richey, and B. R. Ezzell, U.S. Patent 5,037,917 (Aug. 6, 1991).
19. A. P. Kennedy, D. A. Babb, J. N. Bremmer, and A. J. Pasztor, Jr., *J. Polym. Sci. Pt. A: Polym. Chem 33*, 1859–1865 (1995).
20. D. W. Smith, Jr. and D. A. Babb, *Macromolecules 29*, 852–860 (1996).
21. D. A. Babb, N. G. Rondan, and D. W. Smith, Jr., *Polym. Preprints 36*(1), 721–722 (1995).
22. M. T. Bush and P. M. Densen, *Anal. Chem. 20*(2), 121–129 (1948).
23. I. Rico and C. Wakselman, *J. Fluorine Chem. 20*(6), 759–764 (1982).
24. L. Xingya, P. Heqi, and J. Xikui, *Tet. Lett. 25*(43), 4937–4940 (1984).
25. W. J. Pummer and L. A. Wall, *SPE Trans 1963*, 220–224.
26. L. A. Wall and W. J. Pummer, U.S. Patent 3,277,068 (Oct. 4, 1966).

50 David A. Babb

27. J. M. G. Cowie, *Polymers: Chemistry and Physics of Modern Materials, 2nd Ed.*, Chapman and Hall (1991), p. 29.
28. I. M. Salin and J. C. Seferis, *J. Appl. Polym. Sci. 47*(5), 847–856 (1993).
29. F. Carrasco, *Thermochim. Acta 213*(1–2), 115–134 (1993).
30. H. A. Schneider, *J. Therm. Anal. 40*(2), 677–687 (1993).
31. D. J. Perettie, L. D. Bratton, J. Bremmer, D. A. Babb, *Proc. SPIE-Int. Soc. Opt. Eng., 1911: Liquid Crystal Materials, Devices, and Applications II* (1993), pp. 15–20.
32. A. P. Kennedy, L. D. Bratton, Z. Jezic, E. R. Lane, D. J. Perettie, W. F. Richey, D. A. Babb, and K. S. Clement, U.S. Patent 5,246,782 (Sept. 21, 1993).
33. T. A. Tumolillo, Jr. and P. R. Ashley, *Appl. Physl Lett. 82*(24), 3068–3070 (1993).
34. K. S. Clement, D. A. Babb, and B. R. Ezzell, U.S. Patent 5,021,602 (June 4, 1991).
35. K. S. Clement, B. R. Ezzell, D. A. Babb, and W. F. Richey, U.S. Patent 5,037,919 (Aug. 6, 1991).

4

Functional Fluoromonomers and Fluoropolymers

MING-H. HUNG, WILLIAM B. FARNHAM, ANDREW E. FEIRING, and SHLOMO ROZEN

4.1. INTRODUCTION

Fluoropolymers are a special group of polymers that exhibit high thermal stability, excellent chemical resistance, low surface energy, and many other unique properties. Relatively few fluoropolymers containing pendant functional groups are accessible. This is associated with problems arising from: (a) difficult introduction of functional groups into fluoropolymers; (b) maintenance of both functional group effectiveness and fluoropolymer bulk properties, and (c) preparation and polymerization of functional fluoromonomers.

Perfluoropolyethers, which constitute special class of fluoropolymer, are useful as lubricants,[1] elastomers,[2] and heat-transfer fluids under demanding conditions. Several commerical products are available, which are generally prepared by ring-opening polymerization of hexafluoropropylene oxide or by the random copolymerization of tetrafluoroethylene and hexafluoropropylene with oxygen under ultraviolet irradiation.[3] Direct fluorination of hydrocarbon ethers has been reported,[4] but must be done very slowly under carefully controlled

MING-H. HUNG · DuPont Fluroproducts, Wilmington, Delaware 19880-0293.   WILLIAM B. FARNHAM and ANDREW E. FEIRING · DuPont Central Research and Development, Experimental Station, Wilmington, Delaware 19880-0328.   SHLOMO ROZEN · DuPont Central Research and Development, Experimental Station, Wilmington, Delaware 19880-0328. Present address: School of Chemistry, Tel-Aviv University, Tel-Aviv 69978, Israel.

Fluoropolymers 1: Synthesis, edited by Hougham *et al.*, Plenum Press, New York, 1999.

conditions to avoid decomposition of the substrates. A few perfluorinated polyethers have also been prepared by fluorination of partially fluorinated vinyl ether polymers.[5]

In addition, the incorporation of fluoroether units in a regular sequence in the above fluoropolyethers is almost impossible owing to the random nature of the copolymerization process. In this article we describe a novel method for making perfectly alternating fluoropolyethers from functional fluoropolymers that are prepared by base-catalyzed homopolymerization of hydroxy-containing fluoro-monomers.

4.2. RESULTS AND DISCUSSION

Following are several commercially available fluoromonomers (**1–4**) that contain functional groups and have been utilized to make polymeric membranes for ion separations[6–9] or as catalysts for aromatic alkylation and acylation reactions.[10,11] They are also convenient starting precursors, allowing for further functionalization reactions.

$$CF_2\!\!=\!\!CFOCF_2CF(CF_3)OCF_2CF_2\!\!-\!\!X \qquad (\textbf{1}, X = COOMe; \textbf{2}, X = SO_2F)$$

$$CF_2\!\!=\!\!CFO\!\!-\!\!(CF_2)_n\!\!-\!\!X \qquad (\textbf{3}, X = COOMe; \textbf{4}, X = SO_2F; n = 3\!-\!4)$$

Hydroxy-containing fluorovinyl ether monomers (**5,6**) were prepared in excellent yields (80–90%) in a single step from the corresponding esters (**1,3**)[12–14] with sodium borohydride in absolute ethanol.* Protection of the sensitive vinyl ether groups was not required during the reduction. In contrast, the use of a more powerful reducing agent, such as lithium aluminum hydride, resulted in the reduction of the double bond:

$$CF_2\!\!=\!\!CFOCF_2CF(CF_3)OCF_2CF_2\!\!-\!\!CH_2OH \;\textbf{(5)}$$

$$CF_2\!\!=\!\!CFO\!\!-\!\!(CF_2)_n\!\!-\!\!CH_2OH \;\textbf{(6)}$$

During the process of preparing these molecules, we observed that compounds **5** and **6**, formed oligomers when mixed with a strong base (e.g., potassium *t*-butoxide) in polar aprotic solvent. Detailed studies under controlled

---

* The ester precursor, $CF_2\!\!=\!\!CF\!\!-\!\!O\!\!-\!\!CF_2CF(CF_3)O\!\!-\!\!CF_2CF_2\!\!-\!\!COOMe$, is a product of the DuPont Company.

conditions (5 mol% base) revealed that actually a novel base-catalyzed homo-polymerization of the hydroxy-containing vinyl ethers has occurrred:

(1)
$$CF_2=CFOCF_2CF(CF_3)OCF_2CF_2-CH_2OH \textbf{ (5)} \longrightarrow$$
$$-[CF_2CFH-O-CF_2CF(CF_3)OCF_2CF_2-CH_2O]_{n-} \textbf{ (7)}$$

(2)
$$CF_2=CFO-(CF_2)_3-CH_2OH \textbf{ (6)} \longrightarrow$$
$$-[CF_2CFH-O-(CF_2)_3-CH_2O]_n- \textbf{ (8)}$$

The polymerization occurs by ionic addition of OH groups to trifluorovinyl ethers; although additions of nucleophiles to fluorinated olefins are well known, few examples of additions to trifluorovinyl ethers have been reported[15] and no polymerizations by this method have been described.*

This polymerization[16] proceeds quickly at ambient temperatures to give an oligomeric material (7) with M_w of 4000 to 6000 (Table 4.1). Compound 6 was also homopolymerized under similar conditions to afford an oligomer (8) with $M_w \sim 5000.$[†] These new oligomers were characterized by ^{1}H- and ^{19}F-NMR spectroscopy as well as elemental analysis.

The course of the reaction shown in Eqs. (1) and (2) is highly dependent upon reaction conditions, including the nature of the catalyst and the reaction medium, and monomer purity is, of course, of utmost importance in obtaining high-molecular-weight material.[17] We were able to obtain much higher-molecular-weight polymers by catalyst selection. This was done with the aim of minimizing formation of unreactive chain ends. Most of the catalysts found are bases (Table 4.2), but we have not examined the propagating species in detail or determined the various acid/base equilibria that are presumably involved.

Gel permeation chromatography (GPC) studies also indicated that there is always a lower-molecular-weight fraction in the final polymeric product, presumably oligomers of compounds 5 and 6, formed through the cyclization oligomerization. The cyclic oligomers from molecule 5 were isolated and the structures confirmed by NMR spectroscopy.

4.2.1. Linear Polymers

The reaction conditions will determine the relative abundance of linear polymers and cyclic oligomers. To produce linear polymers with the highest molecular weight, reactions were carried out neat to avoid formation of cyclomers.

---

* We have recently described the high-yield addition reaction of phenols to trifluorovinyl esters as a useful process for other polymer systems.
† The alcohols 5 and 6 are also useful as comonomers in free-radical polymerizations.

Table 4.1. Homopolymerization of Compound **5** with *t*-BuOK

Run	Conditions	$M_w{}^a$	M_n	P/D^b	Yield(%)
1	RT 6 h	4100	2700	1.51	53.8
2	RT, 48 h	5920	3410	1.74	58.9
3	50°C, 72 h	4020	2700	1.49	76.1
4	RT, 16 h	4700	2600	1.81	42.1
5	RT, 24 h	4200	1210	3.46	92.0

^{a} Determined by GPC with polymethyl methacrylate (PMMA) as the reference standard.
^{b} Molecular weight distributions.

Although the addition reaction of alcohol to vinyl ether takes place readily at 25°C in ethereal or dipolar aprotic solvents, neat polymerization reactions require substantially higher temperatures. Typical catalyst-screening experiments were carried out at 110–120°C in order to achieve reasonable reaction rates. For the polymerization reactions whose results are summarized in Table 4.2, monomer **5** and the catalyst (typically at 200/1 to 800/1 weight ratio) were combined in screw-cap vials under nitrogen, sealed, and heated in stages to 120°C. Viscosity increases were apparent visually and magnetic stirring was effective only at entry stages of polymerization. The condensation reaction is quite exothermic, and there must be due regard for this factor with reactions on a larger scale.

Table 4.2. Influence of Catalyst/Solvent on the Molecular Weight of Polymer **7**

Catalyst (wt%)	Solvent	T (°C)	$M_n{}^a$	$M_w{}^a$
KH (0.5)	Glyme	8.0	$4,210^b$	$6,530^b$
K$_2$CO$_3$ (3.0)	None	100	6,120	8,570
CsF (0.1) + Cs$_2$CO$_3$ (0.3)	None	120	25,800	52,600
CsF (5) + Cs$_2$CO$_3$ (17)	None	25–80	12,900	22,400
Cs$_2$CO$_3$ (1)	None	120	27,200	57,900
Bu$_4$NCl (1)	None	120	23,400	41,300
Me$_4$NCl (0.2)	None	107	26,700	52,800
PPNCl (0.5)	None	120	20,200	37,800
Ph$_4$PCl (0.5)	None	120	No polymer	
TASCl^c (0.4)	None	120	AC^d	
CsF (0.4)	None	120	16,700	31,100
Cs$_2$CO$_3$ (0.4)	None	120	28,500	65,500
Cs$_2$CO$_3$ (4)	Tetraglyme	25	$5,700^b$	$6,500^b$
KO-*t*-Bu (1.4)	DMF	10	$3,400^b$	$5,920^b$
K$_2$CO$_3$ (3)	DMSO	50	$5,010^b$	$8,870^b$

^{a} GPC analysis.
^{b} Analysis of residue after removing cyclic oligomer.
^{c} TAS = Tris(dimethylamino)sulfonium.
^{d} Apparent decomposition.

The polymerization catalysts that are preferred because of their selectivity are the alkali metal (especially cesium) carbonates, tetraalkylammonium and bis(triphenylphosphoranylidene)ammonium (PPN) chlorides and bicarbonates (Table 4.2). Undesired side reactions are minimized by using relatively low (<5% by weight) catalyst levels. Under these conditions, the fraction of cyclic oligomer was usually 5% or less and was easily removed from the desired polymer by Kugelrohr distillation. Conversions of **5** were essentially quantitative as judged by product weights and lack of detectable amounts of unreacted monomer by GPC.

Using a carefully refined monomer, we obtained polymers with M_n of about 28,500 (M_w 65,500), corresponding to a degree of polymerization of about 72. Alcohol end groups can be detected by NMR, and M_n values calculated from integrals of these signals are in reasonably good agreement with GPC results based upon polystyrene standards. Signals for residual trifluorovinyl groups have not been detected in these samples.

These highly fluorinated homopolymers (**7,8**) derived from hydroxy fluoro-vinyl ethers (**5,6**), in contrast to typtical highly fluorinated compounds and polymers, exhibit unusually good solubility in common organic solvents. The NMR chemical shifts of residual protons in these polymers are highly sensitive to the polarity of the solvent as shown in Table 4.3.

4.2.2. Cyclic Oligomers

At a concentration of about 0.4 M in glyme, **5** reacts with a catalytic amount of potassium hydride to give the cyclic dimer **9** in 55–60% isolated yield. Cyclic trimers (**10**), tetramers, and pentamers constitute a significant fraction of remaining product. The yield of cyclic dimer is diminished in other solvents, such as tetraglyme or DMSO. Although condensation polymerizations are known to feature competitive ring formation and linear chain growth (see, e.g., Semlyen[18]), the substantial fraction of cyclic oligomers obtained from **5** at moderate concentrations is worthy of note. The degree of polymerization for the

9, n = 1   All unspecified
10, n = 2   bonds to fluorine

Table 4.3. Proton Chemical Shifts of Homopolymers **7** and **8**

(Chemical shifts in δ ppm, TMS internal standard)

$\text{\scriptsize \leavevmode}\sim\!\sim\!(CH_2-\underset{\overset{|}{H}\;\textcircled{A}}{CF}-O-CF_2-\underset{\overset{|}{CF_3}}{CF}-O-CF_2-CF_2-\underset{\textcircled{B}}{CH_2}-O-)_n\sim\!\sim\!\sim\!CH_2-\underset{\textcircled{C}}{O}-H$

	DMSO-d_6	Acetone-d_6	THF-d_8	CD$_3$-OD
A	7.18, 7.01	6.88, 6.71	6.76, 6.58	6.70, 6.51
B	4.69	4.72	4.63	4.52
C	3.87	4.06	3.95	3.93
	CD$_3$-CN	CD$_3$-NO$_2$	CDCl$_3$	C$_6$-D$_6$
A	6.48, 6.32	6.43, 6.24	6.04, 5.87	5.26, 5.08
B	4.56	4.58	4.32	3.77
C	3.97	4.17	4.02	3.65

$\text{\scriptsize \leavevmode}\sim\!\sim\!(CF_2-\underset{\overset{|}{H}\;\textcircled{A}}{CF}-O-CF_2-CF_2-CF_2-\underset{\textcircled{B}}{CH_2}-O-)_n\sim\!\sim\!\sim\!CH_2-\underset{\textcircled{C}}{O}-H$

	DMSO-d_6	Acetone-d_6	THF-d_8	CD$_3$-OD
A	7.20, 7.04	6.90, 6.70	6.76, 6.60	6.71, 6.50
B	4.76	4.76	4.60	4.58
C	3.92	4.10	3.92	3.97
	CD$_3$-CN	CD$_3$-NO$_2$	CDCl$_3$	C$_6$-D$_6$
A	6.44, 6.26	6.44, 6.06	6.03, 5.85	5.50, 5.34
B	4.53	4.63	4.37	3.90
C	3.94	4.11	4.04	3.53

linear polymer obtained in solution was limited to modest values, typically in the range 10–15; these oligomers were fully characterized by NMR and GC/MS. A ^{19}F-NMR spectrum of cyclic dimer **9** is shown in Figure 4.1.

4.2.3. Perfluorinated Polymers/Oligomers

The oligomers and polymers described above retain a number of protons that may be undesirable for some demanding applications. The obvious solution is to replace hydrogen with fluorine, and it appears that the only prospect for such transformation is radical fluorination. By the fluorination approach, these F-vinyl ether homopolymers, owing to their unique structures, may provide a new route to novel perfluoropolyethers.

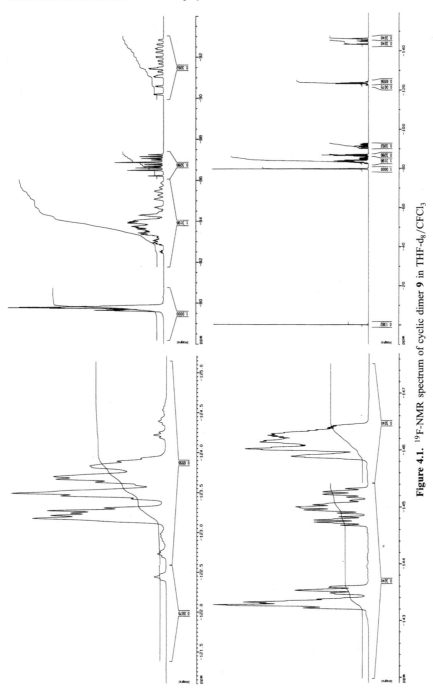

Figure 4.1. ^{19}F-NMR spectrum of cyclic dimer **9** in THF-d_8/CFCl$_3$

By way of illustration, fluorination of homopolymers **7** was studied. It is expected that the resulting perfluorinated polymer chain would have a structure (**11**) with perfectly alternating units of $-(CF_2CF_2O)-$, $[CF_2CF(CF_3)O]-$, and $-(CF_2CF_2CF_2O)-$. Similarly, homopolymers of fluorovinyl ethers (**8**) would afford **12** a perfectly alternating structure of $-(CF_2CF_2O)-$ and $-(CF_2CF_2CF_2CF_2O)-$:

(3) $\quad [-CF_2-CFH-O-CF_2CF(CF_3)-O-CF_2CF_2-CH_2O-]_n$ (**7**) $\rightarrow$

$\qquad\qquad [-CF_2CF_2-O-CF_2CF(CF_3)-O-CF_2CF_2CF_2O-]_n$ (**11**)

(4) $\quad [-CF_2-CFH-O-CF_2CF_2CF_2-CH_2O-]_n$ (**8**) $\rightarrow$

$\qquad\qquad [-CF_2CF_2-O-CF_2CF_2CF_2CF_2O-]_n$ (**12**)

Fluorinations of this type are known in the literature. Lagow has described the fluorination of selected hydrocarbon compounds using elemental fluorine wherein both temperature and fluorine concentration are gradually increased (see, e.g., Huang *et al.*[19]). Adcock developed another process termed "aerosol fluorination," in which the substrate was deposited on solid NaF and then reacted with fluorine.[20] In all cases, however, only relatively small molecules were fluorinated* and it was not obvious that replacing relatively few hydrogens in large chains would proceed smoothly without chain scission. We decided to try an approach based on treating the substrates with relatively concentrated fluorine under irradiation.†

In order to minimize the degradation of macromolecules, the choice of solvent is not a trivial matter. Hydrogen-containing solvents are obviously not suitable, but even halogenated ones such as $CFCl_3$ or CCl_2FCF_2Cl, usually stable to fluorine, can react violently with this element when irradiated at temperatures near $0°C$. We chose two perfluorinated solvents that have been proven safe and suitable for radical fluorination: perfluoro-2-(butyl)-tetrahydro-duran (FC-75 Fluorinert from 3M) and hexafluoropropylene oxide (HFPO) oligomers known as Krytox®.‡

Cyclic dimer **9** was dissolved in FC-75 and treated with a ca. 10-fold excess of fluorine (per hydrogen) while undergoing irradiation with a medium-pressure mercury lamp. Fluorination could be monitored by gas chromatography. After full conversion was achieved, the cyclic product **13** was isolated by distillation in very

---

* For a rare exception, however, see Gerhardt *et al.*[21]
† A similar method has recently been described by Sievert *et al.*[22]
‡ Several oligomers by this name with different molecules weights are commercially available from DuPont.

good yield with no evidence for oligomerization or fragmentation. Its boiling point is very similar to the starting material. Its ^{1}H-NMR spectrum did not show any signal, while the ^{19}F-NMR spectrum showed CF_3 and OCF_2 groups in a narrow region between -80 and -84.5 ppm. The OCF_2CF_2O groups are found at a somewhat higher field (-89 to -90 ppm) and signals for the two "internal" CF_2 groups ($CF_2CF_2CF_2$) appear at a higher field (-129.5 and -129.6 ppm) for the two diastereomers. The tertiary fluorines appear at -145.6 and -146.3 ppm. The high-resolution mass spectrum exhibits characteristic m/z at 829.9426 [$(M-CF_2O)^+$] and other fully fluorinated fragments (see Section 4.4). GPC indicates the molecular weight as ca. 1000. Higher cyclic oligomers were also successfully fluorinated. The cyclic trimer **10** was converted to its fully fluorinated derivative **14** in greater than 80% yield.

13, n = 1   All unspecified
14, n = 2   bonds to fluorine

Fluorination of low-molecular-weight linear oligomers **7** ($n = 7-12$) also proceeded cleanly. In this case Krytox® was found to be a better solvent because of solubility problems encountered with FC-75. The product **11** was obtained in very good yield, and both ^{1}H-NMR (which shows no signals) and ^{19}F-NMR are in accord with essentially complete replacement of all the hydrogen atoms by fluorine. That little fragmentation occurred was shown by GPC analysis ($M_n \approx 3200$).

A similar mixture of polymers **7**, but with higher molecular weight ($n = 20-40$), was also fluorinated to produce perfluoropolyether **15** with the almost identical NMR spectral properties found for the above cases; a representative ^{19}F-NMR spectrum of **15** is shown in Figure 4.2. GPC indicated that cleavage processes were minimal and the average molecular weight exceeds 14,000. The thermal stability of the perfluorinated product in air was compared with the parent hydrogen-containing polymer by thermal gravimetric analysis (TGA) ($10°C/min$, air). While the latter started to lose weight at $340°C$ and practically disappeared at $370°C$, the perfluorinated material degraded completely only around $450°C$.

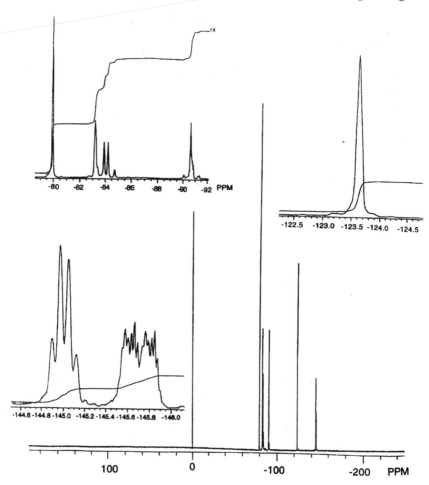

Figure 4.2. ^{19}F-NMR spectrum of perfluoropolyether (**15**) in THF-d$_8$/CFCl$_3$.

Stepwise fluorination of oligomer **7** ($n = 7$–12) in 1,1,2-trichloro-1,2,2-trifluoroethane solvent with 120 pis 25% F$_2$/N$_2$ without photoirradiation at ambient temperature to 80°C was also examined. This slower process allowed us to monitor the rate of fluorination of different kinds of protons in the polymers by ^{1}H-NMR spectra. The results indicated that the —CH$_2$OH group at the polymer chain end was the first fluorinated (the signal quickly disappeared under these conditions). Second-stage fluorination was conducted with 200 psi 25% F$_2$/N$_2$ at 120°C; after this treatment, the —CH$_2$— protons in —CF$_2$CF$_2$—CH$_2$O— were almost gone while the remaining approximately 10%

of the protons, mainly those geminal to a fluorine atom ($-CF_2-CFH-O-$ segment) were still visible. Apparently, the protons in the $-CF_2-CFH-O-$ segments are more resistant to radical fluorination than others, and this is consistent with the higher polarization of these C—H bonds.

The hydroxylated vinyl ether monomers can be further functionalized on the hydroxy group. For example, the conversion of **5** to its various derivatives **16–20** is straightforwad, and thus provides the entry to new difunctional fluorinated monomers.[23]

$$CF_2{=}CFOCF_2CF(CF_3)OCF_2CF_2-CH_2OZ: \quad Z = -C_6H_{13} \text{ (16)};$$

$$Z = -C_{18}H_{37} \text{ (17)}; -C{\equiv}CH \text{ (18)}; -COC(CH_3){=}CH_2 \text{ (19)};$$

$$Z = -CH_2-Ph-CH{=}CH_2 \text{ (20)}$$

4.3. CONCLUSIONS

The two-step process described in this chapter provides a useful new approach to the synthesis of valuable perfluorinated polyethers. Both steps, condensation polymerization and fluorination, occur in high yield under mild reaction conditions. Although most of our work has been conducted with materials derived from the vinyl ether alcohol **5**, since it is the one most readily available to us, preliminary work with **6** suggests that this approach is quite generally applicable. The linear perfluorinated polyethers derived from **5** can be considered as exactly alternating copolymers of tetrafluoroethylene oxide, hexafluoropropylene oxide, and hexafluoro-oxetane and cannot currently be made directly from these monomers. Although a few perfluorinated crown ethers are known, the tedious fluorination process required for their syntheses from the hydrocarbon analogues limits their availability.[2,4] In contrast, our process readily provides multigram quantities of the cyclic dimer **9** in over 50% yield.

The intermediate polyfluorinated species **7**, themselves, show an interesting combination of properties, including high solubility in organic solvents and the ability of the cyclics to complex anions. The strong hydrogen bonds formed by $-CFH-$ and $-CF_2H_2O-$ protons with various acceptors have been noted previously[5,15] and presumably account for the solubility and complexation behavior. Although the reasons for the unusually high yields of cyclic formed in solution oligomerization of **5** is not known, it is tempting to speculate on the existence of a template effect.

The versatility of this synthetic approach can be further extended. Many applications are being explored. First examples are the formation of the macrodiols **21** and macromonomer **22** through the condensation of **5** with fluorinated diols and brominated alcohol, respectively:

(5) $5 + HOCH_2(CF_2)_3CH_2OH \longrightarrow$

$$H[(OCH_2CF_2CF_2CF_2CH_2O)_n(CF_2CHFOCF_2\overset{\displaystyle CF_3}{\overset{\displaystyle |}{C}}FOCF_2CF_2CH_2O)_m]H$$

21

(6) $5 + BrCF_2CFBrOCF_2CF(CF_3)OCF_2CF_2CH_2OH \longrightarrow$

$$ROCF_2CF(CF_3)OCF_2CF_2CH_2O[CF_2CHFOCF_2\overset{\displaystyle CF_3}{\overset{\displaystyle |}{C}}FOCF_2CF_2CH_2O]_nH$$

$$R = BrCF_2CFBr-$$
$$\textbf{22}, R = CF_2{=}CF-$$

These structures are well defined by conducting the polymerization in the presence of appropriate mono- and difunctional reagents. They are of considerable interest for the preparation of segmented block copolymers.[24,25] For instance, the fluorinated macrodiols **21** have already been used to prepare an interesting new series of partially fluorinated segmented polyurethanes,[26] and we are investigating other novel polymers that can be prepared from these intermediates.

Typical experiments for the preparation of hydroxy-containing monomers and their homopolymerizations are described in the Appendix below.

4.4. APPENDIX

4.4.1. Preparation of 9,9-Dihydro-9-Hydroxyperfluoro-(3,6-Dioxa-5-Methyl-1-Nonene) (5)

To a dry flask with a magnetic stirring bar was charged the ester precursor (**1**) (211 g, 0.50 mol) in absolute ethanol (300 nl). Sodium borohydride (11.34 g, 0.30 mol) was added slowly from a solid addition funnel. The reaction was somewhat exothermic and the temperature of the reaction pot was prevented from exceeding 10°C by external cooling. After the addition of sodium borohydride was complete, the reaction mixture was stirred for 1 h at room temperature. The mixture was then dumped into ice water (600 ml)/6N HCl (600 ml). The bottom product layer was separated, washed with water, and distilled to give the desired product as a clear, colorless liquid, b.p. 68–70°C/25 mm. Yield: 168.7 g (85.6%). ^{1}H-NMR (CDCl$_3$): δ400 (*dt*, $J = 1.0$ Hz, 13.5 Hz, 2H), 2.12 (*s*, br, 1H);

^{19}F-NMR (CFCl$_3$ internal standard): -80.4 (s, br, 3F), -84.2 (s, br, 2F), -85.3 (m, br, 2F), -126.6 ($t, J = 14$ Hz, 2F), -145.7 ($t, J = 21.8$ Hz, 1F), $-113.4, -113.7, -113.8, -114.2$ (4s, 1F), $-121.6, -112.1, -122.2, -122.7$ (4$t, J = 5.2$ Hz, 1F), $-135.3, -135.6, -135.9, -136.2$ (4$t, J = 5.8$ Hz, 1F). Anal. Calcd. for C$_8$H$_3$F$_{13}$O$_3$: C: 24.38, H: 0.77, F: 62.67; Found: C: 23.96, H: 0.68, F: 62.78. MS (NCl): [M + F]: Calcd. 412.9859; Found: 412.9843.

4.4.1.1. Cyclic Dimer 9

A mixture of oil-free potassium hydride (117 mg) and glyme (150 ml) was cooled to 0°C and treated dropwise with **5** (23.4 g, 59.4 mmol). A mild exotherm warmed the reaction mixture to ca. 8°C during the addition, and stirring was continued for 4 h at 25°C. The reaction mixture was cooled, treated with ca. 0.2 ml isopropyl alcohol, diluted with ether (200 ml), and washed several times with water. The organic layer was dried and evaporated to give 23.5 g of colorless oil. Kugelrohr distillation (0.2 mm) afforded 12.8 g of colorless oil (b.p. 55–75°C) identified as cyclic dimers **9**, 3.6 g (b.p. 130–165°C), identified as cyclic trimers **10** contaminated with a minor amount of dimers **9**, and 5.6 g of higher-boiling residue ($M_n = 4200$ by GPC analysis). ^{1}H-NMR (THF-d$_8$, 300 MHz) of cyclic dimers **9**: δ6.74–6.70 [five peaks (dd)] (CHF), 4.74 to 4.45 (CH_2F$_2$); ^{19}F-NMR (THF-d$_8$, 470.5 MHz) of **9**: -79.50 to -79.84 (m, CF$_3$), -82.4 to -85.7 (m, overlapping CF$_2$), -86.31 to -86.98 (six lower-field signals for OCF$_2$ AB patterns, $J = 144.3-146.3$ Hz), -90.29 to -92.62 (eight higher-field signals for OCF$_2$ AB patterns, $J = 143.7-146.4$ Hz), -122.5 to -124.4 (m, CF$_2$CH$_2$), -143.3 to -145.42 (m, CF), -146.0 to -146.44 (m, CHF), GC/MS showed for cyclic dimers **9** a parent ion with $m/z = 787.974152$ (calcd. For C$_{16}$H$_6$F$_{26}$O$_6$ = 787.974922); for the cyclic trimer **10** a parent ion with $m/z = 1181.952957$ (calcd. For C$_{24}$H$_9$F$_{39}$O$_9$ = 1181.962383); for the cyclic tetramer a parent ion with nominal $m/z = 1579$.

4.4.1.2. Polymerization of 5 with CsF

Cesium carbonate (10 mg) and **5** (1.0 g) were placed in a vial, sealed, and heated in an oil bath maintained at 120°C. Within 1 h, the mixture had thickened considerably. The reaction mixture was maintained at 120°C for 88 h. ^{1}H-NMR (THF-d$_8$): δ6.65 ($d, J = 52$ Hz, CHF), 4.58 ($t, J = 13$ Hz, internal CF$_2$CH$_2$O), 3.90 ($t, J = 13$ Hz, terminal CH$_2$O); integrated area internal CH$_2$O/terminal CH$_2$O was ca. 55/1. Size exclusion analysis showed the major peak (90%) with $M_n = 27,200$ and $M_w = 57,900$ using polystyrene standards, ^{19}F-NMR (THF-d$_8$) featured internal CF$_2$CH$_2$O (-123.6) and terminal CF$_2$CH$_2$O groups (-125.6) in relative areas of 126/1. No signals for residual trifluorovinyl groups were observed. NMR and size-exclusion analyses were in reasonably good agreement and were consistent with the desired linear condensation polymer [$CF_2CHFOCF_2CF(CF_3)OCF_2CF_2$—$CH_2O]_n$. TGA (20°C/min) of a similarly prepared sample showed onset of thermal decomposition at ca. 300°C (air) and 400°C (N_2). DSC exhibited T_g at −60°C.

4.4.1.3. Polymerization of 5 with R_4NCl

Tetramethylammonium chloride (2 mg) and **5** (1.0 g) were placed in a vial, sealed, and heated in an oil bath at 107°C for 65 h. ^{1}H-NMR analysis of the colorless, viscous grease showed the ratio of signals at 4.6 and 3.90 ppm as ca. 60/1. The small amount of cyclic dimer formed (GC anlaysis) was removed by Kugelrohr distillation (up to 100°C/0.05 mm). ^{19}F-NMR featured the internal/ terminal CF_2CH_2O group ratio as ca. 83/1. Size-exclusion chromatography showed the major peak with $M_n =26,700$ and $M_w =52,800$, consistent with condensation polymer **7**.

Tetrabutylammonium chloride (2 mg) and 1.0 g of **5** were placed in a vial, sealed, and heated to 107°C for 65 h. ^{1}H-NMR analysis showed the ratio of in-chain CH_2O/terminal CH_2OH groups to be ca. 60/1. Size-exclusion analysis showed $M_w = 46,700$; $M_n = 24,000$. $T_g = $ ca. −60°C.

CAUTION. Fluorine is a strong oxidizer and a very corrosive material. An appropriate vacuum line made from copper or Monel in a well-ventilated area should be constructed for working with this element. The reactions themselves were carried out in Teflon® vessels. If elementary precautions are taken, work with fluorine is relatively simple.

4.4.2. General Procedure for Substituting Hydrogens with Fluorine

Mixtures of 25–30% fluorine diluted with nitrogen were used in this work. The gas mixtures were prepared in a secondary container. The appropriate polyether was dissolved either in perfluoro-2-butyl-THF (FC-75) or in Krytox® GPL 100 (a fluorinated oil), which also contained about 5 g of pulverized NaF to absorb the relased HF. The reaction mixture was cooled to −10°C, stirred with the aid of a vibromixer and irradiated with a 450-W medium-pressure mercury lamp. A stream of fluorine in nitrogen (ca. 140 ml/min) was passed into the mixture such that the temperature did not rise above +10°C. The reaction was stopped after 200 mmol of fluorine had been passed through. The mixture was poured into water and the organic layer was washed with sodium bicarbonate solution. The water layer was extracted twice with $CFCl_3$. The combined fluorocarbon fractions were washed with water, dried with $MgSO_4$, and filtered, and the solvent was removed under reduced pressure.

4.4.2.1. Fluorination of Cyclic Dimer **9**

A cold solution ($-10°C$) of 3.66 g of the cyclic dimer **9** in perfluoro-2-butyl-THF (FC-75) was prepared. NaF (5 g) was added and the reaction was performed as described above. The main product was distilled at $50°C/0.25$ mm Hg, using a Kugelrohr and was identified as perfluorinated cyclic dimer **13**, 3.4 g (82% yield). No protons were detectable in the ^{1}H-NMR. ^{19}F-NMR (THF-d_8/CFCl$_3$, 470.5 MHz): -80.94 (apparent $q, J = 9.8$ Hz, CF$_3$), -81.08 (apparent $q, J = 10.2$ Hz, CF$_3$), -81.1 to -85.1 and -89.7 to -90.2 (overlapping OCF$_2$ AB patterns), -129.5 (s) and -129.6 ($d, J = 6.7$ Hz) of equal intensity ($—$CF$_2—$), -145.6 (triplet of doublets, $J = 21.7$, 12.8 Hz, CF), -143.3 (triplet of doublets, $J = 21.7$, 12.8 Hz). MS m/z 829.9426 [(M$—$CF$_2$O)$^+$], 691.9580 {[M$—$CF$_2$O$—$(CF$_2$)$_2$$—$2F]$^+$}, 428.9638 [(M/2$—$F)$^+$], 146.9876 {[CF$_2$CF(CF$_3$)O$—$F]$^+$}.

4.4.2.2. Fluorination of Oligomer Mixture **7** ($n = 7–12$)

The oligometric mixture (2.6 g) was reacted as above using Krytox® GPL 100 as solvent. Upon completion of the reaction, the Krytox® was removed by distillation and the reaction mixture was left at $125°C$ under 2 mm Hg for 4 h. The residue (2.05 g, 78% yield) showed no detectable protons in the ^{1}H-NMR. The ^{19}F-NMR is characteristic of the repeating unit: -82 to -91.5 (m, 13F), -131.5 (m, 2F), -147.3 ppm (m, 1F).

4.4.2.3. Fluorination of Polymer Mixture **7** ($n = 20–40$)

The oligomers (1.05 g) were reacted as before using Krytox® GPL 100 as solvent. Upon completion of the reaction the solvent was distilled and the reaction mixture was left at $150°C$ under 2 mm Hg for 4 h. The residue **15** (0.75 g, 70% yield) showed no protons in the ^{1}H-NMR. The ^{19}F-NMR is characteristic of the repeating unit: -82 to -91.5 (m, 13F), -129.8, -130.5 (m, 2F), -146, -146.7 ppm (m, 1F).

4.5. REFERENCES

1. E. P. Moore, A. S. Milian, and H. S. Eleuterio, U.S. Patent 3,250,808 (1966); W. T. Miller, U.S. Patent 3,342,218 (1966).
2. D. F. Persico, G. E. Gergardt, and R. J. Lagow, *J. Am. Chem. Soc.* **107**, 1197 (1985).
3. K. T. Dishart, U.S. Patent 4,721,578 (1988); D. J. Kalota, J. S. McConaghy, Jr., D. O. Fisher, and R. E. Zielinksi, U.S. Patent 4,871,109 (1989); D. Sianesi and R. Fontanelli, U.S. Patent 3,665,041 (1972); D. Sianesi, A. Pasetti, and G. Belardinelli, U.S. Patent 3,715,378 (1973); D. Sianesi, A. Pasetti, and C. Corti, U.S. Patent 3,442,942 (1969); D. Sianesi and R. Fontanelli, British Patent

1,226,566 (1971); D. Sianesi, G. Bernardi, and G. Moggi, French Patent 1,531,902 (1968); G. Siegemund, W. Schwertfeger, A. E. Feiring, B. Smart, F. Behr, H. Vogel, and B. McKusick, in *Ullmann's Encyclopedia of Industrial Chemistry, Vol. A11* (1988), p. 366.

4. R. J. Lagow, G. E. Gerhardt, U.S. Patent 4,523,039 (1985); D. F. Persico and R. J. Lagow, *Macromolecules 18*, 1383 (1985); G. E. Gergardt and R. J. Lagow, *J. Org. Chem. 43*, 405 (1978); G. E. Gergardt and R. J. Lagow, *J. Chem. Soc. Chem. Commun. 259* (1977); G. E. Gergardt and R. J. Iagow, *J. Chem. Soc. Perkin Trans. 1*, 1321 (1981); R. J. Iagow, T. R. Bierschenk, T. J. Juhlke, and K. Hajimu, in *Synthetic Fluorine Chemistry* (G. A. Olah, R. D. Chambers, and G. K. S. Preakash, eds.), John Wiley and Sons, New York (1992).

5. W. B. Farnham, D. C. Roe, D. A. Dixon, J. C. Calabrese, and R. L. Harlow, *J. Am. Chem. Soc. 112*, 7707 (1990); W. B. Farnham, in *Synthetic Fluorine Chemistry* (G. A. Olah, R. D. Chambers, and G. K. S. Prakash, eds.), John Wiley and Sons, New York (1992).

6. G. K. Kostov and A. N. Atanassov, *Eur. Polym. J. 27*, 1331 (1991).

7. S. J. Sondheimer, N. J. Bunce, and C. A. Fyfe, *J. Macromol. Sci.—Rev. Macromol. Chem. Phys. C26*, 353 (1986).

8. W. G. Grot, C. J. Molnar, and P. R. Resnick, U.S. Patent 4,544,458 (1985).

9. H. Miyake, Y. Sugaya, and M. Yamabe, *Reports Res. Lab. Asahi Glass Co., Ltd. 37*, 241 (1987).

10. F. J. Waller and R. W. Van Scoyoc, *Chemtech 1987*, 438.

11. G. A. Olah, P. S. Iyer, and G. K. S. Prakash, *Synthesis 1986*, 513.

12. D. C. England, U.S. Patent 4,138,426 (1979).

13. M. Yamabe, S. Kumai, and S. Munekata, U.S. Patent 4,275,226 (1981), to Asahi Glass Co., Japan;

14. S. Kumai, S. Munekata, and M. Yamabe, *Reports Res. Lab. Asahi Glass Co., Ltd. 31*, 91 (1981).

15. A. E. Feiring and E. R. Wonchoba, *J. Org. Chem. 57*, 7014 (1992).

16. M.-H. Hung, U.S. Patent 5,093,446 (1992); W. B. Farnham, and M. H. Hung, U.S. Patent 5,134,211 (1992).

17. F. W. Billmeyer, Jr. in *Textbook of Polymer Science*, John Wiley and Sons, New York (1984), Ch. 2.

18. J. A. Semlyen (ed), *Cyclic Polymers*, Elsevier, New York (1986).

19. H.-N. Huang, D. F. Persico, and R. J. Lagow, *J. Org. Chem. 53*, 78 (1988).

20. J. L. Adcock and M. L. Cherry, *Ind. Eng. Chem. Res. 26*, 208 (1987).

21. G. E. Gerhardt, E. T. Dumitru, and R. T. Lagow, *J. Polym. Sci. Polym. Chem. Ed. 18*, 157 (1979).

22. A. C. Sievert, W. R. Tong, and M. J. Nappa, *J. Fluorine Chem. 53*, 397 (1991).

23. M.-H. Hung, U.S. Patent 5,237,026 (1993).

24. B. Boutevin and J. J. Robin, *Adv. Polym. Sci. 102*, 105 (1992).

25. G. M. Cohen, W. B. Farnham, and A. E. Feiring, U.S. Patent 5,185,421 (1993).

26. S. Yang, H. X. Xiao, D. P. Higley, J. Kresta, K. C. Frisch, W. B. Farnham, and M.-H. Hung, *J. Macromol. Sci.—Pure Appl. Chem. A30*, 241 (1992).

5

Use of Original Fluorinated Telomers in the Synthesis of Hybrid Silicones

BRUNO AMEDURI, BERNARD BOUTEVIN, GERARDO CAPORICCIO, FRANCINE GUIDA-PIETRASANTA, ABDELLATIF MANSERI, and AMÉDÉE RATSIMIHETY

5.1. INTRODUCTION

Fluorosilicone polymers have become of great interest for specific applications owing to their outstanding performances.[1,2] They exhibit not only good thermal stability and a high degree of chemical inertness, but also remarkable mechanical and rheological properties. They can be used as lubricants or heat-carrying fluids, as rubbers and resins, and as elastomers in O-rings and gaskets. When dissolved in solvents, fluorosiloxanes can be utilized as stain-resistant coatings on textile fibers, both natural and synthetic. In particular, their surfactant properties (mainly attributable to their fluorine content) allow low-molecular-weight fluorosiloxanes to be used as defoaming agents and as surfactants.

Actually, both Si—O—Si and C—F linkages are efficient for providing rather good thermostability and also confer excellent chemical inertness and low surface energy. On the other hand, while the —$Si(CH_3)_2O$— group offers good properties at low temperatures it usually leads to elastomers that exhibit poor resistance to

BRUNO AMEDURI and BERNARD BOUTEVIN · Ecole Nationale Supérieure de Chimie de Montpellier, F-34296 Montpellier Cédex 5, France · GERARDO CAPORICCIO · Via Filiberto 13, Milan 20149, Italy · FRANCINE GUIDA-PIETRASANTA, ABDELLATIF MANSERI, and AMÉDÉE RATSIMIHETY · Ecole Nationale Supérieure de Chimie de Montpellier, F-34296 Montpellier Cédex 5, France

Fluoropolymers 1: Synthesis, edited by Hougham *et al.*, Plenum Press, New York, 1999.

swelling. As a matter of fact, fluorinated elastomers[3-7] are known to preserve good properties even at high temperatures and to exhibit a high resistance to solvent, oil, UV, and aging, but they have poor properties at low temperatures.

For these reasons, the fluorosilicone elastomers appear to be very promising products as that proposed by the Dow-Corning Company, which commercializes poly(trifluoropropyl methyl siloxane) (PTFPMS) containing —$(CH_3)Si(C_2H_4CF_3)O$— base units. Such a polymer is presently the best compromise for elastomers used in connectics. However, such a product may lead to reversion at high temperatures by undergoing a depolymerization that produces low-molecular-weight cyclic oligomers.

We investigated the synthesis of hybrid fluorinated silicones with high fluorine content, building up fluorinated blocks and grafts via a controlled structure shown in Scheme 1, (where R_F represents a fluorinated group). Actually, since polydimethyl siloxanes still have good properties in the -123 to $+320°C$ range, and Viton elastomers preserve good properties between -30 and $+420°C$, our objective was to obtain a hybrid fluorosilicone that would be stable in the -100 to $+400°C$ range.

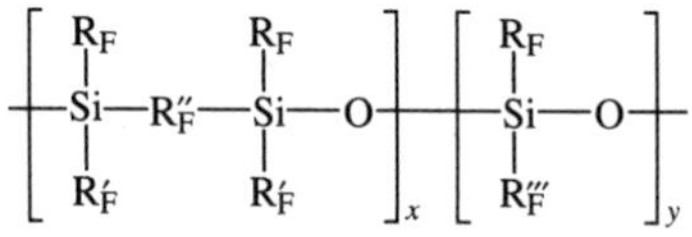

$$\left[\begin{array}{ccc} R_F & & R_F \\ | & & | \\ -Si-R''_F-Si-O- \\ | & & | \\ R_F & & R_F \end{array} \right]_x \left[\begin{array}{c} R_F \\ | \\ -Si-O- \\ | \\ R'''_F \end{array} \right]_y$$

Scheme 1

5.2. RESULTS AND DISCUSSION

Two main routes for the insertion of fluorinated alkyl groups in the silicones are suggested: either by organometallic ① or by hydrosilylation ② methods as shown in Scheme 2. Consequently, research can be pursued in two directions: (a) synthesis of fluorohalides and alkenes, and (b) introduction in siliconated groups and polymerization.

5.2.1. Synthesis of Fluorinated Precursors

Perfluoroalkyl iodides $CF_3(CF_2)_xI$ are commercially available reactants useful for interesting chemistry, offering various functional products such as $CF_3(CF_2)_xC_2H_4I$, $CF_3(CF_2)_xC_2H_4$—G, where G represents hydroxyl, amine, thio, thiocyanate, isocyanate, nitrile, carboxy, or $CF_3(CF_2)_xCH=CH_2$, $CF_3(CF_2)_xCH_2CH=CH_2$.

However, crystallization occurs when $x > 7$ and such behavior is not required in the elastomers. Consequently, it is necessary to synthesize co-oligomers from various fluorinated monomers such as vinylidene fluoride (VDF), trifluoroethylene

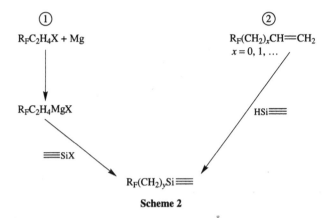

Scheme 2

(VF_3), chlorotrifluoroethylene (CTFE), and hexafluoropropene (HFP), in order to avoid crystallization as described below.

One of the most interesting strategies is telomerization. Such a reaction, introduced for the first time by Hanford in 1942, unlike polymerization, usually leads to low-molecular-weight polymers, called telomers, or even to monoadducts with well-defined end groups. Such products **A** are obtained from the reaction between a telogen or a transfer agent (X-Y), and one or more (n) molecules of a polymerizable compound M (called a taxogen or a monomer) having ethylenic unsaturation under radical polymerization conditions, as follows:

$$X\text{-}Y + n M \longrightarrow \underset{\mathbf{A}}{X(M)_n Y}$$

Telogen X-Y can be easily cleavable by free radicals (formed according to the conditions of initiation) leading to an X radical that will be able to react further with the monomer. After the propagation of the monomer, the final step consists of the transfer of the telogen to the growing telomeric chain. Telomers **A** are intermediate products between organic compounds (e.g., $n = 1$) and macromolecular species ($n = 100$). Hence, in certain cases, end groups exhibit a chemical importance that can provide useful opportunities for further functionalizations.

The scope of telomerization was outlined by Starks in 1974,[8] and further developed by Gordon and Loftus.[9] We recently reviewed such a reaction[10,11] in which mechanisms and kinetics of radical and redox telomerizations were described.

5.2.1.1. Monofunctional Telomers

(*a*) *Telomerization of Fluorinated Monomers with* CF_3CFClI. The addition of [IF] (formed *in situ* from I_2 and IF_5) to CTFE leads to CF_3CFClI (90%)

and $ClCF_2CF_2I$ (10%) in 75% yield.[12,13] CF_3CFClI has been shown to behave as an efficient telogen in the telomerization of CTFE, in contrast to $CF_3(CF_2)_xI,$[11] as well as in that of HFP.[13]

(b) *Synthesis of Viton Cotelomers.* Viton-type cotelomers were synthesized in two ways. $C_4F_9(HFP)_x(VDF)_yI$ was produced in 86% overall yield by stepwise cotelomerization of HFP and VDF with C_4F_9I, unlike the results starting from iC_3F_7I.[14,15] $CF_3(CF_2)_x(VDF)_z(HFP)_tI$ was also produced in a two-step process leading to *ca.* 83–85% overall yield.[15,16] A wide variety of different cotelomers containing VDF, VF_3, CTFE, and HFP base units have been synthesized through such a succession of reactions, as shown in Figure 5.1.

Interestingly, the (co)telomers used behave as further potential telogens for subsequent telomerizations, and all of them can be end-capped with ethylene (E).

These ω-iodinated cotelomers were functionalized into vinyl or allyl type olefins by two different two-step processes as follows[17,18]:

$$R_FI \xrightarrow[\text{CuI}]{CH_2=CH_2} R_FC_2H_4I \xrightarrow{KOH} R_FCH=CH_2$$

$$R_FI + H_2C=CHCH_2OAc \xrightarrow{\text{Peroxide}} R_FCH_2CHICH_2OAc \xrightarrow{Zn} R_FCH_2CH=CH_2$$

The yields obtained were higher for allyl-type alkenes, and the nature of R_F (e.g., the presence of a CF_3 side chain) does not affect the yield.

5.2.1.2. Telechelic Oligomers

The strategy utilized for the synthesis of monofunctional telomers was also successfully adapted to telechelic oligomers. α,ω-Diiodoperfluoroalkanes, which are generally produced by telomerization of tetrafluoroethylene with iodine (see Tortelli and Tortelli[19] and the references therein) are efficient telogens in the thermal telomerization of vinylidene fluoride[20] leading to novel α,ω-diiodofluoroalkanes with various $\overline{DP}_n$ depending on the chain length of the telogen:

$$I(C_2F_4)_nI + H_2C=CF_2 \xrightarrow{\Delta} I(VDF)_x(C_2F_4)_n(VDF)_yI$$

$$n = 1, 2, 3 \qquad\qquad x+y = 1-10$$

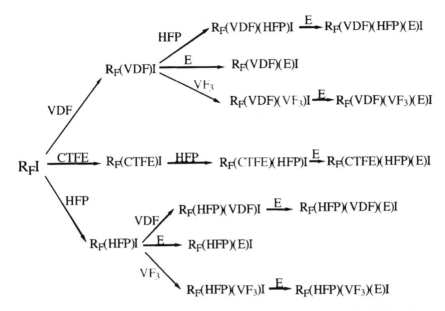

Figure 5.1. Basic strategy of well-architectured cotelomers of vinylidene fluoride (VDF), trifluoroethylene (VF₃), chlorotrifluoroethylene (CTFE), hexafluoropropene (HFP), and ethylene (E).

Apart from IC_2F_4I, which undergoes a β-scission, they also behave nicely with regard to HFP[19]:

$$I(C_2F_4)_pI + C_3F_6 \xrightarrow{\Delta} \begin{cases} I(C_2F_4)_pC_3F_6I \\ I(HFP)(C_2F_4)_p(HFP)I \\ I(HFP)(C_2F_4)_p(C_3F_6)_2I \end{cases}$$

$$p = 2, 3$$

New α,β-diiodofluoroalkanes containing Viton structure were prepared by stepwise telomerization of VDF and HFP (or HPF and VDF) with IR_FI as follows[20,21]:

$$I(C_2F_4)_n(VDF)I + C_3F_6 \longrightarrow I(C_2F_4)_n(VDF)(C_3F_6)I + I(C_3F_6)(C_2F_4)_n(VDF)(C_3F_6)I$$

$$I(VDF)(C_2F_4)_n(VDF)I + C_3F_6 \xrightarrow{\Delta} I(C_3F_6)(VDF)(C_2F_4)_n(VDF)(C_3F_6)I$$

$$I(HFP)(C_2F_4)_p(HFP)I + H_2C=CF_2 \xrightarrow{\Delta} I(VDF)_a(HFP)(C_2F_4)_p(HFP)(VDF)_bI$$

$$p = 2, 3 \qquad a + b = 3{-}20$$

As above, these α,ω-diiodinated cotelomers were functionalized into vinyl or allyl types of nonconjugated dienes as follows[22]:

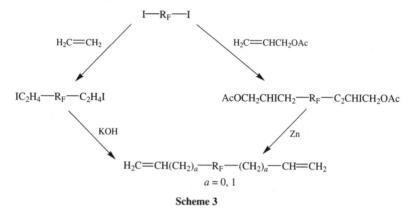

Scheme 3

Diallyls were obtained in higher yields (70–80%) than divinyls (40–55%).

5.2.2. Introduction of Fluorinated Groups in Silanes and Silicones

There are two possible ways to introduce fluorinated groups into silanes and silicones: through the use of fluorinated organometallic reactants (mainly organomagnesians and organolithians) or by hydrosilylation of fluorinated alkenes.

5.2.1.1. Organometallic Derivatives

(*a*) *Organomagnesians.* Several fluorinated alkyl halides were used for the synthesis of these organometallic compounds as follows:

$$R_F X + Mg \longrightarrow R_F Mg\,X + R_F{-}R_F$$
$$(1)$$

where $R_F X$ represents $CF_3C_2H_4Cl$, $C_6F_{13}C_2H_4Y$ (with $Y = Cl, Br, I$), $iC_3F_7C_2H_4I$, and $C_5F_{11}CF(CF_3)C_2H_4I$.

The reactivity series of R_F depends on the nature of X and is $I > Br > Cl$. In addition, the longer the R_F group and the higher the steric hindrance, the lower the yield of **1**.[23] Further, the more hindered R_F, the more dimer produced. In order to prepare tetrafluoroalkyl silanes, such fluorinated magnesians must react with alkyl silane halides as follows:

$$R_F MgX + R_x SiX_{4-x} \longrightarrow (R_F)_y R_x SiX_{4-x-y}$$

The reactivity of these fluorinated magnesian reactants depends on: (a) the chain length and the steric hindrance of the R_F group; (b) the nature of the silane:

$$HSiCl_3 > SiCl_4 > \text{\textbackslash}SiCl_2 > \equiv SiCl;$$ and (c) the nature of the halogen:

$SiF > SiCl > SiOR$. However, it is not possible to synthesize tetrafluoroalkyl silanes as the maximum number of substitutions when $X = Cl$ or F is three.

Moreover, by-products coming from Wurtz reactions have been observed in nonnegligible amounts. Thus, a better alternative involves the use of organolithian intermediates.

(*b*) *Lithian Reactants.* In the same way as organomagnesians, organolithian compounds were produced from fluoroalkyl halides as follows:

$$R_FC_2H_4X + 2Li \longrightarrow R_FC_2H_4Li + LiX + dimer$$

and again, the longer the chain length of R_F and the higher the steric hindrance, the lower the yield of the reaction.

Such fluorinated organolithians have also been shown to be precursors of tetraalkyl silanes:

$$R_FC_2H_4Li + R_1R_2R_3SiX \longrightarrow R_1R_2R_3SiC_2H_4R_F$$

However, the reactivity of such compounds is still low; e.g., when $R_1 = R_2 = R_3 = CH_3$ and $X = Cl$, the silanes produced from $CF_3C_2H_4Li$ and $C_6F_{13}C_2H_4Li$ were obtained in 40 and 20% yields, respectively. As a result of this poor reactivity, this method was abandoned and efforts now center around hydrosilylation reactions involving fluorinated olefins.

5.2.1.2. Hydrosilylation

Various hydrosilylation reactions of hydrogenosilanes were carried out in the presence of Speirs' catalyst for the preparation of fluorinated silanes:

$$R_F(CH_2)_xCH = CH_2 + HSiClR_1R_2 \xrightarrow[100°C/18h]{H_2PtCl_6} R_F(CH_2)_2SiClR_1R_2 \qquad x = 0, 1$$

The yield of the silane produced depends on: (a) the nature of the olefin ($R_FCH_2CH{=}CH_2 > R_FCH = CH_2$); (b) the nature of the fluorinated group (e.g., linear olefins are more reactive than branched ones); (c) the nature of the silane ($HSiCl_3 \sim HSiCl_2R_1 > HSiR_1R_2Cl$).[24]

Nevertheless it was possible to synthesize tetrafluoroalkyl silanes by combining the use of organometallic reactants and the hydrosilylation reaction. A few examples follow:

1. When $R_1 = R_2 = R_3 = R_4 = C_6F_{13}C_2H_4$,

$$C_6F_{13}C_2H_4MgI + HSiCl_3 \longrightarrow HSi(C_2H_4C_6F_{13})_3 \xrightarrow[\text{H}_2\text{PtCl}_6]{\text{H}_2\text{C}=\text{CHC}_6\text{F}_{13}} \underset{(2)}{Si(C_2H_4C_6F_{13})_4}$$

2. For $R_1 = R_2 \neq R_3 = R_4$,

$$C_6F_{13}C_2H_4MgI + CF_3C_2H_4SiF_3 \longrightarrow CF_3C_2H_4SiF(C_2H_4C_6F_{13})_2$$
$$\xrightarrow{CF_3C_2H_4Li} (CF_3C_2H_4)_2Si(C_2H_4C_6F_{13})_2$$

Thus, a large variety of tetrafluoroalkyl silanes were prepared with a wide range of thermal properties. The lowest T_g was obtained for **2** ($T_g = -75°C$) or for $(C_6F_{13}C_2H_4)Si$ $(C_2H_4CF_3)[C_2H_4CF(CF_3)_2]$ $T_g = -67°C$ or for **3** ($T_g = -66°C$), whereas for $Si(C_2H_4CF_3)_4$, $T_m = 131°C$ but no T_g was observed.

Similarly, fluorinated nonconjugated dienes could undergo hydrosilylation with hydrogenochloromethyl silanes:

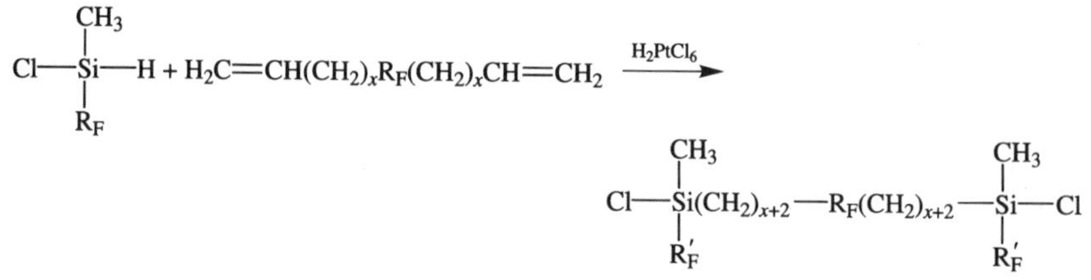

$$\underset{\underset{R_F}{|}}{\overset{\overset{CH_3}{|}}{Cl-Si-H}} + H_2C{=}CH(CH_2)_xR_F(CH_2)_xCH{=}CH_2 \xrightarrow{H_2PtCl_6}$$

$$Cl{-}\underset{\underset{R_F'}{|}}{\overset{\overset{CH_3}{|}}{Si}}(CH_2)_{x+2}{-}R_F(CH_2)_{x+2}{-}\underset{\underset{R_F'}{|}}{\overset{\overset{CH_3}{|}}{Si}}{-}Cl$$

Scheme 4

Indeed, allyl-type dienes were more efficient than divinyl olefins, even for the branched R_F group. Various heptafluoroalkyl disilanes were thus obtained with T_g varying from -42 to $-59°C$.[25]

5.2.1.3. Synthesis of Fluorosilicones

It is well known that silicones can be prepared from dihalogenosilanes that undergo hydrolysis to generate disilanol. These latter intermediates produce cyclic or linear dialkylsiloxanes. The cyclic dialkylsiloxanes can be separated and ring-opening polymerized whereas the linear ones can be condensed to yield silicones. However, it is still difficult to purify cyclic and linear precursors. Thus, in order to avoid cyclization, our objective concerned the polycondensation of fluorohybrid silanes.

(a) *Fluorosilicones without Fluorocotelomers*. Two homopolymers **7a** and **7b**, in Figure 5.2, having different molecular weights, were prepared by copolycondensation of bis-chlorosilane (**4**) (provided by Dow-Corning) and the monomer bis-hydroxysilane (**5**) in the presence of a chain-stopper $(C_6H_5)Si(CH_3)_2Cl$ (**6**).[26]

The monomer silane diol (**5**) was prepared by hydrolysis of the bis-chlorosilane (**4**) in the presence of sodium hydrogenocarbonate $NaHCO_3$ at reflux of diethyl ether for 24 h. The hydrolysis was quantitative. The silane diol (**5**) was characterized by IR, where a narrow band at 3690 cm^{-1} and a wide band between 3650 and 3050 cm^{-1} were observed, respectively, for free $\upsilon_{\equiv SiOH}$ and bonded $\upsilon_{\equiv SiOH}$. Its ^1H- and ^{19}F-NMR spectra exhibited the expected signals and its ^{29}Si-NMR spectrum showed a singlet at +16.4 ppm characteristic for a $\delta_{\equiv SiOH}$ and no trace of the starting chlorosilane at +31 ppm.[26]

A small amount (5%) of oligomeric homopolymer was observed using size-exclusion chromatography (SEC) of silane diol **5**, showing that some polycondensation had already occurred. The copolymerization of **4** and **5** was performed in toluene at 90°C for 19 h. Homopolymers **7a** and **7b**, obtained as oils, were characterized by IR, where no more $\upsilon_{\equiv SiOH}$ band was present, and in ^1H-, ^{19}F-, and ^{29}Si-NMR. Their ^1H-NMR spectra allow us to calculate the value of n by comparing the integration of the aromatic protons to that of the methylene protons linked to the silicon atoms, and thus to determine the average molecular weight $\bar{M}_n$. It is possible to use this NMR method as the values of n that were found, 16.5 and 25.5 for **7a** and **7b** respectively, are not too high. Thus the integrations of the aromatic protons are still accurate. However, for higher values of n, these

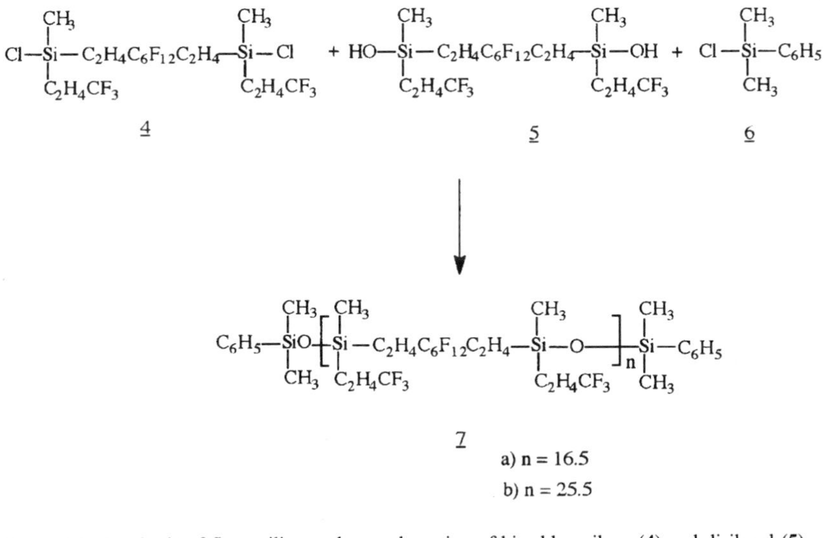

a) n = 16.5

b) n = 25.5

Figure 5.2. Synthesis of fluorosilicones by condensation of bis-chlorosilane (**4**) and disilanol (**5**).

integrations will be more difficult to measure and the values found will become less reliable. $\bar{M}_n$ was also evaluated by SEC using polystyrene standards, and the values found are in good agreement with those obtained from NMR. The results are given in Table 5.1. The ^{29}Si-NMR spectra of the homopolymers exhibited a signal at $+8.90$ ppm for $\delta_{\equiv SiOSi\equiv}$ but not for $\delta_{\equiv SiOH}$ at $+16.4$ ppm.

The homopolymers were also studied by differential scanning calorimetry (DSC), where a glass transition temperature (T_g) was measured, and by thermogravimetric analysis (TGA). The results are summarized in Table 5.1, together with those of a similar hybrid homopolymer prepared by Riley et al.,[1] of the dimethyl disubstituted homopolymer previously synthesized,[27] and of the classic commercially available polytrifluoropropylmethylsiloxane (PTFPMS). It is interesting to note that homopolymers having the same lateral groups have a nearly identical T_g, whatever the length of the linear internal fluorinated chain and that the hybrid homopolymers are much more stable at high temperatures than PTFPMS (cf. the values for 50% weight loss, $T_{-50\%}$). It is also worth noting the influence of the nature of the side chains; e.g., when a methyl group is replaced

Table 5.1. ^{1}H-NMR, SEC, DSC and TGA Data for Homopolymers

Homopolymer		^{1}H-NMR		SEC	DSC 10°C/min	TGA (in N$_2$) 5°C/min
		n	M_n	M_n	T_g, °C	$T_{50\%}$
CH$_3$ | CH$_3$ | —(SiC$_2$H$_4$C$_6$F$_{12}$C$_2$H$_4$SiO)$_n$— **7a** C$_2$H$_4$CF$_3$ C$_2$H$_4$CF$_3$		16.5	11,000	12,500	−29	490
CH$_3$ | CH$_3$ | —(SiC$_2$H$_4$C$_6$F$_{12}$C$_2$H$_4$SiO)$_n$— **7b** C$_2$H$_4$CF$_3$ C$_2$H$_4$CF$_3$		25.5	17,500	16,800	−28	490
CH$_3$ | CH$_3$ **1** —(SiC$_2$H$_4$C$_2$F$_4$C$_2$H$_4$SiO)$_n$— C$_2$H$_4$CF$_3$ C$_2$H$_4$CF$_3$					−27	493
CH$_3$ | CH$_3$ **27** —(SiC$_2$H$_4$C$_6$F$_{12}$C$_2$H$_4$SiO)$_n$— C$_2$H$_4$CF$_3$ C$_2$H$_4$CF$_3$					−53	470
PTFPMS CH$_3$ | —(SiO)$_n$— C$_2$H$_4$CF$_3$					−67	245

Table 5.2. Molecular Weights and Thermal Properties of Fluorinated Hybrid Homopolymers

$$\text{HO} \left[-\underset{\underset{R}{|}}{\overset{\overset{CH_3}{|}}{Si}} C_2H_4(CH_2)_x - R' - (CH_2)_x C_2H_4 \underset{\underset{R}{|}}{\overset{\overset{CH_3}{|}}{Si}} O - \right]_n H$$

			SEC M_n	DSC (10°C/mn)			TGA (5°C/mn)		State
				T_g	T_m	T_c	$T_{50\%}$ (N_2)	$T_{50\%}$ (Air)	
R = CH$_3$	R' = C$_6$F$_{12}$	x = 0	10,000	−53	26	−11	470	380	Solid
		x = 1	10,000	−40	25	−27	465	330	Solid
R = CF$_3$C$_2$H$_4$	R' = C$_6$F$_{12}$	x = 0	40,000	−28			480	410	Oil
		x = 1	14,000	−18			465	360	Oil
R = C$_4$F$_9$C$_2$H$_4$	R' = C$_6$F$_{12}$	x = 0	30,000	−42			490	360	Oil
		x = 1	12,000	−29			470	310	Oil
R = CH$_3$	R' = HFP/C$_4$F$_8$/HFP	x = 1	10,000	−49			425	300	Oil
R = CF$_3$C$_2$H$_4$	R = HFP/C$_4$F$_8$/HFP	x = 1	30,000	−34			445	310	Oil
R = C$_4$F$_9$C$_2$H$_4$	R' = HFP/C$_4$F$_8$/HFP	x = 1	50,000	−38			450	320	Oil
R = C$_4$F$_9$C$_2$H$_4$	R' = C$_2$F$_4$/VF$_2$/HFP	x = 1	9,000	−47			420	315	Oil

by a trifluoropropyl group the T_g increases from -53 to $-29°C$, this loss in T_g being compensated for by an almost equivalent gain of $20°C$ in thermal stability at high temperatures. The influence of the side chains is also seen in the fact that the homopolymers **7a** and **7b** exhibit only a T_g and neither a crystallization (T_c) nor a melting (T_m) temperature as was observed for the dimethyl disubstituted homopolymer.[27]

So, if these hybrid homopolymers exhibit interesting properties at high temperatures, their properties at low temperatures are poorer than those of the classical polysiloxanes, which is why we decided to study fluorinated hybrid polysiloxanes containing the telomers prepared above.

(*b*) *Fluorosilicones Containing Fluorinated (Co)Telomers.* We prepared α,ω-dichlorofluorinated disilanes **8** using the same reaction described in Section 5.2.1.2:

$$\underset{\substack{|\\ R}}{\overset{\substack{CH_3\\ |}}{Cl-Si}}-(CH_2)_{x+2}-R'_F(CH_2)_{x+2}-\underset{\substack{|\\ R}}{\overset{\substack{CH_3\\ |}}{Si}}-Cl$$

(8)

with $R = CH_3$ or $C_2H_4CF_3$ and $R'_F = C_6F_{12}$ or $HFP-C_4F_8-HFP$ or $C_2F_4-VDF-HFP$. Table 5.2 gives the molecular weights and thermal properties of the fluorosilicones we obtained. The influence of the key group in terms of these properties is as follows:

1. The longer the spacer between the fluorinated central group and the silicon atom, the higher the T_g and the less thermostable the fluorosilicon is in air, which confirms a previous investigation performed on model fluorosilicones.[27]
2. Substituting CH_3 by $CF_3C_2H_4$ as the fluorinated side group on the polymeric backbone increases the T_g but a $C_4F_9C_2H_4$ lateral group lowers the T_g of the fluorosilicone containing a $CF_3C_2H_4$ group, and slightly decreases its thermostability in air.
3. The fluorinated central group in the polymeric backbone has a drastic effect on both the T_g and T_{dec}. The introduction of CF_3 side groups coming from HFP base units lowers the T_g (e.g., from $-18°C$ for fluorosilicones where $R = CF_3C_2H_4$, $R' = C_6F_{12}$ and $x = 1$ to $-34°C$ for $R = CF_3C_2H_4$, $R' = HFP/CuF_8/HFP$ and $x = 1$), but does not contribute to a higher thermostability.

5.3. CONCLUSIONS

The wide use of fluorosilicones is the result of their combination of excellent properties: good resistance to heat, solvents, acids, alkalies, and oxidizing media. Gaskets for connectics, hydraulic fluids, and heat-carrier fluids still account for the largest number of applications. After a first generation proposed by Dow-Corning, fluorinated hybrid silicones based on well-architectured fluorooligomers built by controlled stepwise telomerization of various fluoroolefins are presented. Such cotelomers are very interesting precursors of functional or telechelic derivatives (containing hydroxy, carboxy, amine or unsaturated groups).

We also described various strategies for the synthesis of tetrafluoroalkyl silanes or fluorinated disilanes or silicones. Such a series of hybrid polymers compete with those of Dow-Corning. These fluorosilicones are prepared in three steps from original nonconjugated dienes in high yields. Their thermal characteristics depend upon the spacer between the central fluorinated group and the silicon atom, the nature of lateral groups linked to silicon atoms, and the nature of the central fluorinated group (e.g., the presence of the branched CF_3 group).

In spite of their price, fluorinated silicones seem to be high-value-added materials, which should challenge both industrial and academic researchers.

5.4. REFERENCES

1. M. O. Riley, Y. K. Kim, and O. R. Pierce, *J. Fluorine Chem. 10*, 85–110 (1977).
2. B. Boutevin and Y. Pietrasanta. *Prog. Org. Coatings 13*, 297–330 (1985).
3. S. Smith, in *Preparation, Properties and Industrial Applications of Organofluorine Compounds* (R. E. Banks, ed.), Ellis Horwood, Chichester (1982), pp. 235–295.
4. R. E. Uschold, *Polym. J. 17*, 253–263 (1985).
5. M. M. Lynn and A. T. Worm, in *Encyclopedia of Polymer Science and Engineering, Vol. 7* (R. E. Kirk and D. F. Othmer, eds.), John Wiley and Sons, New York (1987), pp. 257–269.
6. A. L. Logothetis, *Prog. Polym. Sci. 14*, 251–272 (1989).
7. D. Cook and M. Lynn, *Rapra Rev. Rep. Shrewsbury 32*, 1–27 (1995).
8. C. M. Starks, *Free Radical Telomerization, 1st Ed*, Academic Press, New York (1974).
9. R. Gordon and R. D. Loftus, in *Encylcopedia of Polymer Science and Technology, Vol. 16* (R. E. Kirk and D. F. Othmer, eds.), John Wiley and Sons, New York (1989), pp. 533–551.
10. B. Boutevin and Y. Pietrasanta, *Comprehensive Polymer Science, Vol. 3* (G. Allen, J. C. Bevington, A. L. Eastmond, and A. Russo, eds.), Pergamon Press, Oxford (1989), pp. 185–193.
11. B. Améduri and B. Boutevin, in *Encyclopedia of Advanced Materials* (D. Bloor, R. J. Brook, R. D. Flemings, and S. Mahajan, eds.), Pergamon Press, Oxford (1994), 2767–2777; in *Topics in Organofluorine Chemistry, Vol. 192* (R. D. Chambers, ed), Springer-Verlag, Heidelberg (1997), 165–233.
12. M. Hauptschein and M. Braid, *J. Am. Chem. Soc. 83*, 2383–2386 (1961).

13. M. P. Amiry, R. D. Chambers, M. P. Greenhall, B. Ameduri, B. Boutevin, G. Caporiccio, G. A. Gornowicz, and A. P. Wright, *Polym. Preprint 34*, 441–442 (1993).
14. J. Balagué, B. Ameduri, B. Boutevin, and G. Caporiccio, *J. Fluorine Chem. 74*, 49–58 (1995).
15. J. Balagué, B. Ameduri, B. Boutevin, and G. Caporiccio (submitted to *J. Fluorine Chem.*)
16. J. Balagué, B. Ameduri, B. Boutevin, and G. Caporiccio, *J. Fluorine Chem. 70*, 215–223 (1995).
17. B. Ameduri, B. Boutevin, M. Nouiri, and M. Talbi, *J. Fluorine Chem. 74*, 191–197 (1995).
18. M. Korota, J. Kvicala, B. Ameduri, M. Hajek and B. Boutevin, *J. Fluorine Chem. 64*, 259–267 (1993).
19. V. Tortelli and C. Tonnelli, *J. Fluorine Chem. 47*, 199–217 (1990).
20. A. Manseri, B. Ameduri, B. Boutevin, R. D. Chambers, G. Caporiccio, and A. P. Wright, *J. Fluorine Chem. 74*, 56–67 (1995).
21. A. Manseri, B. Ameduri, B. Boutevin, R. D. Chambers, G. Caporiccio, and A. P. Wright, *J. Fluorine Chem. 78*, 145–150 (1996).
22. A. Manseri, D. Boulahia, B. Ameduri, B. Boutevin, and G. Caporiccio, *J. Fluorine Chem. 81*, 103–113 (1997).
23. B. Boutevin, G. Caporiccio, F. Guida-Pietrasanta, and A. Ratsimihety, *J. Fluorine Chem. 60*, 211–218 (1993).
24. B. Boutevin, G. Caporiccio, F. Guida-Pietrasanta, and A. Ratsimihety, *J. Fluorine Chem. 68*, 71–79 (1994).
25. B. Boutevin, G. Caporiccio, F. Guida-Pietrasanta, and A. Ratsimihety, *J. Fluorine Chem. 75*, 75–81 (1995).
26. B. Boutevin, G. Caporiccio, F. Guida-Pietrasanta, and A. Ratsimihety, *Macromol. Chem. Phys. 199*, 61–70 (1998).
27. B. Ameduri, B. Boutevin, G. Caporiccio, F. Guida-Pietrasanta, A. Manseri, and A. Ratsimihety, *J. Polym. Sci. Part A: Polym. Chem. 34*, 3077–3090 (1996).

6

Chlorotrifluoroethylene Suspension Polymerization

MILTON H. ANDRUS, JR., ROBERT J. OLSEN,
GILBERT L. EIAN, and RICHARD C. ALLEN

6.1. INTRODUCTION

Polychlorotrifluoroethylene (PCTFE) is ordinarily prepared by emulsion polymerization. A polymer suitable for thermal processing requires coagulation, extensive washing, and postpolymerization workup. Coagulation to provide a filterable and washable solid is a slow, difficult process and removal of surfactant is an important part of it. Complete removal may be extremely difficult depending on the extent of adsorption to the polymer particles. Consequently we set out to develop a suspension polymerization process, which would be surfactant-free and afford an easily isolated product requiring a minimum of postreaction workup.

Early work at Kellog on Kel-F[TM] produced polymers with low zero strength time (ZST) values indicating low molecular weight.[1] These polymers had poor physical properties. Tseng and Young[2] in preliminary experiments in a water/alcohol medium made a suspension polymer that had ZST values approximately those of an emulsion polymer. These results led us to a study of the process variables and of different initiators in the suspension polymerization of chlorotrifluoroethylene. Polymer characterization, including inherent viscosity, GPC molecular weight, and melt flow behavior, was also studied. As a result of this study a suspension process was developed that gives high-molecular-weight product with outstanding physical properties.

MILTON H. ANDRUS, JR., ROBERT J. OLSEN, GILBERT L. EIAN, and RICHARD C. ALLEN · Speciality Materials Division, 3M Company, St. Paul, Minnesota 55144-1000.

Fluoropolymers 1: Synthesis, edited by Hougham *et al.*, Plenum Press, New York, 1999.

6.2. EXPERIMENTAL

6.2.1. Materials

All monomers, surfactants, buffers, and chain transfer agents were used as provided. Water was deionized. Purchased organic peroxides were all reagent grade. Organic peroxides were prepared from the reaction products of acyl chlorides and sodium peroxide. Activity was determined by iodometric titration.

6.2.2. Preparation of Perfluorooctanoyl Peroxide

To a 1-liter three-necked flask fitted with an alcohol thermometer, an addition funnel, and a mechanical stirrer were charged 450 g Freon 113, 6 g sodium hydroxide (0.15 mol) in 45 g of water and 8.5 g 30% hydrogen peroxide (0.075 mol). The solution was stirred and cooled in an ice–salt bath until the temperature reached $-10°C$. Perfluorooctanoyl chloride, obtained from 3 M, (64.9 g, 0.15 mol) was added quickly with rapid stirring, being careful to maintain a batch temperature below 5°C. When all the acyl chloride was added the reaction mixture was stirred for 3–5 min, then poured into a separatory funnel and the lower Freon layer drawn off and filtered through anhydrous sodium sulfate. The dried solution weighed 442 g and contained 0.22 meq peroxide/ml by idometric titration.

6.2.3. Polymerization

Polymerizations were carried out in a jacketed, 1-gal, stirred, pressure tank reactor. Typical reactions were run by adding water, alcohol, or chain transfer agent, phosphate buffer, and persulfate to the reactor. The reactor was pressurized with CTFE monomer. Sulfite solution was fed at a rate to maintain reaction. Copper and iron ions were used at times as catalysts by adding cupric sulfate or ferrous sulfate.[3] The product was filtered, washed with 90 : 10 water methanol followed with deionized water. The product was dried at 110°C.

6.2.4. Molecular Weight Measurements

Inherent viscosities were determined by dissolving PCTFE in refluxing pentafluorochlorobenzene, cooling, and subsequently measuring solution times at 27°C using a Canon-Fenske viscometer. The ability of pentafluorochlorobenzene to dissolve PCTFE while hot but maintain solubility at ambient temperature was discovered by J. Klein at 3M.[4]

Gel permeation chromatography was run by dissolving the polymer in hot 2,5-dichlorotrifluoromethylbenzene and eluting at 130°C with the same solvent as eluent. The chromatograms were run on a Waters 150C HPCL using a Jordi Associates mixed bed Stryrogel column.

6.2.5. Polymer Testing

ZST tests were run according to the procedure described in ASTM D-1430-82. The results are reported as time to break (in minutes) at 250°C.

Melt viscosities were measured using an Instron capillary melt rheometer (Model 3210) using a 0.050-in. diameter capillary ($L : D = 40 : 1$). Corrected viscosities were calculated in the conventional manner. In all cases, samples were preheated for 7 min prior to data acquisition.

Thermal stabilities were assessed by thermogravimetric analysis (TGA). Samples were held at constant temperature (290°C) for 1 h in air in a Perkin-Elmer TGA. Much of the weight loss, particularly for Kel-F 6060, is suspected to be emulsifier used during polymerization.

6.3. RESULTS AND CONCLUSIONS

6.3.1 Polymerizations

A suspension process using redox initiation in a water medium was developed. The redox system is a combination of persulfate/sulfite. Often ferrous or cupric salts were added as a catalyst for the redox reaction. Polymerizations were run in water at low temperature (20–25°C) and low pressure (65–85 psi). Monomer to monomer-plus-water weight ratios of 0.20 to 0.25 were used. Good agitation was required to keep an adequate monomer concentration in the aqueous phase. Yields of up to 100% were obtained with polymer inherent viscosities of 0.4 to 1.5 dl/g in C_6F_5Cl. Reactions were run on both a 1-gal and a 100-gal scale.

A number of different materials were used as chain transfer agents to control molecular weight. These results are shown in Table 6.1. The effect of varying concentration of *t*-butyl alcohol and reaction temperature is shown in Figure 6.1. The results are consistent with normal free radical polymerizations. Polymer output was characterized by inherent viscosity and ZST tests.

The results for suspension polymerizations as well as polymer made by emulsion polymerizations by both 3M and Daikin are shown in Table 6.2. Examination of the results confirms that PCTFE can be prepared by suspension polymerization with molecular weights and physical properties comparable to or better than PCTFE from emulsion polymerization.

Many of the heat-pressured samples prepared for the ZST test showed bubbles and discoloration. This is thought to be due to oligomers formed at the end of the polymerization when the reaction is pushed to high conversion. In

Table 6.1. Chain Transfer Agents

Example	Agent	wt% Chain transfer agent	Inherent viscosity (dl/g)
1	None	—	1.36
2	None	—	1.57
3	Acetone	1.00	1.26
4	*t*-Dodecyl Mercaptan	0.14	1.11
5	Chloroform	0.54	0.99
6	Methanol	1.00	1.10
7	Isopropanol	1.70	0.38

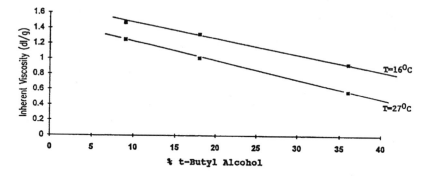

Figure 6.1. Effect of reaction variables.

Table 6.2. Comparison of Zero Strength Time and Inherent Viscosity

Example	Sample	Zero strength time (min)	Inherent viscosity (dl/g in C_6F_5Cl)
8	KEL F^{TM} 6061	337	2.43
9	KEL F^{TM} 6050 (A)	297	1.74
10	KEL F^{TM} 6050 (B)	228	1.58
11	Revised emulsion	100	1.16
12	DAIKIN^{TM} M400H	368	1.77
13	DAIKIN^{TM} M300P	276	1.62
14	ACLON^{TM} 3000	—	—
15	Suspension with 4% *t*-BuOH	5000	1.75
16	Suspension with 9% *t*-BuOH	1473	1.25
17	Suspension with 18% *t*-BuOH	2700	1.01
18	Suspension with 1.7% IPA	—	0.91
19	Suspension with 1.0% MeOH	—	0.38
20	Suspension with MeOH/H_2O wash	1696	1.23
21	Control with HNO$_3$ wash	159	1.23
22	Control with ethylendediamine wash	183	1.05

addition these polymers showed steep viscosity–shear rate responses (Figure 6.3). We attribute this behavior to ionic end groups.[5] The ionic end groups can form associations resulting in a virtual high-molecular-weight polymer. In addition the sulfate end groups can be hydrolyzed followed by loss of HF resulting in the formation of olefin and carboxyl end groups (Scheme 1). These end groups may then be responsible for degradation of the thermal properties.

$$\sim CF_2CFCl\text{-}OSO_3H \quad \xrightarrow{H_2O} \quad \sim CF_2CFCl\text{-}OH \rightarrow \quad \sim CF=CFCL + CF_2\text{-}CO_2H$$

Scheme 1

6.3.2 Organic Peroxides

Both hydrocarbon and fluorocarbon organic peroxides were used to initiate polymerization. The half-lives of several that were used are shown in Table 6.3. The perfluoro-organic peroxides were prepared at temperatures below 0°C by the reaction of the corresponding acyl chloride and sodium peroxide (Scheme 2). Sodium peroxide was formed from an aqueous mixture of sodium hydroxide and hydrogen peroxide.

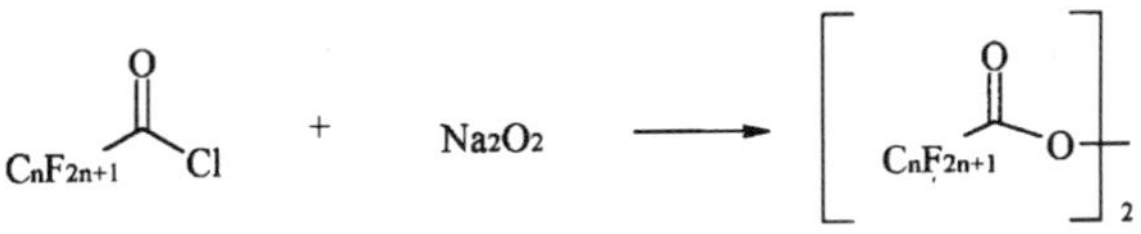

Scheme 2

The results of polymerizations using these various initiators are shown in Table 6.4. The best results were obtained with FC-8; thus a number of aqueous suspension polymerizations were run under a variety of conditions using FC-8 dissolved and fed to the reaction in several different solvents. The results are shown in Table 6.5. The low yields are probably due to precipitation, premature decomposition in the feed line, or hydrolysis of the FC-8 peroxide prior to

Table 6.3. Low-Temperature Initiators

Initiator	Trade name	$T_{1,2}$(min) at 27°C	Reference
$[Cl_3CCO_2]_2$		< 10	6
$[C_3F_7CO_2]_2$	FC-4	209	7
$[C_7F_{15}CO_2]_2$	FC-8	159	7
$[(CH_3)_3CCH_2C(CH_3)(CN)N=]_2$	VAZO 33	1300	8
$[(CH_3)_3CC_6H_{12}\text{-}O\text{-}CO_2]_2$	Percadox 16N	120	9
$[C_4F_9OC_2F_4OCF_2CO_2]_2$	FC-4-2-2		

Table 6.4. New Initiators

Example	Initiator	Yield (%)	Inherent viscosity (dl/g in C_6F_5Cl)
23	Bu_3B/H_2O	Tr.	<0.1
24	Vazo 33	0	
25	$[Cl_3CCO_2]_2$	5	
26	Percadox 16N	10	0.4
27	Percadox 16N	15	0.94
28	FC-8	5	0.5
29	FC-8	55	1.3
30	FC-4-2-2	8	1.0
31	$[C_3F_7CO_2]_2$	< 5	

initiation. A high-molecular-weight polymer was obtained as indicated by the high viscosities shown in Table 6.4.

6.3.3 Gel Permeation Chromatography

Gel permeation chromatography was run in 2,5-dichlorotrifluoromethylben-zene at 130°C. The results of samples from different sources are shown in Table 6.6. The results indicate, based on the styrene equivalent molecular weights, that a moderately high-molecular-weight polymer is formed and that the molecular weight of the suspension polymer (Examples 2 and 27) is comparable to that formed by the emulsion polymerization (Example 9). The chromatograms (Figure 6.2), although of low quality, indicate that both the emulsion polymer and the redox suspension polymer and the redox suspension polymer contain some high-molecular-weight material although in slightly different ratios. Daikin polymer, with very good thermal properties, has very little of the high-molecular-weight fraction (Example 12). We attribute the high-molecular-weight fractions to association of ionic end groups.[5]

Table 6.5. FC-8 Initiator $[C_7F_{15}CO_2]_2$

Example	Initiator (%)	Solvent[a]	Temperature (°C)	Yield (%)	Inherent viscosity (dl/g C_6F_5Cl)
32	0.22	F-113	13	7	0.80
33	0.20	FC-80	27	55	0.62
34	0.05	FC-75	27	37	0.74
35	0.20	FC-75	17	43	0.72
36	0.10	FC-75	19	46	1.26

[a] FC-75 and FC-80 are inert fluorochemical fluids from 3M.

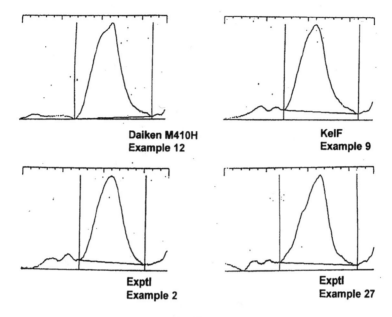

Figure 6.2. Chromatograms.

Table 6.6. Gel Permeation Chromatography[a]

Example	Sample	Zero strength time	Inherent viscosity (dl/g C_6F_5Cl)	M_n	M_w	P
12	Daikin M400H	368	1.78	3.56E4	7.80E4	2.19
9	Kel-F 6050	297	1.74	2.85E4	7.39E4	1.92
2	Suspension–redox	711	1.57	3.23E4	7.19E4	2.22
27	Suspension–C_8 Peroxide	116	0.94	3.45E4	6.74E4	1.96

[a] Conditions: eluent: 2,5-dichlorotrifluoromethyl benzene, temperature 130°C (Run by Mike Stephens, CRL Analytical).

6.3.4 Capillary Rheometry

A similar variety of samples was tested for thermal stability by capillary rheometry and TGA. Figure 6.3 shows the viscosity–shear rate dependence for PCTFE homopolymers and one copolymer (Alcon 3000). All materials, save one, showed virtually identical viscosity relationships despite large changes in inherent viscosity. Only the polymers from runs initiated by fluorochemical peroxides (FCP) showed a dependence of molecular weight (as measured by inherent viscosity) upon melt viscosity.

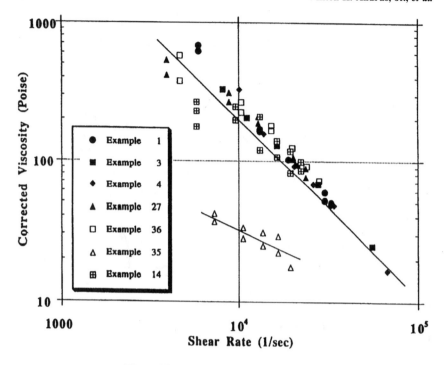

Figure 6.3. Corrected viscosity vs. shear rate.

Thermal stabilities were assessed by the time-dependent change of melt viscosity at a constant temperature and shear rate (290°C, 50 s^{-1}, respectively). Figure 6.4 shows that three of the six resins showed a significant drop in viscosity as a function of time at 290°C. The average decrease in viscosity for Kel-F 6050, Alcon 3000, and an experimental suspension is 37%.

TGA results (Figure 6.5) confirm the improved stability of suspension polymers, especially when organic peroxides are used.

6.4. SUMMARY

Suspension polymerization was shown to give high yields of controlled high-molecular-weight polymers. The presence of bubbles in the thermally pressed samples, especially the high-yield polymers, suggest that low-molecular-weight oligomers are formed at the end of the polymerization. The perfluoroorganic peroxides gave high-molecular-weight polymers albeit with low yields. These polymers exhibited "standard" viscosity–shear rate response compared to the

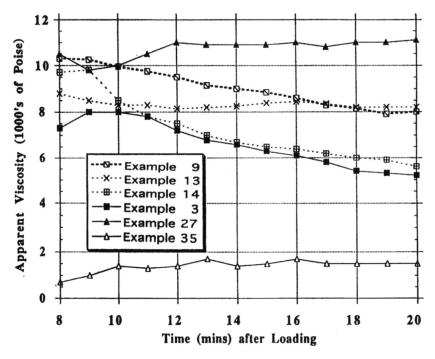

Figure 6.4. Apparent viscosity vs. time after loading.

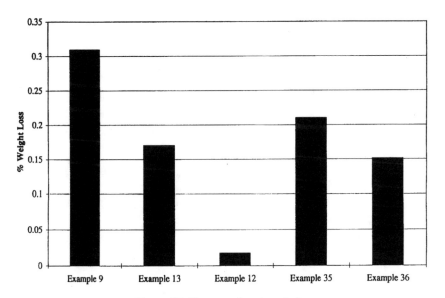

Figure 6.5. Thermogravimetric analysis.

redox initiated products. The thermal stability of the suspension polymer was shown to be better than the emulsion polymer.

ACKNOWLEDGMENTS: The authors would like to acknowledge Sue Andrews and Becky Anderson for running inherent viscosities, Don Klein for running ZSTs, and Scott Anderson for running thermal analysis. We also acknowledge helpful discussions with Cesar Garcia-Franco on rheology and Gary Kwong on polymerization reactions. We also thank Division and 3M management for their support.

6.5. REFERENCES

1. A. N. Bolstad, Private communication (1989).
2. C. M. Tseng and C. I. Young, Internal 3M Report (1989).
3. S. Chandrasekaran, U.S. Patent 4,469,864 (1984).
4. J. A. Klein, Internal 3M Report (1988).
5. A. L. Logothetis, *Prog. Polym. Sci. 14*, 251–296 (1989).
6. V. R. Kokatnur and M. Jelleng, *J. Am. Chem. Soc. 63*, 1432 (1941).
7. Z. Chengxue *et al.*, *J. Org. Chem. 47*, 2009–2013 (1982).
8. DuPoint Bulletin, VAZO Free Radical Sources, *Bulletin* E-93156.
9. AKZO Product Catalog, Initiation for Free Radical Polymerizations, Bulletin 88-57.

7

Fluorinated Polymers with Functional Groups

Synthesis and Applications. Langmuir–Blodgett Films from Functional Fluoropolymers

BORIS V. MISLAVSKY

7.1. INTRODUCTION

Fluorinated polymers, especially polytetrafluoroethylene (PTFE) and copolymers of tetrafluoroethylene (TFE) with hexafluoropropylene (HFP) and perfluorinated alkyl vinyl ethers (PFAVE) as well as other fluorine-containing polymers are well known as materials with unique inertness. However, fluorinated polymers with functional groups are of much more interest because they combine the merits of perfluorinated materials and functional polymers (the terms "functional monomer/polymer" will be used in this chapter to mean monomer/polymer containing functional groups, respectively). Such materials can be used, e.g., as ion exchange membranes for chlorine–alkali and fuel cells, gas separation membranes, solid polymeric superacid catalysts and polymeric reagents for various organic reactions, and chemical sensors. Of course, fully fluorinated materials are exceptionally inert, but at the same time are the most complicated to produce.

There are two principal ways to synthesize functional fluoropolymers: (a) polymerization of functional fluoromonomers or copolymerization of fluoroolefins with functional monomers, and (b) modification of common

BORIS V. MISLAVSKY · Institute of Chemical Physics, Russian Academy of Science, 117977 Moscow, Russia.

Fluoropolymers 1: Synthesis, edited by Hougham *et al.*, Plenum Press, New York, 1999.

fluoropolymers by graft copolymerization of functional monomers. Chemical modification of common fluoropolymers (mainly fluoro- or perfluoroolefins) is also possible. However because of their unique chemical inertness there is a very limited range of reagents that can affect them and so only a very limited range of functional groups can be attached to them in this way.

In this chaper, the two principal methods for their synthesis and some of the main applications of functional fluoropolymers will be reviewed. However, as there are many patents and papers as well as several reviews on these subjects, the review will be brief and the primary focus will be on selected applications and specific properties of functional fluoropolymers that can be achieved in highly ordered Langmuir–Blodgett (LB) films of these polymers.

7.2. SYNTHESIS OF FUNCTIONAL FLUOROPOLYMERS

7.2.1. Polymerization of Functional Fluoromonomers and Copolymerization of Functional Monomers with Fluoroolefins

Since PTFE was first synthesized more than 50 years ago, fluoropolymers have been produced by radical polymerization and copolymerizaton processes, but without any functional groups, for several reasons. First, the synthesis of functional vinyl compounds suitable for radical polymerization is much more complicated and expensive in comparison with common fluoroolefins. In radical polymerization of one of the simplest possible candidates—perfluorovinyl sulfonic acid (or sulfonyl fluoride)—there was not enough reactivity to provide high-molecular-weight polymers or even perfluorinated copolymers with considerable functional comonomer content. Several methods for the synthesis of the other simplest monomer—trifluoroacrylic acid or its esters—were reported,[1] but convenient improved synthesis of these compounds as well as radical copolymerization with TFE induced by γ-irradiation were not described until 1980.[2]

Radical copolymerization of TFE with hydrocarbon functional monomers has also not been widely used, owing perhaps to the high activity in the reaction with the C—H bond with its high probability of chain transfer to the monomer and the polymer, which is a feature of growing perfluoroalkyl radicals, and to poor chemical stability of the copolymers.

One of the first functional fluoropolymers was poly-1,2,2-trifluorostyrene. On one hand, it has much better oxidation and chemical resistance in comparison with common hydrocarbon polymers and, on the other hand, a wide range of functional groups can be attached to the aromatic ring. A sulfonated polymer was successfully used as a membrane for fuel cells by General Electric Co.[3]

The first commercially successful example of a perfluorinated functional polymer was achieved in 1960 at DuPont.[4] It was a random copolymer of TFE and functional PFAVE with sulfonyl fluoride group[5]; many other functional fluoro-monomers were patented as well.[6] In spite of the fact that ion exchange membranes made from polymers with sulfonic groups demonstrated high performance in water electrolysis and fuel cells, it appeared that their efficiency in chlorine–alkali electrolysis was not as good. In the 1970s PFAVE with carboxylic groups were suggested by Japanese firms; membranes for chlorine–alkali electrolysis from these polymers showed higher selectivity in that process.[7] All modern ion exchange membranes for chlorine–alkali electrolysis usually contain at least two layers: one made of a copolymer with sulfonic groups and another made of a copolymer with carboxylic groups.

The structure of such copolymers was intensively investigated in the decade from 1970 to 1980. Although many models have been proposed to describe the structural organization of these polymers, they fall basically into two groups. The first set of models suggests that all sulfonic groups are in an aqueous phase consisting of well-defined clusters and channels.[8] Another set of models assumes the presence of regions intermediate between the fluorocarbon backbone and the aqueous phase that would include the side ether chains and possibly some fluorocarbon backbone as well as some of the ionic groups and associated counterions.[9]

Two comprehensive reviews including functional PFAVE synthesis, copolymerization of functional PFAVE with fluoroolefins (mainly TFE), investigations of the structure of the copolymers as well as their ion-exchange membrane properties were prepared by leading specialists from Asahi Glass Co.[10] and by Russian specialists from the Membrane Processes Laboratory in the Karpov Research Institute of Physical Chemistry.[11]

7.2.2. Radiation Grafting of Functional Monomers onto Fluoropolymers

As was noted above, functional fluoropolymers produced by copolymerization of fluoroolefins with functional PFAVE have several unique properties, with the main disadvantage of these materials being the extremely high cost of functional monomers and the resulting high cost of the functional polymers produced from them. The fact that they are so expensive limits their wider industrial application in other fields such as catalysis and membrane separation, except for chlorine–alkali electrolysis and fuel cells, where the only suitable materials are fully fluorinated polymers because of the extreme conditions associated with those processes.

From this point of view attaching of functional groups to common inexpensive fluoropolymers seemed to be a very attractive way to improve the chemical and thermal stability of the final functional polymer materials in comparison with common hydrocarbon functional polymers, while keeping the costs quite reasonable. This can be accomplished either by chemical modification of fluoropolymers or by grafting polymerization of functional monomers onto fluoropolymers.

As noted above, chemical modification has not been widely used because of the high chemical inertness of fluoropolymers. The range of chemicals that can react with perfluorinated materials is very limited, being mainly organic complexes of alkali metals. In addition, the reaction usually takes place in thin surface layer of fluoropolymer so the achievable degree of modification is relatively low. This method is generally used to increase the adhesion of perfluoropolymer materials to other surfaces. Polychlorotrifluoroethylene (PCTFE), better known as Kel-F polymer, is close to perfluorinated polymers but can be treated with organometallic compounds such as organolithium or organomagnesium as well.[12] Modified polymer is suggested for high-performance liquid chromatography. Of course, common hydrogen-containing fluoropolymers, e.g., polyvinylidene fluoride (PVDF) or copolymer of vinylidene fluoride (VDF) with HFP can react with a wider range of chemicals, primarily dehydrofluorinating agents with the formation of double bonds in the main fluoropolymer chains that can then be converted to functional groups including hydrophilics.

Most of the information regarding attachment of functional groups to fluoropolymers deals with radiation grafting of functional monomers onto fluoro-polymers. The starting point of this method is the well-known property of fluoropolymers of stocking up radical sites in the matrix, which are quite stable over time and are suitable for initiating radical polymerization of vinyl monomers. For instance, irradiated PTFE can still be used for grafting polymerization several years (!) after it has been exposed. Evident advantages of this method are low cost and easy availability of fluoropolymers, functional vinyl monomers, and irradia-tion, as well as the simplicity of the procedure.

Investigations of radiation grafting of functional monomers onto fluoropoly-mers started in the late 1950s. Since that time several hundred papers and patents have been published, as have several reviews,[13-16] and some specific aspects have been considered, but a broad overview has not yet appeared in print.

Many vinyl monomers were reported to have been grafted onto fluoropoly-mers, such as (meth)acrylic acid and (meth)acrylates, acrylamide, acrylonitryl, styrene, 4-vinyl pyridine, N-vinyl pyrrolidone, and vinyl acetate. Many fluoro-polymers have been used as supports, such as PTFE, copolymers of TFE with HFP, PFAVE, VDF and ethylene, PCTFE, PVDF, polyvinyl fluoride, copolymers of VDF with HFP, vinyl fluoride and chlorotrifluoroethylene (CTFE). The source of irradiation has been primarily γ-rays and electron beams. The grafting can be carried out under either direct irradiation or through the use of preliminary irradiated fluoropolymers. Ordinary radical inhibitors can be added to the reaction mixture to avoid homopolymerization of functional monomers.

Some important specific features of radiation grafting should be pointed out. In contrast to chemical modification, the final grafted copolymers can be obtained with functional monomer uniformly distributed in the fluoropolymer matrix. The degree of grafting can achieve several parts per one part of the initial fluoropolymer (but perhaps in this case the final material could be formally named as a functional hydrocarbon polymer modified by a fluoropolymer). The structure of final grafted copolymer generally depends on the ratio between the rate of polymerization and the rate of diffusion of the functional monomer into the fluoropolymer matrix, and once polymerization is started, the upper layer with the grafted copolymer usually promotes the diffusion of functional monomer into the lower layers of the fluoropolymer matrix. Thus the final structure of the grafted copolymers could be varied over a wide range by modulating reaction conditions such as intensity and total dose of irradiation, temperature, functional monomer concentration, and solvent nature.

Some monomers can promote the grafting copolymerization of other monomers that have low rate of grafting or cannot be grafted individually. With this approach a very wide range of functional monomers can be grafted by using a comonomer that can promote grafted copolymerization.

It is a well-known fact that the mechanical properties of fluoropolymers, especially perfluoropolymers, degrade dramatically under irradiation. Nevertheless a considerable improvement of the mechanical properties of the final grafted copolymers was observed in comparison with mechanical properties of the initial irradiated fluoropolymer. Thus it is possible to minimize or completely avoid the degradation of mechanical properties of the final grafted composites in comparison with the initial fluoropolymers by choosing appropriate reaction conditions.

Though the composites with grafted hydrogen-containing monomers usually have better chemical and thermal stability in comparison with hydrocarbon functional polymers, they are inferior to perfluorinated polymers in thermal and chemical stability. From this point of view the composites based on graft copolymers of polyfluoroolefins and perfluoromonomers with functional groups could be the most interesting ones. Even in case of grafting of expensive functional PFAVE, for some applications, e.g., for catalysis, synthesizing the graft copolymers retains all the advantages of perfluorinated acid catalysts, and also significantly increases the effectiveness of the functional groups, as the final composite has a "core-shell" type of structure and most of the functional groups are located in the surface layer. Thus we can decrease the expensive functional fluoromonomer content without prejudice to the catalytic activity of the copolymers (i.e., to increase specific catalytic activity substantially). While there have been very few articles on this subject, successful grafting of trifluoroacrylates onto fluoropolymers has been reported.[17,18] We also reported previously[19] on radiation-induced graft copolymerization of perfluorinated alkyl vinyl ethers with sulfonyl fluoride groups (PFAVESF) onto fluoropolymers. The properties of the materials obtained in comparison with other functional polymers and catalysts will be analyzed below.

7.3. APPLICATIONS OF FUNCTIONAL FLUOROPOLYMERS

7.3.1. Chlorine–Alkali Electrolysis

Chlorine–alkali electrolysis is the largest application of such materials as these are the only materials that can be used successfully in this process. As this process provides alkali of better quality than the conventional diaphragm process, and is much more attractive environmentally than the mercury process, its part in industrial world manufacturing of alkali is expanding rapidly.

Modern membranes usually consist of at least two layers: one from sulfonic type copolymer and another from carboxylic type copolymer. The membranes are usually reinforced by fluoropolymer fabric to provide better mechanical properties and long lifetimes. The most important properties are considered in detail in the reviews mentioned above[10,11] and in a basic text by Seko et al.[6]

7.3.2. Fuel Cells

It was the development of fuel cells that initiated the development of functional fluoropolymers; perfluorinated ion exchange materials (such as Nafion) provided the best results because of very good thermal and oxidative stability. Properties of hydrogen–oxygen solid polymer electrolyte (SPE) fuel cells as well as their applications in space and for specific ground and undersea power sources were widely reviewed.[20–23] It was reported that they also could be used as power source for vehicles.[24] In hydrogen–halogen fuel cells charge and discharge can be carried out in the same unit, which enables their use for off-peak electric power storage systems.[25,26]

7.3.3. Water Electrolysis

Hydrogen will possibly play a major role among prospective energy carriers, and the most suitable method for industrial hydrogen production is water electrolysis. Membrane cells provide much better efficiency in comparison with other methods.[27]

7.3.4. Polymer Catalysts

Cross-linked polystyrene and its functional derivatives are widely used in organic syntheses as polymeric reagents and catalysts.[28] However, thermal and chemical stability of such materials has to be better. Some improvement in these properties can be achieved by the grafting of styrene with the following chemical modification or grafting of other functional monomers.

However, only perfluorinated ion exchangers can be used at substantially higher temperatures—up to 220–230°C—and so cover the main temperature range of acid catalyzed organic reactions. Besides, Nafion-type resin is a stronger acid than polystyrenesulfonic acid (it is a polymeric analogue of trifluoromethane sulfonic acid so it can be considered a superacid), and is much more stable in any aggressive environment. Application of Nafion-type catalysts in different classes of organic reactions such as isomerization and disproportionation of alkylated benzenes, alkylation and acylation of aromatic rings, electrophilic substitutions, and polymerization are considered in detail in several reviews.[29–31]

Unfortunately Nafion materials have not found commercial application as catalysts because of their extremely high cost. There were several attempts to use supported catalysts made by applying of low-molecular-weight Nafion polymer from solutions onto inert supports. However, such catalysts could only be used in very few reactions between nonpolar reagents; in other cases the surface catalytic layer was easily washed away from the surface.

The only way to apply chemically bonded thin perfluorosulfonic acid layer onto the surface of an inert support is to graft perfluorinated functional monomers onto perfluorinated polymers. Some features of radiation-induced graft copolymerization of PFAVESF onto fluoropolymers were investigated.[19] The studies showed that neither irradiation of a fluoropolymer–PFAVESF mixture (direct grafting) or interaction of PFAVESF with previously irradiated fluoro-polymers (preirradiation grafting) yielded the grafted copolymers. It was assumed that this is connected with the low activity of PFAVE in radical polymerization. A special method has been developed for the synthesis of grafted copolymer. Previously irradiated fluoropolymer powders were used to prevent waste of PFAVESF.

As was shown, the rate of graft polymerization and the composition of grafted copolymers depend on the monomer concentration, temperature, and the composition of fluorpolymer support. The former also depends on the dose of previous irradiation of the fluoropolymer support. It was assumed that the structure of the composites obtained is close to the core-shell type.

Catalytic properties of the active acid form of the composites obtained in comparison with random copolymers of tetrafluoroethylene and PFAVESF (Nafion-type) were investigated in esterification, oligomerization, and aromatic compounds alkylation reactions.

7.3.4.1. Esterification of Acetic Acid by Primary Alcohols

The results obtained in the reaction at 70°C are shown in Table 7.1. Even in the case of a "small" molecule of ethanol the specific rate (per functional group) with the grafted copolymer is seven times higher than the one with the Nafion catalyst, and this difference increases with the size of the alcohol molecule. This

Table 7.1. Specific Rates in the Esterification of Acetic Acid by Primary Alcohols

Alcohol			n		
$C_nH_{2n+1}OH$	2	3	4	8	10
Nafion (W_{Nf})	36.1	28.5	16.1	10.6	8.0
Grafted copolymer (W_{gc})	245	194	131	103	100
W_{gc}/W_{Nf}	6.8	6.8	8.1	9.7	12.5

reaction with ethanol was also carried out in a flow reactor at 120°C. With a contact time of 6–9 s the conversion to ethyl acetate was 98% for both the Nafion catalyst and for the grafted copolymer, with the content of functional groups in the copolymer 35 times lower than in Nafion.

7.3.4.2. Alkylation of Phenol by Styrene

The catalysis of this reaction by grafted copolymer in comparison with well-known catalysts such as *p*-toluene sulfoacid, and cross-linked sulfonated polystyrene was investigated. The yield of mono- and disubstituted products is shown in Table 7.2. The specific catalytic activity of grafted copolymers is 100 times or even higher than the same values for the other catalysts. This also can be related to the much higher level of acidity of perfluoroalkyl sulfoacid.

Specific catalytic activity of the composites obtained was at least several times higher than the same value for the random copolymer Nafion (even in an esterification reaction considered to be a diffusion-uncontrolled reaction). For the oligomerization reaction of decene-1 with strong diffusion control, the specific catalytic activity of the composites was 35 times higher than that for the random copolymer. Esterification of acrylic acid and alkylation of mesitilene by a substituted phenol were also performed using the composite catalyst.

The advantages of the grafted copolymers obtained in comparison with the other catalysts can be summarized as follows:

- Versus the Nafion catalyst (random copolymer): higher specific catalytic activity and surface area, relative simplicity of the synthetic method.
- Versus the supported catalysts based on Nafion: impossibility of washing off the catalytic active layer and substantial simplicity of the production method.
- Versus low-molecular-weight sulfoacids: multiple use, simplicity of separation by filtering, higher acidity, possibility of use in a flow reactor.
- Versus cross-linked sulfonated polystyrene: higher temperature (up to 220°C), multiple regeneration, higher acidity.

Table 7.2. Yield of Products in Alkylation Reaction of Phenol by Styrene

Catalyst[a]	Catalyst content (wt %)	Products[b]					
		A			B		
		Mono-	Di-	Yield	Mono-	Di-	Yield
1	1.5	78.2	10.8	89	16.3	72.2	88.5
2	10	51	15.2	66.2	12.5	50.5	63
3	10	66.8	11.2	78	18.4	65	83.4
4	5	55.3	17.1	72.4	12.3	50.2	62.5

[a] 1–grafted copolymer, ion exchange capacity (IEC) = 0.026 meq/g; 2,3–sulfonated polystyrenes, IEC = 2.5 meq/g; 4–toluene sulfoacid, (IEC) = 6.2 meq/g.
[b] A–phenol/styrene = 1 : 1, reaction time 60 min; B–phenol/styrene = 1 : 2, reaction time 90 min.

As the initial sulfonyl fluoride groups can be easily modified by the reaction with corresponding amino derivatives, e.g., those of crown ethers, the composites obtained can be used as polymeric reagents for a wide range of organic reactions.

7.3.5. Permselective Membranes

Grafting of functional monomers onto fluoropolymers produced a wide variety of permselective membranes. Grafting of styrene (with the following sulfonation), (meth)acrylic acids, 4-vinylpyridine, N-vinylpyrrolidone onto PTFE films gave membranes for reverse osmosis,[32–34] ion-exchange membranes,[35–39] membranes for separating water from organic solvents by pervaporation,[40–42] as well as other kinds of valuable membranes.

7.3.6. Selective Adsorbents

Grafted copolymers of fluoropolymers with functional vinyl monomers can give selective adsorbents for different metal ions. For example, amidoxime-containing grafted copolymers produced by grafting of acrylonitrile followed by treatment with hydroxylamine were found to be selective for copper ions.[16] Such species could be also used for the extraction of uranium from seawater. Complexation constants of copper and rare earth metal ions were found to be higher for grafted acrylic acid than for polyacrylic acid, and much higher than for low-molecular-weight carboxylic acid. This phenomenon was explained assuming that the concentration of ligands is higher in the polymer domain, especially in the surface domain of the grafted polymer, and therefore once a metal ion is attached to one functional group on the polymer chain, the other ligands coordinate more readily.

7.3.7. Ion Selective Electrodes

Ionomer-film-modified electrodes (especially those modified with fluoro-polymers) can be used in electroanalytical systems as chemical sensors as they show remarkable affinities for large organic cations relative to simple monovalent or divalent inorganic cations.[43,44] Such ionomer films can extract and preconcentrate large organic cations from a contacting aqueous medium. The extent of the pre-concentration effect depends on the magnitude of the ion-exchange selectivity coefficient, and these values have been reported for a variety of organic cations in Nafion ionomer[45]; it was shown that Nafion film preconcentrates these cations and that the equilibrium film concentrations are orders of magnitude higher than the concentrations in solution. It was also found that Nafion has much higher selectivity in comparison with conventional ion-exchange materials such as sulfonated polystyrene, which was explained by assuming a non-cross-linked structure for Nafion and a stronger hydrophobic interaction of the perfluorinated chain with nonpolar parts of the exchanging ions.

Because Nafion film can preconcentrate organic cations at a substrate electrode surface, Nafion-coated electrodes can be used as sensitive and selective sensors for electroactive organic cations. The film-entrapped ion is then either oxidized or reduced and the resulting current is recorded and related to the concentration of ions in the aqueous solution (qualitatively similar to anodic stripping voltammetry). It was shown that ion-exchange voltammetric determination of the dication methyl viologen at a Nafion-film-coated electrode results in an improvement of three orders of magnitude in detection limit relative to an uncoated electrode.[46]

This Nafion-coated-microelectrode technique can be applied for *in vivo* determination of dopamine and other neurotransmitters to investigate brain activity. As these transmitters are amines, they are protonated at physiological pH values and so can be ion-exchanged into Nafion, while the other principal components in the brain environment are anions at physiological pH, should be rejected by the Nafion film. It was demonstrated that Nafion-coated carbon microelectrodes show very little anion interference. This differentiation of the neurotransmitters from metabolites and ascorbate has been sorely missed in *in vivo* electrochemistry.[47] Nafion should also preconcentrate these cations, and ion-exchange voltammetry at a Nafion-coated electrode should give an improved detection limit; in fact this detection limit for dopamine cations was found to be two orders of magnitude lower than the corresponding detection limit at a naked electrode.[46]

The maximum preconcentration advantage for an ionomer-film-modified electrode is obtained with ionomer-film–analyte solution equilibrium is achieved. As the time required for establishment of equilibrium depends on the size, charge, structure, and concentration of the analyte counterions and on the ionic strength of the analyte solution, relatively small counterions (e.g., dopamine cations) can reach equilibrium within quite a reasonable time (several minutes). Both a reduction in the thickness of the Nafion film and improvements in the measurement techniques have been enabled a shortening of the response time,[48] but larger counterions {e.g., $[Ru(bpy)_3]^{2+}$} might require time periods that are 100-fold longer, which, from an analytical viewpoint, is unacceptable. It was suggested that using Nafion-impregnated PTFE porous film instead of a Nafion cast layer might improve the transport properties of the electrode coating; diffusion coefficients of relatively large $[Ru(NH_3)_6]^{3+}$ and $[Ru(NH_3)_6]^{2+}$ cations in impregnated membrane were found to be about three orders of magnitude higher than the analogous values in Nafion.[45]

It was pointed out[45] that while ionomer-modified electrodes have greater selectivity than naked electrodes, ionomers provide only a general form of selectivity (charge type and mass or hydrophobicity selectivity), and special methods were suggested to improve specific selectivity. For example, it was thought that applying an additional cellulose acetate layer over the Nafion film would make the electrode selective with respect to dopamine specifically (in contrast to other neurotransmitters). To make the electrode system specifically selective to o-nitrophenol (in contrast to p-isomer), a complexation with cyclo-dextrin was performed, which increased the selectivity ratio one order of magnitude with respect to the o-isomer.[50]

Research was done with redox systems incorporated into Nafion-modified electrodes[51] because of interesting possible applications of such systems, e.g., for electron-transfer catalysis.

In conclusion, the advantage of Nafion polymer as an electrode modifier can be summarized as follows. Applying a Nafion polymer layer is usually simpler compared with other electrode-coating techniques (electrosorption, glow-discharge polymerization, covalent binding, electropolymerization). Final coat-ings appeared attractive because of their strong adhesion to the electrode surface, their good swelling in aqueous solutions, their water insolubility, their reasonable ion-exchange capacity, and their high chemical and mechanical stability. These factors are responsible for a lot of research and valuable applications of Nafion polymer as an electrode modifier.[52–59] Humidity-sensitive and water-resistant polymeric materials based on 4-vinyl pyridine grafted onto PTFE have also been reported.[60]

7.3.8. Mixed Conductivity Polymers

The main characteristic of mixed conductivity polymers (MCP) is superposition of the electronic conducting properties of the polyaromatic skeleton, such as polypyrrole or polyaniline, and the ionic conducting properties of an ion-exchange polymer, such as perfluorinated ion-exchanger. Mixing on the molecular level can be achieved either by chemical or electrochemical inclusion of the electronically conducting polymer in the ionomer membrane or by electrosynthesis in a solution containing dispersed ionomer and monomer precursor of the aromatic polymer. In the case of oxidizing polymerization of an aromatic monomer, if Nafion film is used as a separator between the solutions of monomer and oxidizer, a structure with gradually changing electronic conductivity through the thickness of the film can be obtained.[61] Such materials might have potential for specific electronics applications.

7.4. LANGMUIR–BLODGETT FILMS FROM FUNCTIONAL FLUOROPOLYMERS

The recently demonstrated[62–64] poor stability of Langmuir–Blodgett (LB) films of conventional amphiphilic low-molecular-weight compounds with respect to temperature and chemical and mechanical treatment is a serious drawback for their practical utilization in microelectronics, where their defined structure, controllable thickness, and high homogeneity would be very attractive. Numerous attempts have been made to resolve these problems using different polymerizable monomers or polymer materials for LB technology.[65–70] For example, polyimides have been found to have great potential for electronics application because of their high-dielectric properties and thermal, chemical, and mechanical stability.[71–73]

However, fluorocarbon compounds might be of considerable interest for LB-layer fabrication. Their dielectric and mechanical characteristics and thermal and chemical stability are not inferior to those of polyimides, and highly developed synthesis technology makes it possible to create systems with various predictable properties. Such films have been found to demonstrate a high degree of perfection and excellent dielectric characteristics.[69,70]

Unlike the approach in which LB multilayers are formed from polymerizable fluoromonomers followed by UV-polymerization,[69] we started to use functional fluoropolymers for the preparation of LB films. These films seem to be better, because in the case of monomeric LB film the polymerization process causes contraction and hence possible defects in the final cured film. Two different kinds of functional amphiphilic fluoropolymers were used for LB-film preparation. One type (referrred to in Section 7.4.2), which has large fluoroalkyl groups as side groups on the hydrocarbon polymer backbone could also be considered as a functional fluoropolymer as it contains substantially hydrophobic fluorinated fragments and functional hydrophilic ester and carboxy groups, and could be used for the same purposes as, say, polyimide LB films. Another type (referred to in Section 7.4.1) has a very specific structure, unusual for common electronics applications of LB films, and could be suitable for specific applications in sensors and nanoelectronic devices.

7.4.1. Multilayered Structures with Metal Cations Based on LB Films of Perfluorinated Sulfoacid Polymer (Nafion-Type)

As was noted above, the structure and behavior of Nafion polymers were intensively investigated and structural models were suggested. It is obvious that during the transformation from precursor (nonionic form) to ionic form the internal structure of the polymer is reorganized into the ordered type. But this order has a "random" nature, and all earlier investigations were peformed with such "randomly" ordered films.

The reason for Nafion LB-film fabrication was the wish to obtain the highly ordered systems from perfluorinated ion exchange polymer with multilayered structure, where the ionic layers (conductors) would alternate with fluorocarbon polymer layers (insulators), and to investigate the properties of such films.[74] This polymer contains a hydrophobic fluorocarbon polymeric chain and hydrophilic ionic groups, so it is sufficiently amphiphilic; it has a comblike structure that makes it a suitable polymer for LB-film deposition.

The polymer forms a stable monolayer on the surface of water. The surface pressure–area diagram of the monolayer is shown in Figure 7.1, where the surface area S is calculated per ionic side group. The collapse pressure of the monolayer is equal to 54 mN/m.

The film has noticeable planar conductivity, which depends on the number of monolayers as shown in Figure 7.2. The conductivity of the film is detectable for two monolayers, but the value is small for very thin films (two to six monolayers). From six monolayers the conductivity begins to increase linearly with the number of monolayers, a feature that is also found in LB films made of charge-transfer salts, and is perhaps a function of the imperfection in continuity of the first monolayers on the metal electrode–quartz substratum boundary. This imperfection came about during the deposition process as a result of different hydrophilic properties of metal and quartz surfaces.

The conductivity of the film was calculated for 30 monolayers. The film was deposited onto a Ag microelectrode array with a 1-mm distance between fingers. The thickness of the monolayer was taken to be 2×10^{-7} cm. For an air humidity value of 60% the conductivity equals 1.3×10^{-6} $(\Omega/cm)^{-1}$. The current through the film has an ionic character, and there is apparently layered solid electrolyte

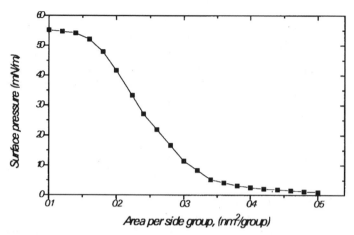

Figure 7.1. Surface pressure–area plot for polymer monolayer measured at 20°C and at a compression rate 2×10 nm/s for each side group.

structuring. When the polymer monolayer is being deposited by LB technology, cations from sulfonic groups SO_3^- and Li^+ are packed in the layer, which also contains water molecules from the ion aqueous cover. It is known that ordinary ion-exchange membranes from such polymers contain from 4 to 20 water molecules per ionic group.[11] Thus the LB monolayer consists of the layer of fluorocarbon polymer and an "electrolyte" layer, containing sulfonic groups, Li^+

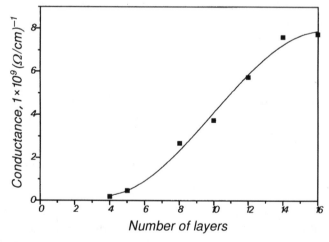

Figure 7.2. Dependence of the film conductivity on the number of LB monolayers deposited onto an Al microelectrode array with a 0.1-mm distance between fingers.

cations, and water molecules. These are "electrolytic" layers, which conduct current in the planar direction. Obviously, when voltage is applied, there is electrolysis of the water that is included in the electrolytic layers. This is the analogue of the electrolysis that takes place in ordinary membranes of such polymers.[11]

The perpendicular conductivity of the film was found to be much smaller than the planar one and was determined to be less than 1×10^{-9} $(\Omega/cm)^{-1}$. Thus the difference between planar and perpendicular conductivity is not less than three orders of magnitude.

As was shown, the planar conductivity of the film can be increased by immersing the substratum with the film in the ethanol–water $(1:1)$ solution of $LiNO_3$ (0.1 mol/liter) for a short time. Then the film should be washed in water and allowed to dry. After such treatment the conductivity becomes 500 times greater and reaches the value 6×10^{-4} $(\Omega/cm)^{-1}$. This increase may be due to the fact that in considering the second general model of the structure of this polymer[9] it could be assumed that some additional quantity of Li^+ cations might be absorbed into the ionic sphere of SO^- groups, so that the total amount of Li^+ in the "electrolytic" layers increases, and the conductivity then also increases.

As the film is conductive owing to electrolytic layers that contain water, the conductivity should depend on the quantity of water in these layers, and thus on the humidity in the air. This dependence for the Li^+ form of LB film is shown in Figure 7.3. The conductivity increases within three orders of magnitude when the humidity rises from 0 to 100%. So, this film can be used in humidity sensors.

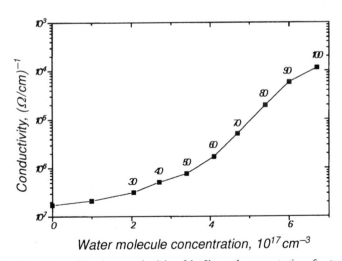

Figure 7.3. Dependence of the planar conductivity of the film on the concentration of water molecules; number on the curve show relative humidity corresponding to 20°C.

Another conductivity mechanism could be suggested for LB films of this polymer with Ag^+ cations. Such cations can accept or release electrons easily, so in the layer of such cations the conductivity could be caused by electron transitions between the ions with different degrees of oxidation. With tunneling microscopy an anomaly in the $dI/dV(V)$ curves near zero bias was discovered for the LB films in Ag form with an odd number of layers: there was a conductivity peak some 150–200 mV wide (Figure 7.4, Curves 1, 3) but no anomaly for these same films with an even number of layers (Figure 7.4, Curve 2). For LB films with an odd number of layers the ordered superstructure of the scale $11.5 \times 11.5 \times 10^{-8}$ cm has been found in a conductivity $dI/dV(x, y)$ measurement regime. The scale of such a structure corresponds to 3×2 surface reconstruction (Figure 7.5).

Cations other than Li^+ can be introduced into the film by a simple ion-exchange procedure, and this may be a simple method for obtaining multilayered structures with any cation. Organic cations, e.g., cyanine dyes, can also be introduced into the film by this procedure. Perhaps this approach for the application of highly ordered thin LB films onto the surface of electrodes could yield some new results, such an improvement in response time, as was discussed in Section 7.3.7.

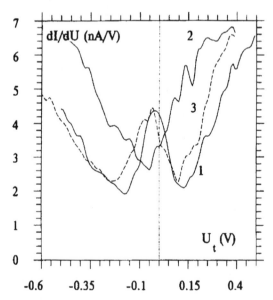

Figure 7.4. dI/dV plots for LB films with different numbers of layers: (1) one layer; $U = -10$ mV, $I = 0.8$ nA, $U_{mod} = 50$ mV; (2) two layers; $U = -10$ mV, $I = 0.8$ nA, $U_{mod} = 50$ mV; and (3) three layers; $U = -15$ mV, $I = 0.8$ nA, $U_{mod} = 50$ mV.

Figure 7.5. Photo of the surface of LB film of three layers in the conductivity $dI/dV(x, y)$ measurement regime; dimension is $70 \times 74 \times 10^{-8}$ cm; gray range of brightness corresponds to relative units; $U = -50\,\text{mV}$, $I = 0.8\,\text{nA}$, $U_{\text{mod}} = 50\,\text{mV}$.

7.4.2. LB Mono- and Multilayers of Fluorocarbon Amphiphilic Polymers and Their Application in Photogalvanic Metal–Insulator–Seminconductor Structures

A_3B_5-type semiconductors, especially GaAs[75] have always been of great interest. These materials have unique electrical and optical properties, but there are some difficulties with technology. In particular, insulating layer fabrication on the surface of such semiconductors and, therefore, creation of metal–insulator–semiconductor (MIS) structures based on them is still a problem, but LB technology might represent a solution.[68,75]

Several polymer and random copolymers of fluoroalkyl acrylates have been studied,[76] including: (1) Poly-1,1-dihydroperfluoroheptylacrylate (PFHA), (2) copolymer of 1,1-dihydroperfluoroheptylacrylate (PFHA) with acrylic acid (PFHA-AA, 1 : 1), and (3) copolymer of the monoester of 1,1-dihydroperfluoroheptyl alcohol and maleic acid (PFHM) with vinyl acetate (PFHM-VA, 1 : 1). The average degree of polymerization was found to be about ten for all the polymers tested. The tendency of polymers to form well-ordered two-dimensional structures on the water surface was demonstrated. Polymers were deposited as multilayers onto different kinds of substrates, e.g., quartz slides, monocrystalline silicon, GaAs, and metals such as Ag, Au, Pt, by the conventional LB method.

The feasibility of the fabrication of comblike fluorocarbon polymer LB films has been shown. These films can be deposited onto different kinds of substrates as y-type layers by the usual LB technique. In this case the deposition procedure is much simpler than the one for polyimide LB films, but the temperature, chemical and mechanical stability, and dielectric properties of the fluorocarbon polymers are not inferior to those of polyimides. The fluorocarbons are more hydrophobic than ordinary hydrocarbons, hence shorter hydrophobic chains can be used and thinner monolayers can be prepared (the PFHA-AA LB monolayer thickness investigated was 16.5×10^{-8} cm).

The Ag/LB film/GaAs structures based on fluorocarbon (PFHA-AA) LB films display the behavior of stable MIS diodes and have high photogalvanic efficiency, demonstrating the attractiveness of these films for creating A3B5-type MIS solar cell insulator layers. The high strength of these monolayers makes it possible to use them as two-dimensional polymer matrices incorporating different functional units. This leads to the expectation that a method of cross-linking of such polymers in LB films on solid substrates can be developed and that two-dimensional polymer networks can be created.

7.5 REFERENCES

1. D. C. England, L. Solomon, and C. G. Krespan, *J. Fluorine Chem.* *3*, 63–89 (1973).
2. T. Watanabe, T. Momose, I. Ishigaki, Y. Tabata, and J. Okamoto, *J. Polym. Sci.: Polym. Phys. Ed.* *19*, 599–602 (1981).
3. R. B. Hodgdon, *J. Polym. Sci. A-1* *6*, 171–191 (1968).
4. W. Grot, *Chemistry and Industry, Oct. 7 1985*, 647–649.
5. D. J. Connolly and W. F. Gresham, U.S. Patent 3,282,875.
6. M. Seko, S. Ogawa, and K. Kimoto, in *Perfluorinated Ionomer Membranes*, ACS Symposium Series 180, American Chemical Society., Washington, D.C. (1982), pp. 365–410.
7. H. Ukihashi, *Chemtech 10*, 118–120 (1980).
8. T. D. Gierke, G. E. Munn, and F. C. Wilson, *J. Polym. Sci.: Polym. Phys. Ed.* *19*, 1687–1704 (1981).
9. H. L. Yeager, in *Perfluorinated Ionomer Membranes*, ACS Symposium Series 180, American Chemical Society, Washington, D.C. (1982), pp. 41–64.
10. H. Ukihashi, M. Yamabe, and H. Miyake, *Prog. Polym. Sci. 12*, 229–270 (1986).
11. Yu. E. Kirsch, S. A. Smirnov, Yu. M. Popkov, and S. F. Timashev, *Uspekhi Khim. 59*, 970–994 (1990).
12. N. D. Danielson, R. T. Taylor, J. A. Huth, R. W. Sienglek, J. G. Galloway, and J. B. Paperman, *Ind. Eng. Chem. Prod. Res. Dev. 22*, 303–307 (1983).
13. A. Chapiro, *Radiat. Phys. Chem. 9*, 55–67 (1977).
14. A. Chapiro and A.-M. Jendrychowska-Bonamour, *Polym. Eng. Sci. 20*, 202–205 (1980).
15. D. O. Hummel, *J. Polym. Sci.: Polym. Symp. 67*, 169–184 (1980).
16. J. Okamoto, *Radiat. Phys. Chem. 29*, 469–475 (1987).
17. H. Omichi and J. Okamato, *J. Polym. Sci.: Polym. Chem. Ed. 20*, 521–528 (1982).
18. H. Omichi and J. Okamoto, *J. Polym. Sci.: Polym. Chem. Ed. 20*, 1559–1568 (1982).

19. B. V. Mislavsky and V. P. Melnikov, *Polym. Preprints 34*, 377–378 (1993).
20. E. A. Ticianelli, C. R. Derouin, A. Redondo, and S. Srinivasan, *J. Electrochem. Soc. 135*, 2209–2214 (1988).
21. G. A. Eisman, *J. Power Sources 29*, 389–398 (1990).
22. R. Baldwin, P. Pham, A. Leonida, J. McElroy, and T. Nalette, *J. Power Sources 29*, 399–412 (1990).
23. K. Prater, in *Proceedings of the Fourth Canadian Hydrogen Workshop*, Toronto, Canada, Nov. 1–2, 1989, pp. 1–20.
24. R. A. Lemons, in *Proceedings of the Grove Anniversary Fuel Cell Symposium, London, Sept. 18–21* (D. G. Lover, ed.), Elsevier Applied Science, London and New York (1989), pp. 251–264.
25. R. S. Yeo and J. McBreen, *J. Electrochem. Soc. 126*, 1482–1687 (1979).
26. R. S. Yeo and D. T. Chin, *J. Electrochem. Soc. 127*, 549–555 (1980).
27. P. W. T. Lu and S. Srinivasan, *J. Appl. Electrochem. 9*, 269–283 (1979).
28. P. Hodge and D. C. Sherrington (eds.), *Polymer-Supported Reactions in Organic Synthesis*, John Wiley and Sons, Chichester (1980).
29. B. Kipling, in *Perfluorinated Ionomer Membranes*, ACS Symposium Series 180, American Chemical Society, Washington, D.C. (1982), pp. 475–488.
30. S. J. Sondheimer, N. J. Bunce, and C. A. Fyfe, *J. Macromol. Sci—Rev. Macromol. Chem. Phys. C26*, 353–413 (1986).
31. G. A. Olah, P. S. Iyer, and G. K. S. Prakash, *Synthesis 1986*, 513–531.
32. S. Munari, F. Vigo, M. Nicchia, and P. Canepa, *J. Appl. Polym. Sci. 20*, 243–253 (1976).
33. A. Chapiro, G. Bex, A.-M. Jendrychowska-Bonamour, and T. O'Neill, *Adv. Chem. Ser. 91*, 560–573 (1969).
34. H. Ali-Miraftab, A. Chapiro, and T. LeDoan, *Europ. Polym. J. 14*, 639–646 (1978).
35. S. Munari, F. Vigo, G. Teadlo, and C. Rossi, *J. Appl. Polym. Sci. 11*, 1563–1570 (1967).
36. E.-S. A. Hegazy, I. Ishigaki, and J. Okamoto, *J. Appl. Polym. Sci. 26*, 3117–3124 (1981).
37. I. Ishigaki, N. Kamiya, T. Sugo, and S. Machi, *Polym. J. 10*, 513–519 (1978).
38. E.-S. A. Hegazy, I. Ishigaki, A. Rabie, A. M. Dessouki, and J. Okamoto, *J. Appl. Polym. Sci. 28*, 1465–1479 (1983).
39. E.-S. A. Hegazy, A. M. Dessouki, A. M. Rabie, and I. Ishigaki, *J. Polym. Sci.: Polym. Chem. Ed. 22*, 3673–3685 (1984).
40. F. Aptel, J. Cuny, J. Jozefowicz, G. Morel, and J. Neel, *J. Appl. Polym. Sci. 16*, 1061–1076 (1972).
41. F. Aptel, J. Cuny, J. Jozefonwicz, G. Morel, and J. Neel, *J. Appl. Polym. Sci. 18*, 351–364 (1974).
42. G. Morel, J. Jozefonvicz, and P. Aptel, *J. Appl. Polym. Sci. 23*, 2397–2407 (1979).
43. C. R. Martin and H. Freiser, *Anal. Chem. 53*, 902–904 (1981).
44. C. R. Martin, I. Rubinstein, and A. J. Bard, *J. Am. Chem. Soc. 104*, 4817–4824 (1982).
45. M. W. Espenscheid, A. R. Ghatak-Roy, R. B. Moore, R. M. Penner, M. N. Szentirmay, and C. R. Martin, *J. Chem. Soc.: Faraday Trans. I 82*, 1051–1070 (1986).
46. M. N. Szentirmay and C. R. Martin, *Anal. Chem. 56*, 1898–1902 (1984).
47. G. Nagy, G. A. Gerhardt, A. F. Oke, M. E. Rice, and R. N. Adams, *J. Electroanal. Chem. 188*, 85–94 (1985).
48. E. W. Kristensen, W. G. Kuhr, and R. M. Wightman, *Anal. Chem. 59*, 1752–1757 (1987).
49. J. Wang and P. Tuzhi, *Anal. Chem. 58*, 3257–3261 (1986).
50. T. Matsue, U. Akiba, and T. Osa, *Anal. Chem. 58*, 2096–2097 (1986).
51. N. Oyama, T. Ohsaka, and T. Okajima, *Anal. Chem. 58*, 979–981 (1986).
52. N. Oyama, T. Ohsaka, K. Sato, and H. Yamamoto, *Anal. Chem. 55*, 1429–1431 (1983).
53. I. Rubinstein and A. J. Bard, *J. Am. Chem. Soc. 102*, 6641–6642 (1980).
54. I. Rubinstein and A. J. Bard, *J. Am. Chem. Soc. 103*, 5007–5013 (1981).
55. D. A. Buttry and F. C. Anson, *J. Am. Chem. Soc. 104*, 4824–4829 (1982).
56. D. A. Buttry and F. C. Anson, *J. Am. Chem. Soc. 105*, 685–689 (1983).
57. H. S. White, J. Leddy, and A. J. Bard, *J. Am. Chem. Soc. 104*, 4811–4817 (1982).

58. H. L. Yeager and A. Steck, *Anal. Chem. 51*, 862–865 (1979).
59. A. Steck and H. L. Yeager, *Anal. Chem. 52*, 1215–1218 (1980).
60. Y. Sakai, Y. Sadaoka, and H. Fukumoto, *Sensors and Actuators 13*, 243–250 (1988).
61. P. Albert, P. Audebert, M. Armand, G. Bidan, and M. Pineri, *J. Chem. Soc.: Chem. Commun. 1986*, 1636–1638.
62. M. Sugi, *J. Mol. Electron. 1*, 3–17 (1985).
63. N. R. Couch, C. M. Montgomery, and R. Jones, *Thin Sol. Films 135*, 173–182 (1986).
64. I. R. Peterseon, *J. Mol. Electron. 2*, 95–99 (1986).
65. G. Leiser, B. Tieke, and H. Wegner, *Thin Sol. Films 68*, 77–90 (1980).
66. R. H. Tredgold and C. S. Winter, *Thin Sol. Films 99*, 81–85 (1983).
67. A. J. Vickers and R. H. Tredgold, *Thin Sol. Films 134*, 43–48 (1985).
68. R. H. Tredgold, *Thin Sol. Films 152*, 223–230 (1987).
69. A. Laschewsky, H. Ringsdorf, and G. Schmidt, *Thin Sol. Films 134*, 153–172 (1985).
70. S. Sha, T. Hisatsune, T. Moriizumi, K. Ogawa, H. Tamura, N. Mino, Y. Okahata, and K. Ariga, *Thin Sol. Films 179*, 277–282 (1989).
71. M. Kakimoto, M. Suzuki, T. Konishi, Y. Imai, M. Iwamoto, and T. Hino, *Chem. Lett. 1986*, 823–826.
72. M. Uekita, H. Fwaji, and M. Murata, *Thin Sol. Films 160*, 21–32 (1988).
73. S. Baker, A. Seki, and J. Seto, *Thin Sol. Films 180*, 263 (1989).
74. B. V. Mislavsky, R. G. Yusupov, N. S. Maslova, Yu. N. Moiseev, S. V. Savinov, D. A. Znamensky, and A. Yu. Rozhkov, in *Technical Papers, Soc. Plastic Engrs. 40*, 2174–2178 (1994).
75. D. A. Znamensky, B. N. Levonovich, S. N. Maksimovskii, P. P. Sidorov, P. A. Todua, and V. N. Ulasyuk, *Phys. Rapid Commun. 8*, 18–20 (1990).
76. D. A. Znamensky, R. G. Yusupov, and B. V. Mislavsky, *Thin Sol. Films 219*, 215–220 (1992).

8

Synthesis of Fluorinated Poly(Aryl Ether)s Containing 1,4-Naphthalene Moieties

FRANK W. MERCER, MATILDA M. FONE,
MARTIN T. McKENZIE, and ANDY A. GOODWIN

8.1. INTRODUCTION

Linear aromatic polymers have long been known for their usefulness in meeting the high-performance requirements for structural resins, polymer films, and coating materials needed by the aerospace and electronics industries. Aromatic polyimides[1,2] and poly(aryl ether ketone)s[3-6] are the polymers of choice for these applications because of their unique combination of chemical, physical, and mechanical properties. Poly(aryl ether ketone)s are especially desirable because they are economically accessible by both nucleophilic and electrophilic routes. In addition, because of the flexibilizing ether and ketone groups present in the polymer backbone, poly(aryl ether ketone)s are generally more easily processed than polyimides. In recent years, poly(aryl ether ketone)s have been prepared containing 2,6-naphthalene and 1,5-naphthalene units. Ohno[4] and Hergenrother[5] reported that polyetherketones containing naphthalene moieties in the main chain possess higher T_g's but low solubility in ordinary organic solvents.

Considerable attention has also been devoted to the preparation of fluorine-containing polymers because of their unique properties and high-temperature

FRANK W. MERCER, MATILDA M. FONE, and MARTIN T. McKENZIE · Corporate Research and Development, Raychem Corporation Menlo Park, California 94025. ANDY A. GOOD-WIN · Department of Materials Engineering, Monash University, Clayton, Victoria, Australia.

Fluoropolymers 1: Synthesis, edited by Hougham *et al.*, Plenum Press, New York, 1999.

performance. Among the high-performance fluorinated polymers being studied for use in aerospace and electronics applications are the fluorinated aromatic poly-ethers prepared containing hexafluoroisopropylidene (HFIP) units.[7] Polymers containing HFIP units have been studied for applications as films, coatings for optical and microelectronics devices, gas separation membranes, and as a matrix resin in fiber-reinforced composites. Frequently the incorporation of HFIP units into the polymer backbone leads to polymers with increased solubility, flame resistance, thermal stability, and glass transition temperature, while also resulting in decreased color, crystallinity, dielectric constant, and moisture absorption.

Recently the synthesis and characterization of novel fluorinated poly(aryl ether)s containing perfluorophenylene moieties[8–10] was also reported. These fluorinated polyethers were prepared by reaction of decafluorobiphenyl with bisphenols. These polymers exhibit low dielectric constants, low moisture absorption, and excellent thermal and mechanical properties. Tough, transparent films of the polymers were prepared by solution-casting or compression-molding. The fluorinated poly(aryl ether)s containing perfluorophenylene moieties are good candidates for use as coatings in microelectronics applications.

As part of an effort to develop high-performance, high-temperature-resistant polymers for microelectronics applications, we also recently described a series of both partially fluorinated and nonfluorinated poly(aryl ether ketone)s containing amide, amide–imide, cyano oxadizole, or pyridazine groups and characterized their thermal and electrical properties.[11]

On the basis of the above studies reported thus far, we have designed and synthesized a series of novel poly(aryl ether)s containing both hexafluoroisopro-pylidene and 1,4-naphthalene moieties. We found these new polyarylethers to have good solubility, high T_g's, and excellent thermal stability. We report herein the synthesis and characterization of these poly(aryl ether)s containing fluorinated 1,4-naphthalene moieties.

8.2. EXPERIMENTAL

8.2.1. Starting Materials

N,N-Dimethylacetamide (DMAc), 4-fluorobenzoic acid, 4-fluorobenzoyl chloride, aluminum chloride, 1-bromonaphthalene, nitrobenzene, ferric chloride, dimethyl sulfone, 4,4′-dihydroxybiphenyl (DHB), and potassium carbonate were obtained from Aldrich and used without purification. 4,4-(Hexafluoroiso-propylidiene)-diphenol (6F-BPA), 9,9-bis(4-hydroxyphenyl)fluorene (HPF), and 1,1-bis(4-hydroxyphenyl)-1-phenylethane (Bisphenol AP) were obtained from Ken Seika Corporation and used without purification. 4,4′-Dihydroxydiphenyl sulfone (DHDS) was obtained from Nachem Incorporated and used without purification.

8.2.1.1. 2,2-Bis[4-(1-Naphthoxy)Phenyl]Hexafluoropropane (**1**)

A 500-ml three-neck round-bottom flask equipped with nitrogen inlet, thermometer, overhead stirrer, and condenser was charged with 6F-BPA (24.5 g, 0.073 mol), 1-bromonaphthalene (31.0 g, 0.15 mol), potassium carbonate (21.9 g, 0.162 mol), copper (I) iodide (0.8 g), and DMAc (250 ml). The reaction mixture was purged with nitrogen for 10 min and then heated, with stirring, at about 150°C under nitrogen for 5 days. The mixture was allowed to cool to room temperature and poured slowly into 500 ml of water, resulting in the formation of a water-insoluble oil. The water/DMAc mixture was carefully decanted from the oil. The oil was dissolved in 200 ml of a 25/75 (by volume) mixture of hexane/toluene and was washed once with 100 ml of 5 wt% NaOH in water and twice with 100 ml of water. The hexane/toluene mixture was dried over magnesium sulfate, filtered, and concentrated under reduced pressure to yield a brown oil (38.4 g), which partially crystallized on standing overnight at room temperature. The product was suspended in a 50/50 (by weight) mixture of ethanol/methanol and filtered to yield an off-white crystalline solid (m.p. 94–95°C). The yield was 34.2 g (79%). GC/MS $m^+/e = 588$. Calculated for $C_{49}H_{28}O_4F_8$: C, 71.43; H, 3.77; F, 19.37. Found: C, 71.54; H, 3.63; F, 19.51.

8.2.1.2. 2,2-Bis[4-(4-{4-Fluorobenzoyl}-1-Naphthoxy)Phenyl]Hexafluoropropane (**2**)

This compound was prepared as depicted in Scheme 1 using the following procedure: To a 250-ml round-bottom flask was added 10.00 g (0.0170 mol) of 2,2-bis[4-(1-naphthoxy)phenyl]hexafluoropropane (**1**), 5.71 g (0.0360 mol) of 4-fluorobenzoyl chloride, 2.3 g (0.0246 mol) of dimethyl sulfone, and 132 g of dichloromethane. The mixture was stirred under nitrogen until the solids dissolved, then cooled in an ice bath, and 14.5 g (0.109 mol) of aluminum chloride was added. The mixture was stirred for 1 h at ice-bath temperature, 24 h at room temperature, and 1 h at reflux. The mixture was allowed to cool to room temperature and poured into methanol. The resulting solid was filtered, washed with methanol and water, dried, and recrystallized from DMAc to yield 11.6 g (82% yield). m.p. = 261–263°C. Calculated for $C_{49}H_{28}O_4F_8$: C, 70.67; H, 3.39; F, 18.25. Found: C, 70.74; H, 3.36; F, 18.54.

8.2.2. Polymerizations

8.2.2.1. Homopolymerization of **1**: Preparation of 6FNE

As depicted in Scheme 2, homopolymerization of **1** to form the aromatic poly(aryl ether) 6FNE utilized the Scholl reaction in nitrobenzene with anhydrous ferric chloride at room temperature. A 100-ml round-bottom flask was charged with **1** (3.6 g, 0.0061 mol) and dry nitrobenzene (25 ml). Anhydrous ferric chloride (2.96 g, 0.0183 mol) was added portionwise with stirring over 15 min. The mixture was stirred under nitrogen for 6 h. The mixture was poured into 300 ml of methanol acidified with 1 ml of concentrated HCl. The precipitate was collected by filtration, washed three times with methanol and once with deionized water, and dried to yield 3.2 g (89%) of a tan powder.

8.2.2.2. Poly(Aryl Ether Ketone) Synthesis

Aromatic poly(aryl ether ketone)s containing 1,4-naphthalene moieties were prepared by the reaction of a bisphenol and **2** in the presence of potassium carbonate in DMAc at 160°C as depicted in Scheme 3. A typical polymerization was carried out as follows: To a 100-ml round-bottom flask was added 8.32 g (0.010 mol) of **2**, 3.36 g (0.010 mol) of 4,4′-(hexafluoroisopropylidene) diphenol, 51.2 g of DMAc, and 3.1 g (0.022 mol) of potassium carbonate. The mixture was heated to 160°C with stirring under nitrogen for 18 h. The mixture was allowed to cool to room temperature. The polymer was precipitated by pouring the reaction mixture into a blender containing about 100 ml of water, filtered, washed three times with water and dried to yield 8.1 g (92% yield) as a white powder.

8.2.3. Polymer Films

6FNE was spin-coated from diethylbenzene at 17 wt% solids onto glass substrates. Solutions of polymers **3–7** (15–20 wt% solids) in xylene/NMP mixture (1 : 4 by weight) were spin-coated onto glass substrates. All the coatings were dried for 1 h at 100°C and for 45 min at 200°C. The films, about 10 μm thick, were released from the glass substrates by placing the substrate in deionized water after cooling to room temperature. All six poly(aryl ether)s yielded tough, flexible, creasable films.

8.2.4. Measurements

Dielectric constants were measured using the previously described fluid displacement method.[12] The capacitance of the films was measured using circular gold electrodes (1 in. diameter) mounted in a brass dielectric cell held to a constant 25°C and a GenRad Precision LC Digibridge (Model 1688) at 10 KHz. Percent relative humidity was measured using a General Eastern dewpoint hygrometer (System 1100DP). Glass transition temperatures (T_g) reported in this chapter were determined in air using differential scanning calorimetry (DSC) with a heating rate of 10°C/min. Thermal gravimetric analyses (TGA) were determined in air using a heating rate of 20°C/min. Both DSC and TGA were performed on a Seiko SSC 5200 System DSC 220C TGA/DTA 320. IR spectral analysis was performed on a Bio-Rad FTS-60A FTIR.

The molecular-weight distributions were measured using a Waters GPC in the dual-detector mode (DRI and UV). The UV detector was operated at 254 nm. The samples were prepared by dissolving 2 mg of polymer in 10 ml of THF. The injection volume was 200 µl. Separations were effected using two Polymer Labs 10-µm PL mixed-B columns. THF was used as the mobile phase. The molecular-weight distributions were calculated relative to narrow polystyrene standards ranging from 10^2 to 4×10^6 M_W.

Proton-decoupled ^{13}C-NMR spectra were recorded on a Varian XL-300 operating at 75.4 MHz. Approximately 250 mg of the sample was dissolved in 3 ml of deuterated chloroform. ^{13}C chemical shifts were referenced internally to $CDCl_3$ (77 ppm). A delay of 200 s was used to ensure relaxation of all the carbon nuclei and 1000 transients were collected to assure a good signal-to-noise ratio.

Low-temperature dynamic mechanical analysis (DMA) studies were carried out on a Perkin-Elmer DMA7 operating in the penetration mode. Samples were scanned at 3°C/min and 1 Hz between −140°C and ambient temperature. The temperature was calibrated using high-purity indium and n-octane standards. Above ambient temperature studies were carried out using a Rheometric Scientific DMTA Mk II operating in dual-cantilever bending mode. Samples were scanned from ambient, through the glass transition, at 2°C/min over the frequency range 0.1–30 Hz. Samples were prepared by pressing together solvent cast films at 300°C. The principal parameter determined was tan δ.

8.3. RESULTS AND DISCUSSION

The syntheses of **1** utilized the Ullmann ether synthesis.[13] Reaction of 2 mol of 1-bromonaphthalene with 4,4-(hexafluoroisopropylidene)diphenol afforded the desired product **1**. The reaction was carried out in DMAc at 160°C in the presence of potassium carbonate as the base and copper (I) iodine as the reaction catalyst to yield **1**, as depicted in Scheme 1. The reaction proceeded slowly but in good yield with easy isolation of the desired compound. Acylation of **1** with 4-fluorobenzoyl chloride to prepare **2** was carried out under modified Friedel–Crafts reaction conditions[14] using dimethyl-sulfone as catalyst moderator. Both **1** and **2** were easily recrystallized to yield high-purity monomers suitable for polymerizations.

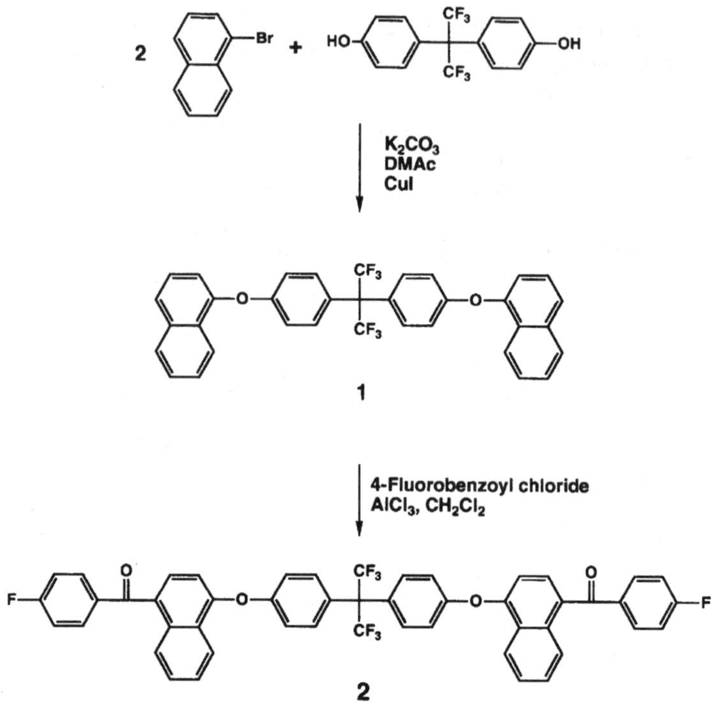

Scheme 1. Synthesis of monomer **2**.

Compounds **1** and **2** were identified by FTIR and ^{13}C-NMR. The ^{13}C proton decoupled spectra for **1** and **2** are dominated by signals ranging from 62 to 195 ppm. The ^{13}C chemical shift assignments were made based on comparisons with 4,4′-(hexafluoroisopropylidene)diphenol and from calculations based on substituted benzenes and naphthalenes.[15] The ^{13}C-NMR spectrum clearly showed that the Friedel–Crafts acylation of **1** by 4-fluorobenzoyl chloride yielded the 1,4-addition product exclusively. The ^{13}C chemical shifts for **2** are listed in Table 8.1. The key structural features in the FTIR spectrum of **2** include the following absorptions: aromatic C—H, 3074 cm^{-1}, ketone C=O, 1658 cm^{-1}, aromatic ether Ar—O—Ar, 1245 cm^{-1}, and C—F, 1175 cm^{-1}.

The homopolymerization of **1** consists of a room-temperature reaction of the monomer dissolved in nitrobenzene in the presence of anhydrous ferric chloride. Polymerizations were carried out under a stream of dry nitrogen. As depicted in Scheme 2, the homopolymerization of **1** to form 6FNE takes place by means of the Scholl reaction. The mechanism of the Scholl reaction was assumed to proceed through a radical-cation intermediate derived from the single-electron oxidation of the monomer and its subsequent electrophilic addition to the nucleophilic monomer. The reaction releases two hydrogens, both as protons, to form the

Table 8.1. [13]C Chemical Shifts (ppm) of 2,2-Bis[4-(4-{4-Fluorobenzoyl}-1-Naphthoxy)-Phenyl]Hexafluoropropane (**2**)

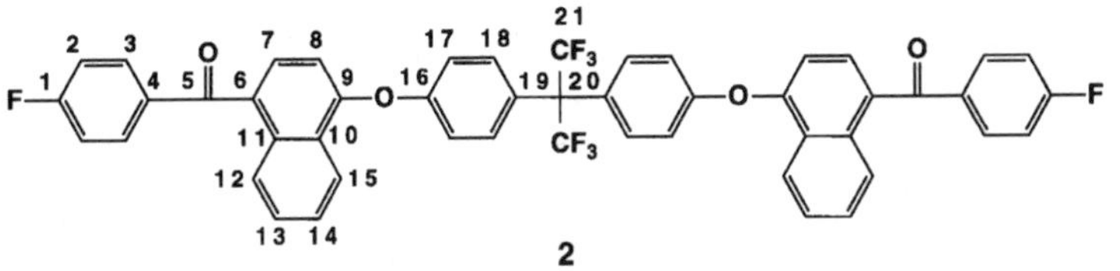

C1 = (167.4, 164.1)	C11 = 134.8
C2 = 115.7	C12 = 129.0
C3 = 132.0	C13 = 132.8
C4 = 133.0	C14 = 126.6
C5 = 195.6	C15 = 132.0
C6 = 131.6	C16 = 157.3
C7 = 128.3	C17 = 118.6
C8 = 115.4	C18 = 132.9
C9 = 155.1	C19 = 128.6
C10 = 126.8	C20 = (64.6, 64.3, 64.0, 63.7, 63.4, 63.1, 62.8)
	C21 = (118.6, 115.4, 122.2, 125.8)

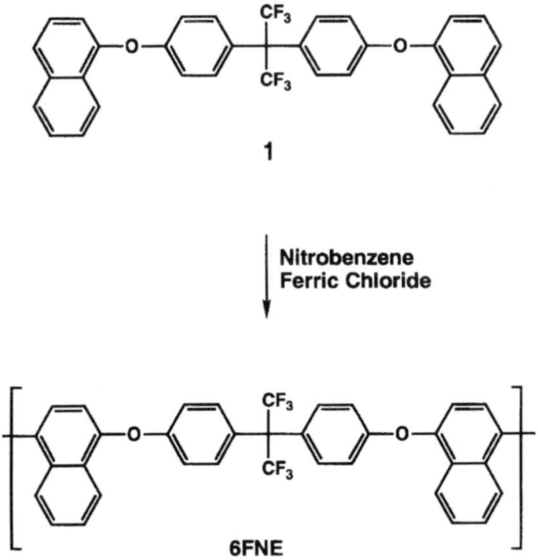

Scheme 2. Homopolymerization of **1**.

neutral polymer.[16] More recently, Percec[17] described the use of the Scholl reaction to prepare poly(ether ketone)s and poly(ether sulfone)s containing 4,4'-disubstituted-1,1'-dinaphthyl units.

The homopolymerization of **1** was performed with a molar ratio of [FeCl$_3$]/[monomer] equal to 3. The isolated polymer was thoroughly washed with methanol and water to remove the residual nitrobenzene and iron salts. 6FNE was soluble in a broad range of solvents, including NMP, DMAc, xylene, and THF. 6FNE displayed a T_g of 247°C and a TGA onset of decomposition in air at 523°C. Films of 6FNE were tough, flexible, and creasable. The dielectric constant of 6FNE was 2.72 at 0% relative humidity (RH) and rose to only 2.79 at 83% RH. The FTIR spectrum of **3** is shown in Figure 8.1.

Figure 8.2 shows the α-relaxation of 6FNE, which is related to the glass transition process. In the first scan, a low-temperature shoulder is present on the main peak. Similar peaks have been observed in other rigid amorphous polymers.[18] It is claimed that this is not a true relaxation, but is due rather to the presence of defects that collapse on heating.[19] The peak is indeed diminished on the second scan, which suggests that it may be related to the thermal history of the glassy polymer. The T_g of 6FNE, defined at the temperature corresponding to the peak maximum, increases from 242°C on the first scan to 247°C on the second. In addition, the intensity of the α-peak increases significantly. An analysis of the frequency dependence of the T_g, using an Arrhenius equation, shows that the apparent activation energy remains constant on both the first and second scans, at

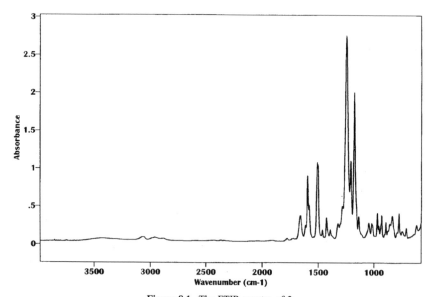

Figure 8.1. The FTIR spectra of **3**.

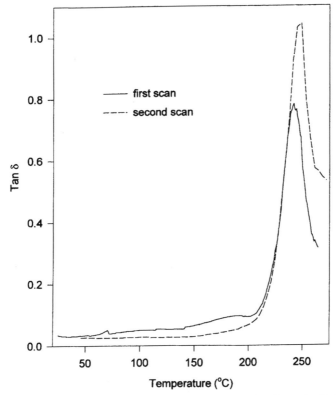

Figure 8.2. Temperature dependence on tan δ at 1 Hz in the α-relaxation region for 6FNE.

800 kJ/mol. This high value of the activation energy is a reflection of the rigid chain structure of 6FNE.

Figure 8.3 shows the temperature dependence of tan δ in the β-relaxation region of 6FNE. The β-peak is typically broad (centered around −80°C) and weak and arises from the localized motions of the polymer chains. The origins of secondary relaxations in rigid aromatic polymers have been attributed to superpositions of two motional processes: a low-temperature component reflecting local intrachain motions and a high-temperature cooperative component attributed to the local motions in ordered amorphous regions.[20] The mechanism of the process is suggested to be flips of the phenyl rings. The β-process is known to be sensitive to water content, aging history, and morphology.

Polycondensation reactions of **2** with diphenols were carried out in DMAc at 160°C using an excess of potassium carbonate to yield viscous solutions of the desired poly(aryl ether ketone)s. Judging by the viscosity increase, the polymerization reaction was near completion after only about 8 h at 160°C. Aqueous workup of the reaction mixture yielded the 1,4-naphthalene-containing polymers as white powders. The isolated polymers were thoroughly washed with water to remove the residual DMAc and potassium salts.

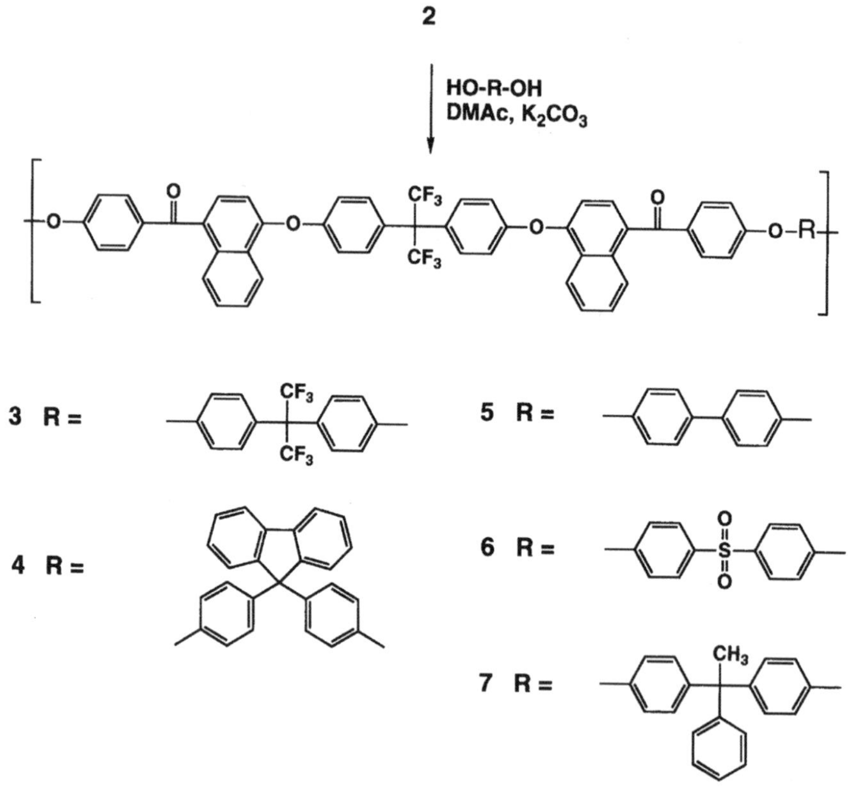

2

Scheme 3. Synthesis of fluorinated poly(aryl ether ketone)s **3–7**.

Poly(ether ketone)s **3**, **4**, **5**, **6**, and **7** were soluble in polar aprotic solvents such as DMAc and NMP and in chlorinated solvents such as chloroform. The improved solubility of these fluorinated poly(ether ketone)s can be explained by the presence of both the flexible hexafluoroisopropylidene groups and the bulky 1,4-naphthalene moieties, which inhibit polymer crystallization and facilitate the penetration of solvent molecules between the polymer chains.

The T_g's for the poly(ether ketone)s prepared from **2** and 6F-BPA, HPF, DHB, DHS, and Bisphenol AP were 194, 230, 200, 205 and 198°C, respectively. The T_g for **3** is higher than for the fluorinated polyetherketone **8**, reported to be 175°C by R. N. Johnson *et al.*[21] The higher T_g of **3** compared to **8** is attributed to the presence of the bulky 1,4-naphthalene moieties in **3**. The high T_g of **4** is

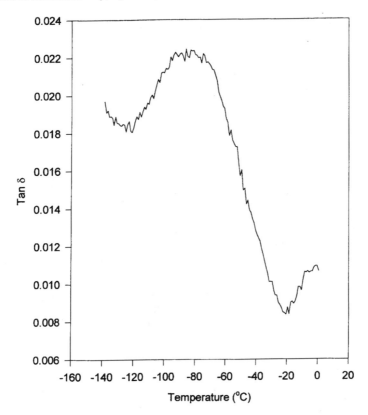

Figure 8.3. Temperature dependence on tan δ at 1 Hz in the β-relaxation region for 6FNE.

attributed to the incorporation of the cardo group from the 4,4′-dihydroxyphenyl-fluorene. Korshak *et al.*[22] previously reported that the incorporation of cardo groups in polyarylates endows them with enhanced glass transition temperatures, improved thermal stability, and increased solubility. TGA of **3**, **4**, **5**, **6**, nd **7** reveals that the polymers exhibit initial weight loss in air at about 500°C (scan rate = 20°C/min). The broad temperature range between the T_g and the thermal decomposition temperature makes these polymers attractive for thermoforming processes. Table 8.2 presents the characterization data of the poly(ether ketone)s.

At 0% RH, **3**, **4**, **5**, **6**, and **7** display dielectric constants (measured at 10 KHz) of 3.32, 3.23, 3.36, 3.63, and 3.32, respectively. At about 80% RH, the dielectric constants of **3**, **4**, **5**, **6**, and **7** increased to 3.88, 3.52, 3.87, 4.18, and 3.47, respectively. The relationships of dielectrics constant to relative humidity for polymers **3**, **4**, **5**, **6**, and **7** are depicted graphically in Figure 8.4. For comparison, the fluorinated poly(aryl ether ketone) **8**, made by the reaction of 6F-BPA and 4,4′-

difluorobenzophenone, displayed a dielectric constant of 2.94 at 0% RH and increased to 3.25 at 58% RH.

8

All of the poly(ether ketone)s containing 1,4-naphthalene moieties described in this chapter have dielectric constants greater than that of **8**. Since the dielectric constant of a polymer is a function of the polymer's total polarizability, α_T[23,24] the higher dielectric constants of the poly(ether ketone)s **3**, **4**, **5**, **6**, and **7**, compared to that of **8**, are attributed to the more polar nature of the naphthalene moieties present in **3**, **4**, **5**, **6**, and **7**. In addition, polymer **6** displays the highest dielectric constant of the poly(ether ketone)s described in this chapter. The higher dielectric constant observed in **6** is attributed to the presence of the polarizable sulfone groups in the polymer. Polymer **6** also displays the greatest increase in dielectric constant with increasing relative humidity. The larger increase in the dielectric constant of **6** with increasing relative humidity, compared to **3**, **4**, **5**, and **7**, is attributed to higher levels of moisture absorption facilitated by hydrogen-bonding of water to the sulfone moiety in **6**.

Size-exclusion chromatography of **3–7** showed that the polymers were of high molecular weight and displayed very high polydispersity. A low-molecular-weight fraction, which we attribute to cyclic oligomers, was observed for all five polymers tested. However, the presence of the oligomeric species did not have an especially negative effect on the polymer properties since all the polymers yielded flexible, creasable films showing good mechanical integrity. Molecular-weight

Table 8.2. Properties of Fluorinated Poly(Aryl Ether)s Containing 1,4-Naphthalene Moieties

Polymer	Diphenol	T_g (°C) (DSC)	T_g (°C) onset[a]	Dielectric constant		Molecular weight	
				0% RH	80% RH	M_n	M_w
6FNE	—	247	523	2.72	2.79	18,100	52,800
3	6F-BPA	194	524	3.32	3.88	6,900	51,000
4	HPF	230	513	3.23	3.52	6,200	45,700
5	DHB	200	510	3.36	3.87	13,800	58,500
6	DHDS	205	502	3.63	4.18	6,700	34,300
7	Bisphenol AP	198	503	3.32	3.74	5,900	38,600

[a] Onset of decomposition characterized by 5% wt loss in air.

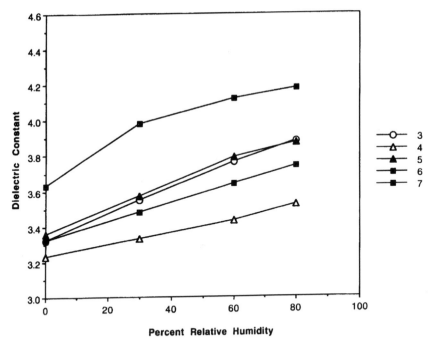

Figure 8.4. Effect of humidity on the dielectric constant of polyetherketones **3–7**.

distributions for **3** and **7** are depicted in Figure 8.5 and GPC results for **3–7** are tabulated in Table 8.2.

8.4. CONCLUSIONS

Six novel fluorinated poly(aryl ether)s containing 1,4-naphthalene moieties were synthesized in high yield using 2,2-bis[4-(1-naphthoxy)phenyl]hexafluoropropane (**1**). Oxidative coupling of **1** yielded a polymer with high T_g, low moisture absorption, and low dielectric constant that could be cast into flexible films. The low dielectric constant and low moisture absorption of 6FNE may make it useful as a dielectric insulator in microelectronics applications.

Reaction of **1** with 4-fluorobenzoyl chloride yielded the difluoro-containing monomer **2**, which is readily polymerized with a bisphenol using potassium carbonate in *N,N*-dimethylacetamide to yield poly(ether ketone)s. The five poly-(ether ketone)s prepared were soluble in polar aprotic solvents and were cast into flexible, creasable films showing good thermal stability. We have demonstrated that 1-phenoxy-substituted naphthalene moieties undergo a Friedel–Crafts acyla-

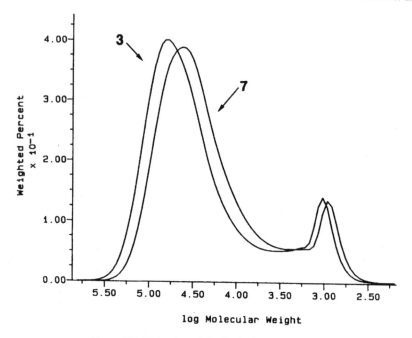

Figure 8.5. Molecular weight distributions for **3** and **7**.

tion reaction at the 4-position when reacted with 4-fluorobenzyl chloride to yield the 1,4-substituted product in high yield. The appropriately substituted monomers were easily prepared and subjected to displacement polymerizations with various bisphenols to yield polymeric products. High molecular weight was readily achieved, and structural variety was introduced by selection of the appropriate bisphenol.

ACKNOWLEDGMENT: The assistance at Raychem Corporation of Gloria Merlino in obtaining TGA/DSC results is greatly appreciated.

8.5. REFERENCES

1. A. H. Frazer, in *Polymer Reviews*, John Wiley and Sons, New York (1968), pp. 159–175.
2. K. Mittal, Ed., *Polyimides: Synthesis Characterization, and Applications, Vols. 1 and 2*, Plenum Press, New York (1984).
3. T. E. Attwood, P. C. Dawson, J. L. Freeman, L. R. H. Hoy, J. B. Rose, and P. A. Staniland, *Polymer* 22, 1096 (1981).
4. M. Ohno, T. Takata, and T. Endo, *J. Polym. Sci. Pt. A: Polym. Chem. 33*, 2647 (1995).
5. P. M. Hergenrother, B. J. Jensen, and S. J. Havens, *Polymer 29*, 358 (1987).
6. G. L. Tullos, P. E. Cassidy, and A. K. St. Clair, *Macromolecules 24*, 6059 (1991).

7. M. Bruma, J. W. Fitch, and P. E. Cassidy, *J. Macromol. Sci.—Rev. Macromol. Chem. Phys. C36*, 119 (1996).
8. F. W. Mercer and T. D. Goodman, *Polym. Preprints 32* (2), 189 (1991).
9. F. W. Mercer, T. D. Goodman, J. Wojtowicz, and D. Duff, *J. Polym. Sci. Pt. A: Polym. Chem. 30*, 1767 (1992).
10. J. Irvin, C. J. Neef, K. M. Kane, and P. E. Cassidy, *J. Polym. Sci. Pt. A: Polym. Chem. 30*, 1675 (1992).
11. F. W. Mercer, M. T. McKenzie, G. Merlino, and M. M. Fone, *J. Appl. Polym. Sci. 56*, 1397 (1995).
12. F. W. Mercer and T. D. Goodman, *High Perf. Polym. 3* (4) 297 (1991).
13. J. Lindley, *Tetrahedron 40*, 1433 (1984).
14. P. J. Horner, V. Jansons, and H. C. Gors, European Patent 0 178 184 (1986).
15. D. F. Ewing, *Org. Magn. Reson. 12*, 499 (1979).
16. R. G. Feasey, A. Turner-Jones, P. C. Daffurn, J. L. Freeman, *Polymer 14*, 241 (1973).
17. V. Percec, J. H. Wang, Y. Oishi, and A. E. Feiring. *J. Poly. Sci. Pt. A: Polym. Chem. 29*, 965 (1991).
18. T. Sagusa and M. Hagiwara, *Polymer 26*, 501 (1985).
19. L. David, and S. Etienne, *Macromolecules 26*, 4489 (1993).
20. L. David and S. Etienne, *Macromolecules 25*, 4302 (1992).
21. R. N. Johnson, A. G. Farnham, R. A. Clendinning, W. F. Hale, and C. N. Merriam, *J. Polym. Sci. A-1 5*, 2375 (1967).
22. V. V. Korshak, S. V. Vinogradova, and Y. S. Vygodskii, *J. Macromol. Sci.—Rev. Macromol. Chem. C11* (1), 45 (1974).
23. G. Hougham, G. Tesoro, and J. Shaw, *Polym. Mat. Sci. Engr. 61*, 369 (1989).
24. G. Hougham, G. Tesoro, A. Viehbeck, and J. Chapple-Sokol, *Polym. Preprints 34* (1), 375 (1993).

9

Synthesis and Properties of Fluorine-Containing Aromatic Condensation Polymers Obtained from Bisphenol AF and Its Derivatives

SHIGEO NAKAMURA and YUKO NISHIMOTO

9.1. INTRODUCTION

Fluorine-containing polymers exhibit unique chemical and physical properties and high performance that are not observed with other organic polymers. They possess high thermal stability, high chemical stability, a low coefficient of friction, low adhesion, water and oil repellency, low refractive index, and outstanding electric insulation. In addition, there have recently been new expectations of selective permeability, piezoelectricity, and biocompatibility.

These characteristics of fluorine-containing polymers are imparted by the highest electronegativity (4.0), the third smallest van der Waals radius (0.135 nm), and low polarizability ($\alpha = 1.27 \times 10^{-24}$ cm^3) of the fluorine atom, strong bonding energy (472 kJ/mol) and small atomic distance (0.138 nm) of the C—F bond, and the weak intermolecular cohesive energy of the polymer.

SHIGEO NAKAMURA · Faculty of Engineering, Kanagawa University, Rokkakubashi, Kanagawa-ku, Yokohama 221–8686, Japan.   YUKO NISHIMOTO · Faculty of Science, Kanagawa University, Tsuchiya, Hiratsuka 259-1293, Japan.

Fluoropolymers 1: Synthesis, edited by Hougham *et al.*, Plenum Press, New York, 1999.

Research on fluorine-containing condensation polymers is rather limited compared to that on fluorine-containing addition polymers. This fact is attributed to the difficulty in synthesis and the high cost of fluorine-containing condensation monomers. Recently, 2,2-bis(4-hydroxyhpenyl)-1,1,1,3,3,3-hexafluoropropane (Bisphenol AF) with a hexafluoroisopropylidene unit, $HOC_6H_4C(CF_3)_2$-C_6H_4OH, was produced commercially from hexafluoroacetone and phenol, and now Bisphenol AF and its derivatives are available as condensation monomers.

Replacement of isopropylidene units in polymer backbones by units of hexafluoroisopropylidene is known to enhance solubility, water and oil repellency, thermal stability, and glass transition temperature (T_g) and to decrease crystallinity and water absorption.[1]

This chapter deals with the synthesis and properties of fluorine-containing condensation polymers from Bishpenol AF and its derivatives. Much of the work reported has been conducted in our laboratory. An extensive review has already been published by Cassidy and co-workers[1] on the synthesis of fluorine-containing condensation polymers.

9.2. POLY(CARBONATE)S

9.2.1. Synthesis

A fluorine-containing poly(carbonate) was first synthesized by Knunyants and co-workers from Bisphenol AF and phosgene.[2] However, detailed properties of this polymer other than the softening temperature of 170°C have not been reported.

Commerical poly(carbonate)s have been produced mainly through a low-temperature solution polycondensation of 2,2-bis(4-hydroxyphenyl)propane (Bisphenol A) using phosgene as a carbonylation agent.[3] However, phosgene vapor is extremely toxic, and the preparation using phosgene is not usually suitable for a laboratory-scale synthesis of poly(carbonate)s. Disphosgene (trichloromethyl chloroformate), a liquid, and triphosgene [bis-(trichloromethyl)-carbonate], a solid, react similarly to phosgene as carbonylation agents but can be handled more easily and safely than phosgene.

It has been reported that high-molecular-weight poly(carbonate)s can be synthesized using N,N'-carbonyldiimidazole as a condensing agent under mild conditions.[4,5] However, the reagent is very sensitive to moisture and must be handled with extreme care.

Bisphenol-AF-derived poly(carbonate) (2) has been synthesized by the two-phase transfer-catalyzed polycondensation of Bisphenol AF (1) with trichloro-methyl chloroformate (TCF) in organic-solvent–aqueous-alkaline solution systems with a variety of quaternary ammonium salts at room temperature (Scheme 1).[6]

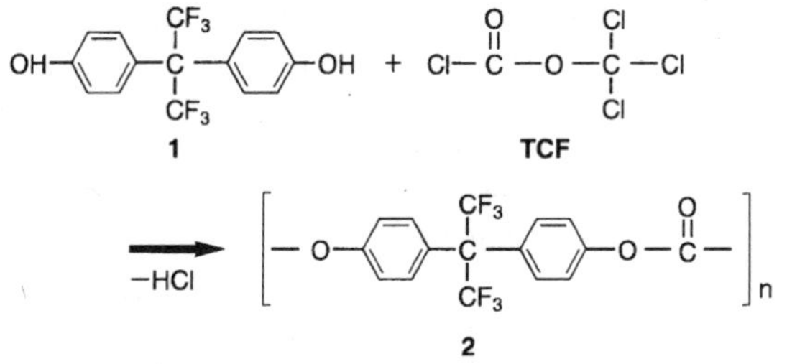

Scheme 1

The following quaternary ammonium salts are used as phase transfer catalyst: *tetra-n*-butylammonium chloride (TBAC), *tetra-n*-butylammonium bromide (TBAB), benzyltriethylammonium chloride (BTEAC), and benzyltriethylammonium bromide (BTEAB). Chlorinated hydrocarbons, such as dichloromethane (DCM), chloroform (CF), tetrachloromethane (TCM), 1,2-dichloromethane (DCE), and nitrobenzene (NB) are used as solvents. The effects of phase-transfer catalyst and solvent on the yield and reduced viscosity are summarized in Table 9.1.

When TBAB is used as a phase-transfer catalyst, sodium hydroxide as a base, and DCE as a solvent, both the molecular weight and yield of the poly(carbonate) are relatively high. Bisphenol AF-derived poly(carbonate) (**2**) having reduced viscosity of 0.35 dl/g is obtained in a 84% yield at ambient temperature under the

Table 9.1. Effects of Phase-Transfer Catalyst and Organic Solvent on the Yield and Reduced Viscosity of Bisphenol-AF-Derived Poly(Carbonate)[6]

No.	Reaction conditions		Polymer	
	Catalyst	Solvent	Yield (%)	η_{red} (dl/g)
1	TBAB	DCM	73	0.41
2	TBAB	CF	74	0.30
3	TBAB	TCM	83	0.14
4	TBAB	DCE	84	0.35
5[a]	TBAB	DCE	87	0.05
6[b]	TBAB	DCE	66	0.05
7[c]	TBAB	DCE	71	0.37
8	TBAB	NB	76	0.37
9	TBAC	DCM	84	0.25
10	BTEAB	DCM	60	0.13
11	BTEAC	DCM	63	0.35

[a] 5.00 mmol of TCF was used.
[b] 14.3 mmol NaOH was used.
[c] KOH (28.5 mmol) was used instead of NaOH.

optimum reaction conditions, in which 7.50 mmol of Bisphenol AF is reacted with 7.50 mmol of TCF using 3.15 mmol of TBAB, 30 ml of 1 *M* aqueous sodium hydroxide and 37.5 ml of DCE. Although polymers with slightly higher molecular weights are obtained using DCM and NB as organic solvents, the yields are lower. This rather lower molecular weight of Bisphenol AF poly(carbonate) (**2**) compared to Bisphenol A-derived poly(carbonate) (**3**) is attributed to trifluoromethyl groups having a higher electron-withdrawing property than methyl groups, which cause a decrease in reactivity of phenolic hydroxyl groups.

Random copoly(carbonate)s (**5**) are obtained by the reaction of mixtures of Bisphenol AF (**1**) and Bisphenol A (**4**) with TCF (Scheme 2).[6]

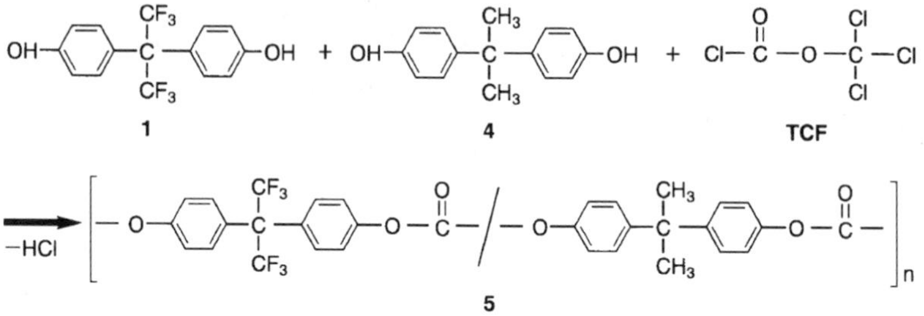

Scheme 2

Table 9.2 shows the results of a preparation of a series of copoly(carbonate)s having a wide range of hexafluoroisopropylidene unit content. Reduced viscosity of copoly(carbonate)s (**5**) decreases gradually with increasing feed ratio of Bisphenol AF (**1**) to Bisphenol A (**4**). This is attributed to the lower nucleophilicity of Bisphenol AF induced by strongly electron-withdrawing trifluoromethyl groups as described above.

Table 9.2. Preparation of Fluorine-Containing Poly(Carbonate)s[6]

Bisphenol AF/Bisphenol A (unit ratio)		Yield (%)	η_{red} (dl/g)
Feed	Polymer		
100/0	100/0	84	0.35
80/20	81/19	79	0.19
60/40	54/46	81	0.27
50/50	47/53	80	0.28
40/60	41/59	85	0.51
20/80	19/81	75	0.54
0/100	0/100	85	0.73

9.2.2. Properties

Aromatic poly(carbonate) from 2,2-bis(4-hydrozyphenyl)propane (Bisphenol A) and phosgene is utilized as one of the high-performance engineering thermoplastics and finds versatile applications owing to its excellent characteristics, including high thermal stability, high impact strength, high refractive index, high transparency, and self-extinguishing property.

The solubility and thermal stability of polymers are usually improved and the surface energy is decreased by substituting fluorine atoms for the hydrogen atoms in the isopropylidene units connecting phenylene units in the main chain.

It has been determined from X-ray diffraction measurements that poly(carbonate)s containing Bisphenol AF moiety are all amorphous.[6] The (T_g) of poly(carbonate)s increases with an increase in hexafluoroisopropylidene unit from 149°C for Bisphenol A poly(carbonate) (3) to 169°C for Bisphenol AF poly(carbonate) (2) (Table 9.3).[6] Thermooxidative stability is also improved by the introduction of fluorine atoms into the isopropylidene units. The 10% weight-loss temperature (DT_{10}) increases from 429 to 460°C and the residual weight (RW) at 500°C goes from 37 to 57% by perfluorination of the isopropylidene units.

The solubility is generally improved by the introduction of fluorine atoms into aromatic condensation polymers. Poly(carbonate)s containing hexafluoroisopropylidene units are much more soluble than Bisphenol A poly(carbonate) (3). All of the hexafluoroisopropylidene-unit-containing poly(carbonate)s become soluble in acetone, ethyl acetate, chloroform, and dimethyl sulfoxide (DMSO) in addition to the solvents of Bisphenol A poly(carbonate) (3). Colorless, transparent, and flexible films are prepared from hexafluoroisopropylidene-unit-containing poly(carbonate)s by casting or pressing.

The contact angle (θ) by water at 25°C in air is 84° for Bisphenol A poly(carbonate) (3) film (Table 9.4).[6] Introduction of 19% of Bisphenol AF unit increases the value of θ to 90° and from that point it is almost constant irrespective of fluorine content. This abrupt increase in θ is attributed to the migration and

Table 9.3. Thermal Properties of Fluorine-Containing Poly(Carbonate)s[6]

Bisphenol AF/Bisphenol A (unit ratio)	T_g (°C)	DT_{10} (°C)	RW (%)
100/0	169	460	57
81/19	163	442	48
54/46	160	437	50
47/53	158	436	48
41/59	156	439	41
19/81	152	434	45
0/100	149	429	37

Table 9.4. Water Contact Angle of Fluorine-Containing
Poly(Carbonate)s[6]

Bisphenol AF/Bisphenol A (unit ratio)	θ_w (degree)
100/0	91
81/19	91
54/46	92
47/53	91
41/59	91
19/81	90
0/100	84

concentration of trifluoromethyl groups to the surface of the polymer film owing to their lower surface energy as observed by ESCA.[7] The critical surface tension of Bisphenol AF poly(carbonate) (2) is determined to be 20 mN/m at 25°C from the Zisman plot using *n*-butanol–water mixtures as wetting liquids.[6] This value is much lower than the 45 mN/m at 25°C reported for Bisphenol A poly(carbonate) (3).[8]

The high refractive index of Bisphenol A poly(carbonate) (3) is decreased by the incorporation of the Bisphenol AF moiety. The refractive index of Bisphenol A poly(carbonate) (3) is 1.585,[9] whereas that of Bisphenol AF poly(carbonate)s (2) is 1.426.[6] Poly(carbonate)s containing a small amount of Bisphenol AF moiety can be used as sheaths of optical fibers of Bisphenol A poly(carbonate) (3).

The tensile strength and tensile modulus decrease and the elongation increases by the introduction of fluorine atoms into isopropylidene units of Bisphenol A poly(carbonate) (3),[6] i.e., poly(carbonate) becomes more flexible by the introduction of hexafluoroisopropylidene units. The increased flexibility is attributed to the weaker intermolecular interaction induced by fluorine atoms.

9.3. POLY(FORMAL)S

9.3.1. Synthesis

Hay *et al.*[10,11] have prepared high-molecular-weight a Bisphenol-A-derived poly(formal) (6) using a phase-transfer catalyst in DCM. A Bisphenol-AF-derived poly(formal) (7) is also synthesized by solution polycondensation of Bisphenol AF (1) with DCM in highly polar cosolvents in the presence of potassium hydroxide (Scheme 3).[12] Aprotic polar solvents such as *N,N*-dimethylformamide

(DMF), *N,N*-dimethylacetamide (DMAc), DMSO, hexamethylphosphoric triamide (HMPA), and *N*-methyl-2-pyrrolidone (NMP) may be used as cosolvents.

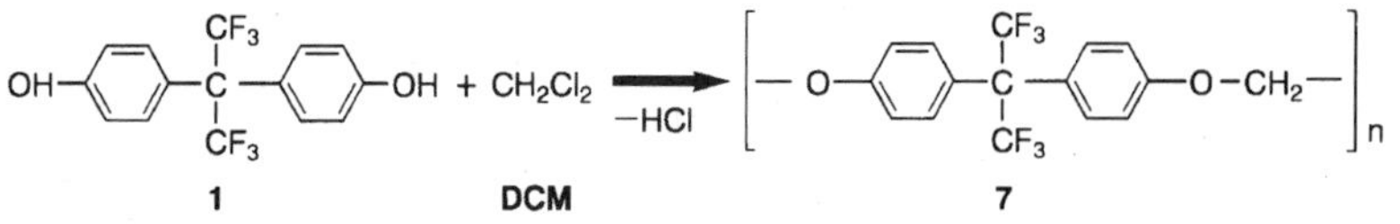

Scheme 3

A large excess of DCM to Bisphenol AF (**1**) is required for production of high-molecular-weight Bisphenol A poly(formal) (**7**) because DCM acts as the reactant as well as the cosolvent. The most suitable molar amount of DCM is about 12 times that of Bisphenol AF (**1**).

The effects of cosolvents on the reduced viscosity and yield are summarized in Table 9.5. DMAc and NMP lead to the formation of high-molecular-weight Bisphenol AF poly(formal) (**7**) in a high yield. The optimum reaction conditions are 48 mmol of DCM, 14 mmol of potassium hydroxide, and 5 ml of NMP for 5 mmol of Bisphenol AF, resulting in the formation of Bisphenol AF poly(formal) (**7**) with reduced viscosity of 4.62 dl/g in a 87% yield at 75°C.[12]

Fluorine-containing copoly(formal)s (**8**) are synthesized in a similar manner by reacting the mixtures of Bisphenol AF (**1**) and Bisphenol A (**4**) with DCM (Scheme 4).[12]

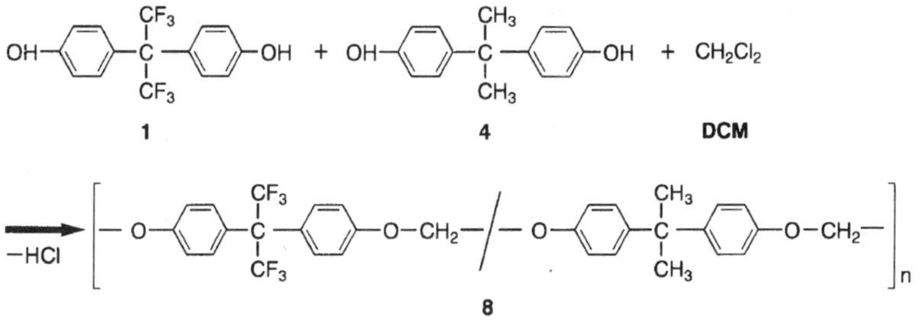

8
Scheme 4

Copoly(formal)s (**8**) with high reduced viscosities are readily obtained in high yields irrespective of the feed ratio of Bisphenol AF (**1**) and Bisphenol A (**4**) (Table 9.6).

9.3.2. Properties

Bisphenol-A-derived poly(formal) (**6**) shows poor solubility and is only soluble in dichloromethane, chloroform, THF, HMPA, and NMP. However, poly(formal)s containing the Bisphenol AF moiety are easily soluble in a wide variety of organic solvents, such as acetone, ethyl acetate, benzene, toluene,

Table 9.5. Effect of Cosolvent on the Yield and Reduced
Viscosity of Bisphenol-AF-Derived Poly(Formal)[12]

	Polymer	
Solvent	Yield (%)	η_{red} (dl/g)
DMF	49	0.11
DMAc	89	4.40
DMSO	92	1.96
HMPA	92	1.40
NMP	87	4.62

m-cresol, and most aprotic polar solvents in addition to the solvents for Bisphenol
A poly(formal) (**6**).[12]

Resistance to acids is improved by the introduction of fluorine atoms.
Bisphenol A poly(formal) (**6**) decomposes rapidly with significant coloration in
strong acids, whereas Bisphenol AF poly(formal) (**7**) is stable in concentrated
sulfuric acid.[12]

Poly(formal)s containing Bisphenol AF moiety are all amorphous.[12] The T_g
is 88°C for Bisphenol A poly(formal) (**6**) increases with increasing Bisphenol AF
content, and rises to 123°C for Bisphenol AF poly(formal) (**7**) (Table 9.7).[12]

The thermooxidative stability is improved by increasing the hexafluoroisopropylidene unit content.[12] The DT_{10} in air is raised from 363°C for Bisphenol A
poly(formal) (**6**) to 398°C for Bisphenol AF poly(formal) (**7**), and the RW at
500°C is increased from 48 to 73%.[12]

A contact angle by water is 86° for Bisphenol A poly(formal) (**6**), is
increased to 93° by the introduction of 28 mol% of hexafluoroisopropylidene
unit, and then becomes almost constant irrespective of fluorine content.[12] As

Table 9.6. Preparation of Fluorine-Containing
Poly(Formal)s[12]

Bisphenol AF/Bisphenol A (unit ratio)		Yield (%)	η_{red} (dl/g)
Feed	Polymer		
100/0	100/0	87	4.62
75/25	75/25	90	1.44
50/50	52/48	88	2.05
25/75	28/72	91	5.31
0/100	0/100	92	3.54

Table 9.7. Thermal Properties of Fluorine-Containing
Poly(Formal)s[12]

Bisphenol AF/Bisphenol A (unit ratio)	T_g (°C)	DT_{10} (°C)	RW (%)
100/0	123	398	73
75/25	114	389	60
52/48	108	375	60
28/72	100	359	56
0/100	88	363	48

observed for poly(carbonate)s, this is attributed to the concentration of hydrophobic trifluoromethyl groups on the surface of the films.

9.3.3. Thermal Degradation

In the thermal degradation (TG) curves of poly(formal)s, the onset temperatures of weight loss shift to higher temperatures with increasing Bisphenol AF content.[13] The weight loss of Bisphenol A poly(formal) (**6**) occurs in a single step, whereas the weight of hexafluoroisopropylidene-unit-containing poly(formal)s decreases in two stages and the inflection at about 500°C becomes more definite with increasing Bisphenol AF content. The typical TG–DTA (differential thermal

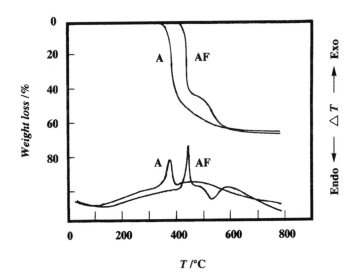

Figure 9.1. TG–DTA curves for Bisphenol A-derived poly(formal) (A) and Bisphenol AF-derived poly(formal) (AF).

analysis) curves are shown in Figure 9.1 for Bisphenol A poly(formal) (**6**) and Bisphenol AF poly(formal) (**7**). The first stage of the degradation is exothermic and the second one is endothermic in the DTA curve of Bisphenol AF poly(-formal) (**7**).

The FTIR spectra of the gas mixture evolved in thermal decomposition of Bisphenol AF poly(formal) (**7**) at various temperatures suggest the existence of benzene rings, C—O—C bonds, and C=C bonds. In a pyrogram of pyrolysis gas chromatography (Py-GC) of Bisphenol A (**3**), α-methylstyrene, phenol, p-cresol, 4-hydroxy-α-methylstyrene, and isopropyl phenol are observed as major peak products. The cleavage reactions shown in Scheme (5) is suggested for the formation of phenol and 4-hydroxy-α-methylstyrene from Bisphenol A (**3**).

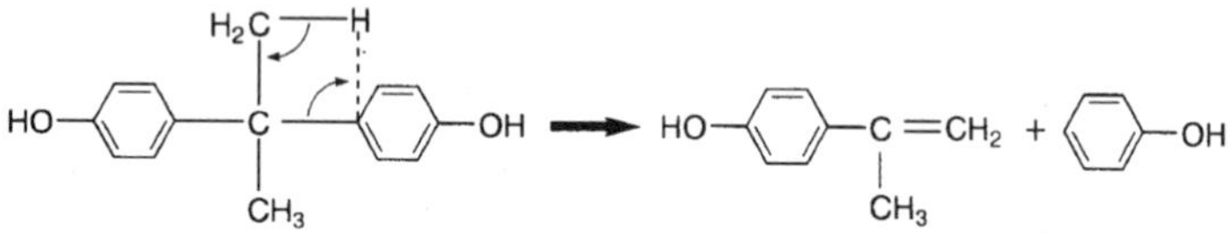

Scheme 5

In a pyrogram of Bisphenol A poly(formal) (**6**), the peak products are identified as α-methylstyrene, phenol, 4-hydroxy-α-methylstyrene, and isopropyl phenol by Py-GC/MS. These products are identical with the degradation products from Bisphenol A. In addition to the decomposition products of Bisphenol A, 4-isopropenyl anisole is also identified as a product. The pyrograms of Bisphenol AF poly(formal) (**7**) contain only two major species, pentafluoroisopropenyl benzene (product **1′**) and pentafluoroisopropenyl anisole (product **2′**). They correspond to α-methylstyrene, 4-hydroxy-α-methylstyrene from Bisphenol A poly(formal) (**6**) and are produced by the cleavage of phenylene–oxy bonds and oxy–methylene bonds according to (Scheme 6).

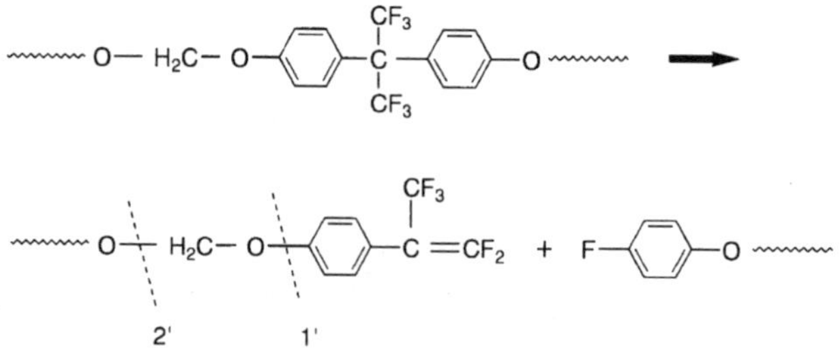

Scheme 6

The bond dissociation energy of phenylene–oxy bonds is not very different from that of oxy–methylene bonds. Accordingly, the two-step decrease in weight of hexafluoroisopropylidene-unit-containing poly(formal)s, especially Bisphenol AF poly(formal) (**7**), is not brought about by a different reaction mechanism from that of Bisphenol A poly(formal) (**6**), but results rather from the difference in vapor pressures, (i.e., the different rates of vaporization) between product **1'** and product **2'** themselves or their polymerized products. Specifically, product **2'** is produced by Scheme (5) and the cleavage of oxy–methylene bonds, and then product **1'** is produced by the secondary degradation of product **2'**. The more volatile **1'** is evaporated in the first stage and then the less volatile **2'** in the second stage, which is why the weight loss of hexafluoroisopropylidene-unit-containing poly(formal)s occurs in two steps in TG curves.

9.4. POLY(KETONE)S

9.4.1. Synthesis

Hexafluoroisopropylidene-unit-containing aromatic poly(ether ketone)s were first synthesized from an alkaline metal salt of Bisphenol AF (**1**) and 4,4'-difluoro-benzophenone.[14] Cassidy and co-workers prepared hexafluoroisopropylidene-unit-containing poly(ether ketone)s by condensing 2,2-bis[4-(4-fluorobenzoyl)-phenyl]-1,1,1,3,3,3-hexafluoropropane (**9**) and 2,2-bis[4-(4-fluorobenzoyl)-phenyl]propane (**10**) with Bisphenol AF (**1**) or Bisphenol A (**4**) (Scheme 7).[15] The reactions are nucleophilic aromatic displacements and were conducted in DMAc at 155–160°C with an excess of anhydrous potassium carbonate. After 3 to 6 h of reaction, high-molecular-weight poly(ketone)s are obtained in high yields.

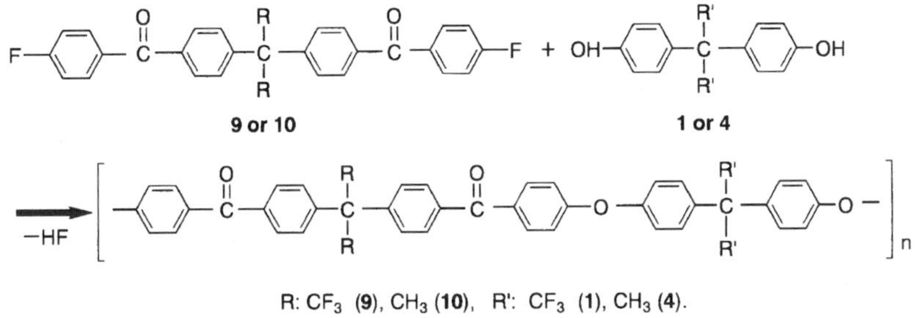

9 or 10   **1 or 4**

R: CF$_3$ (**9**), CH$_3$ (**10**),   R': CF$_3$ (**1**), CH$_3$ (**4**).

Scheme 7

A powerful and efficient method for the preparation of poly(ketone)s is the direct polycondensation of dicarboxylic acids with aromatic compounds or of aromatic carboxylic acids using phosphorus pentoxide/methanesulfonic acid (PPMA)[16] or polyphosphoric acid (PPA)[17] as the condensing agent and solvent. By applying both of these reagents to the synthesis of hexafluoroisopropylidene-unit-containing aromatic poly(ketone)s, various types of poly(ketone)s such as poly(ether ketone) (**11**), poly(ketone) (**12**), poly(sulfide ketone) (**13**), and poly-

(sulfone ketone) (**14**) have been prepared by the reaction of 2,2-bis(4-carbox-yphenyl)-1,1,1,3,3,3-hexafluoropropane (**15**) with diphenyl compounds such as diphenyl ether, diphenyl ketone, diphenyl sulfide, and diphenyl sulfone.[18] The structures of the resulting polymers are confirmed by ^{1}H-NMR and IR.

The reaction of **15** with diphenyl ether and diphenyl sulfide yields poly(ether ketone) (**11**) and poly(sulfide ketone) (**13**) with exclusively *para*-substituted benzene rings according to Scheme 8.

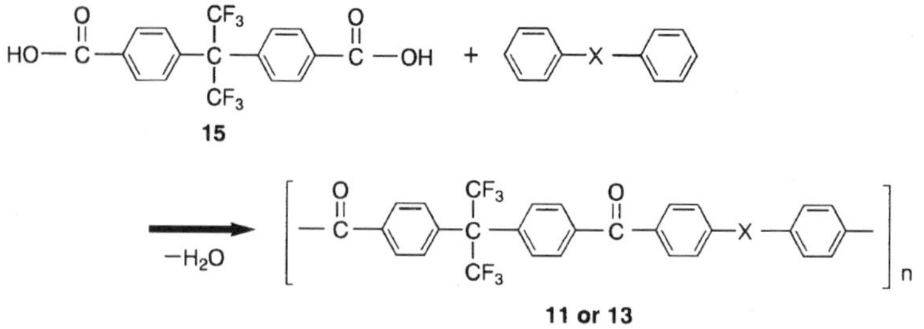

Scheme 8

Poly(ketone) (**12**) and poly(sulfone ketone) (**14**) are produced according to Scheme (9). They contain significant numbers of phenylene units substituted in *p*- and *o*-positions in addition to the *m*-phenylene units.

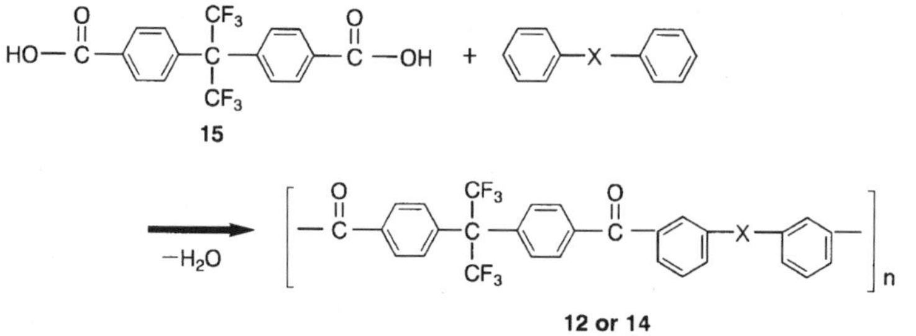

Scheme 9

Reaction temperatures below 100°C and high monomer concentrations are essential for preparing polymers with higher molecular weights in high yields. All of the polymerizations proceed in homogeneous solutions. However, higher monomer concentration, i.e., one molar monomer solution in PPMA causes precipitation of the polymers during the reaction and restricts the molecular weight of the polymers to a low value.

The molecular weights of poly(ketone)s obtained from diphenyl ketone and diphenyl sulfone are lower than those from diphenyl ether and diphenyl sulfide. The former monomers possess highly electron-withdrawing carbonyl or sulfone groups, which lower the electron density of the rings.

Hexafluoroisopropylidene-unit-containing copoly(ether ketone)s and copoly-(sulfide ketone)s are also synthesized by reacting mixtures of 2,2-bis(4-carboxy-phenyl)-1,1,1,3,3,3-hexafluoropropane   (**15**)   and   2,2-bis(4-carboxyphenyl)-propane (**16**) with diphenyl ether or diphenyl sulfide in PPMA, respectively (Scheme 10).[18]

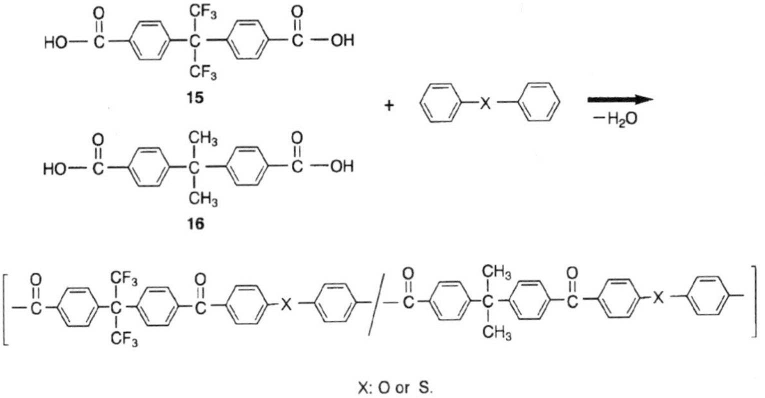

X: O or S.

Scheme 10

Irrespective of the feed ratio of **15** and **16**, the reaction proceeds homogeneously, and quantitative yields are attained with reduced viscosities between 0.13 and 0.37 dl/g.

9.4.2. Properties

Four poly(ether ketone)s obtained from 2,2-bis[4-(4-fluorobenzoyl)-phenyl]-1,1,1,3,3,3-hexafluoropropane (**9**) or 2,2-bis[4-(4-fluorobenzoyl)-phenyl]propane (**10**) with Bisphenol AF (**1**) or Bisphenol A (**4**) are all soluble in chloroform, benzene, THF, and aprotic polar solvents such as DMF, DMAc, and NMP.[15] Poly(ether ketone) from **9** and **1**, which has the highest fluorine content, dissolves easily in ethyl acetate.

The T_g of the poly(ether ketone) from **9** and **1** is 180°C.[15] This is the highest among the four poly(ether ketone)s and higher by 30°C than that of PEEK. The T_g of poly(ether ketone) from **10** and **4** having no fluorine atom shows the lowest T_g of 169°C because of the smaller steric hindrance of methyl groups to the free rotation of the polymer backbone compared to more bulky trifluoromethyl groups. Thermal stability of these poly(ether ketone)s is high. The DT_{10} of the poly(ether ketone) from **9** and **1** is 485°C in air and 536°C in nitrogen, respectively. A membrane from poly(ether ketone) from **9** and **1** shows high selectivity ratios for H_2/CH_4 and CO_2/CH_4.

Poly(ketone)s   obtained   from   2,2-bis(4-carboxyphenyl)-1,1,1,3,3,3-hexa-fluoro-propane **15** with diphenyl compounds dissolve readily in aprotic polar

Table 9.8. Thermal Properties of Poly(Ketone)s
Having Hexafluoroisopropylidene Units[18]

Polymer	T_g (°C)	T_m (°C)	DT_5 (°C)	RW (%)
11	171	—	> 500	98
12	158	202	> 500	96
13	134	—	338	78
14	142	—	338	74

solvents such as DMF, DMAc, DMSO, and NMP and even in *m*-cresol, sulfolane, trichloroethane, and THF.[18]

Crystallinity of these hexafluoroisopropylidene-unit-containing poly(ketone)s is low except for poly(sulfide ketone) (**13**). The water contact angle for the fluorine-containing poly(ketone) films is high, being 98° for poly(ether ketone) (**11**), from 2,2-bis(4-carboxyphenyl)-1,1,1,3,3,3-hexafluoropropane (**15**) and 96° for poly(sulfide ketone) (**13**) from **15**, whereas it is 78° for poly(ether ketone) from 2,2-bis(4-carboxy-phenyl)propane (**16**) and 74° for the poly(sulfide ketone) from **16**. This result indicates that the substitution of isopropylidene units of poly-(ketone)s with hexafluoroisopropylidene units has a remarkable effect on the surface properties of poly(ketone) films.

The T_g of hexafluoroisopropylidene-unit-containing poly(ether ketone) (**11**) is 171°C, which is higher than that of the hexafluoroisopropylidene-unit-containing poly(sulfide ketone) (**13**) by 13°C (Table 9.8).[18] The T_g for the former (**11**) is remarkably higher than those of 144 and 155°C for PEEK and PEK, respectively. The T_g values of poly(ether ketone) and poly(sulfide ketone) having no fluorine atoms from **16** and dipheny; ether or diphenyl sulfide are 134 and 142°C. Hexafluoroisopropylidene-unit-containing poly(sulfide ketone) (**13**) shows a melting point of 202°C, whereas other hexafluoroisopropylidene unit-containing poly(ketone)s are amorphous.

Thermooxidative stability of the fluorine-containing poly(ether ketone) (**11**) and poly(sulfide ketone) (**13**) from **15** is very high. The 5% weight-loss temperatures (DT_5) are 391 and 436°C for poly(ether ketone) and poly(sulfide ketone) analogues having no fluorine atoms, whereas those of poly(ether ketone) (**11**) and poly(sulfide ketone) (**13**) are higher than 500°C.

9.5. POLY(AZOMETHINE)S

9.5.1. Synthesis

Many approaches have been attempted to obtain soluble or fusible high-molecular-weight poly(azomethine)s, including the introduction of flexible bonds, spacers, or pendant bulky groups to the polymer backbone and ring substitutions.

Novel fluorine-containing aromatic poly(azomethine)s are prepared by the reaction of hexafluoroisopropylidene-unit-containing aromatic diamines, 2,2-bis[4-(4-aminophenoxy)phenyl]-1,1,1,3,3,3-hexafluoropropane (**17**) and 2,2-bis(4-aminophenyl)-1,1,1,3,3,3-hexafluoropropane (**18**) with terephthalaldehyde (**19**) or isophthalaldehyde (**20**)[20] (Scheme 11).

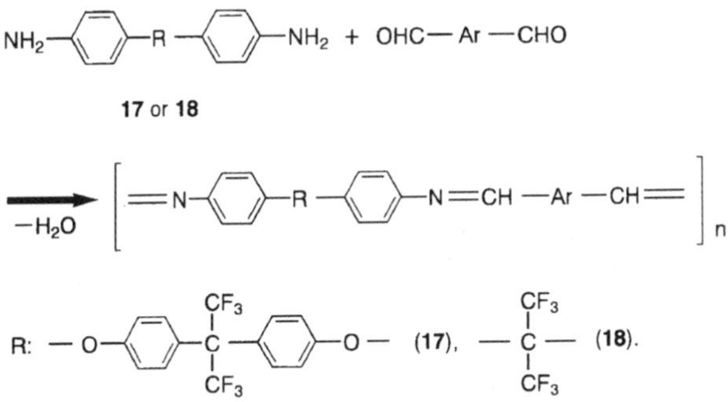

Scheme 11

Room-temperature solution polycondensation is used for the preparation of hexafluoroisopropylidene-unit-containing poly(azomethine)s. At the end of the reaction, the water liberated by the reaction is thoroughly taken off as an azeotrope by vacuum distillation to allow the reaction to go to completion. Among DMF, DMSO, HMPA, NMP, and m-cresol used as reaction solvents, m-cresol yields a polymer with higher reduced viscosity in higher yield. The reaction proceeds rapidly and is essentially completed in 30 min.

Hexafluoroisopropylidene-unit-containing copoly(azomethine)s are also prepared under similar reaction conditions using an equimolar mixture of 2,2-bis[4-(4-aminophenoxy)phenyl]-1,1,1,3,3,3-hexafluoropropane (**17**) and 2,2-bis[4-(4-aminophenoxy)phenyl]propane (**21**) or of 2,2-bis(4-aminophenyl)-1,1,1,3,3,3-hexafluoropropane (**18**) and 2,2-bis(4-aminophenyl)propane (**22**) as the diamine components with terephthalaldehyde (**19**) or isophthalaldehyde (**20**) (Scheme 12).[20]

9.5.2. Properties

Wholly aromatic poly(azomethine)s possess remarkable thermal stability, and high strength and modulus. However, owing to their limited solubility and infusibility, it is difficult to obtain poly(azomethine)s having sufficiently high molecular weight and useful processability.

The incorporation of fluorine atoms improves the solubility of aromatic condensation polymers without causing them to lose their high thermal stability and modifies the processability. Hexafluoroisopropylidene-unit-containing poly-(azomethine)s and copoly(azomethine)s are readily soluble in highly polar solvents such as DMAc, HMPA, and NMP, and they also dissolve completely in dichloromethane, chloroform, and THF, whereas poly(azomethine)s derived from **21** and **22** and having no fluorine atom are insoluble in these solvents.[20] Accordingly, the solubility of aromatic poly(azomethine)s is remarkably improved by substituting isopropylidene units with fluorine atoms.

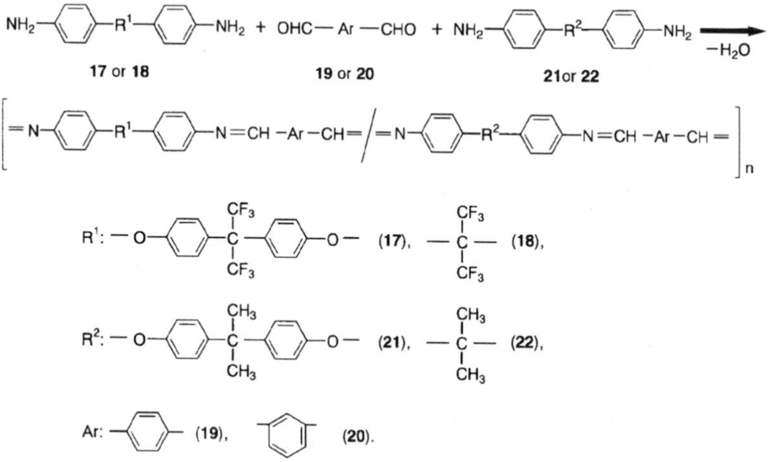

Scheme 12

The cast films of poly(azomethine)s are transparent but pale to deep yellow in color. The water contact angles of fluorine-containing poly(azomethine) films from diamine (**17**) are 80° for terephthalaldehye (**19**) and 75° for isophthalalde-hyde (**20**) as dialdehyde component (Table 9.9).[20] These low values of water contact angles are attributed to the lower fluorine content in these polymers.

None of the poly(azomethine)s show glass transition, and they are amorphous irrespective of the presence or absence of fluorine atoms.[20] All the poly(azomethine)s are thermooxidatively stable at temperatures as high as 400°C. Hexafluoroisopropylidene-unit-containing poly(azomethine)s are more stable to thermooxidation than those having no fluorine atom (Table 9.9). This is expected because the methyl group is more susceptible to oxidation than aromatic rings.

Table 9.9. Preparation and Properties of Poly(Azomethine)s Having Hexafluoroisopropylidene Units[20]

Polymer		Contact angle[a]	DT_5
Diamine	Dialdehyde	(degree)	(°C)
17	19	80	467
	20	75	490
18	19	—	443
	20	—	486
11	19	—	452
	20	71	479
12	19	—	420
	20	—	475

[a] Water contact angle at 20°C in air.

9.6. POLY(AZOLE)S

9.6.1. Synthesis

Several routes have been reported for the synthesis of aromatic poly(azole)s such as poly(benzimidazole), poly(benzoaxazole), and poly(benzthiazole): melt polycondensation of dicarboxylic acid diphenyl esters with tetramines[21] and high-temperature solution polycondensation of dicarboxylic acids or their derivatives with tetramine hydrochlorides in PPA.[22] PPA acts as condensing agent and solvent. Ueda et al.[23] developed a modified method for the synthesis of polyazoles with the use of PPMA.

Fluorine-containing aromatic poly(benzimidazole)s are synthesized by direct polycondensation of 2,2-bis(4-carboxyphenyl)-1,1,1,3,3,3-hexafluoropropane (**15**) with 3,3'-diaminobenzidine tetrahydrochloride (**23**) (Scheme 13) and 1,2,4,5-benzenetetramine tetrahydrochloride (**24**) (Scheme 14) in PMMA or PPA.[24]

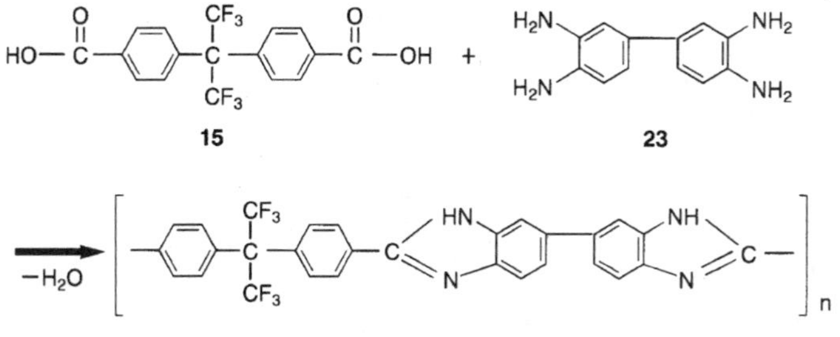

Scheme 13

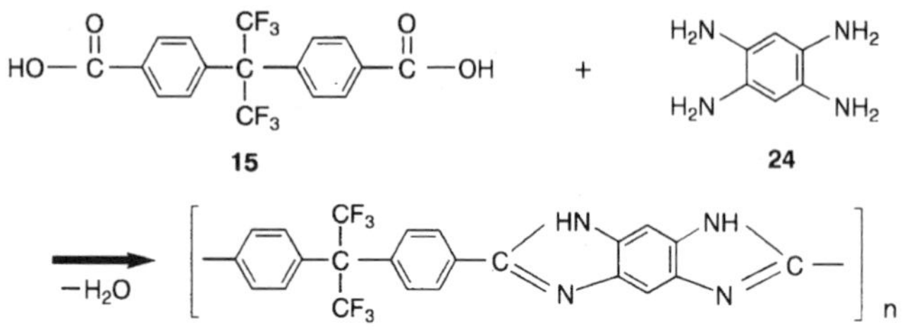

Scheme 14

The effects of various reaction conditions are examined in detail, in which 1 mmol of tetramine (**23**) or (**24**) is reacted with 1 mmol of dicarboxylic acid (**15**). As for the amount of solvent, 5 ml of PPMA is appropriate for the 1-mmol scale reaction. A higher concentration makes it difficult for the reaction to proceed homogeneously, whereas a lower concentration reduces the rate of reaction. The reduced viscosity markedly increases with increasing temperature, and the polycondensation of **15** and benzidine (**23**) at 140°C results in a sufficiently high reduced viscosity of 0.90 dl/g in 24 h. The reaction of 2,2-bis(4-carboxyphenyl)-1,1,1,3,3,3-hexafluoropropane (**15**) with tetramines occurs slowly, requiring more than 24 h for completion, because (**15**) has the highly negative hexafluoroisopropylidene unit.

The reduced viscosity of poly(benzimidazole) from **15** and tetramine (**24**) is not high enough under the reaction conditions described above in PPMA. The reaction at 200°C for 24 h in PPMA affords only a low-molecular-weight polymer in a low yield owing to low solubility in PPMA and sublimation of dicarboxylic acid (**15**). However, an improved reduced viscosity of 0.59 dl/g is attained by a two-step method, first at 140°C for 24 h and then at 200°C for 24 h.

High-molecular-weight fluorine-containing aromatic poly(benzoxazole)s have not been obtained either by the direct solution polycondensation in PPA at 200°C or by the low-temperature solution polycondensation in DMAc at 0 to 5°C from 2,2-bis(3-amino-4-hydroxyphenyl)-1,1,1,3,3,3-hexafluoropropane and aromatic diacid derivatives because the fluorine-containing monomer has low nucleophilicity owing to the presence of the electron-withdrawing hexafluoroisopropylidene unit.

Maruyama et al.[25] have obtained high-molecular-weight poly(benzoxazole)s by the low-temperature solution polycondensation of N,N',O,O'-tetrakis(trimethylsilyl)-substituted 2,2-bis(3-amino-4-hydroxyphenyl)-1,1,1,3,3,3-hexafluoropropane (**25**) with aromatic diacids and subsequent thermal cyclodehydration of the resulting poly(o-hydroxy amide)s *in vacuo*. In this method, aromatic diamines with low nucleophilicity are activated more positively through the conversion to the N-silylated diamines, and the nucleophilicity of the fluorine-containing bis(o-aminophenol) can be improved by silylation.

The low-temperature solution polycondensation of **25** with dichloride (**26**) of 2,2-bis(4-carboxyphenyl)-1,1,1,3,3,3-hexafluoropropane (**15**) is carried out in

DMAc at 0 to 5°C for 8 h and fluorine-containing poly(o-hydroxy amides) (27) having an inherent viscosity of 0.40 dl/g is obtained (Scheme 15).[25]

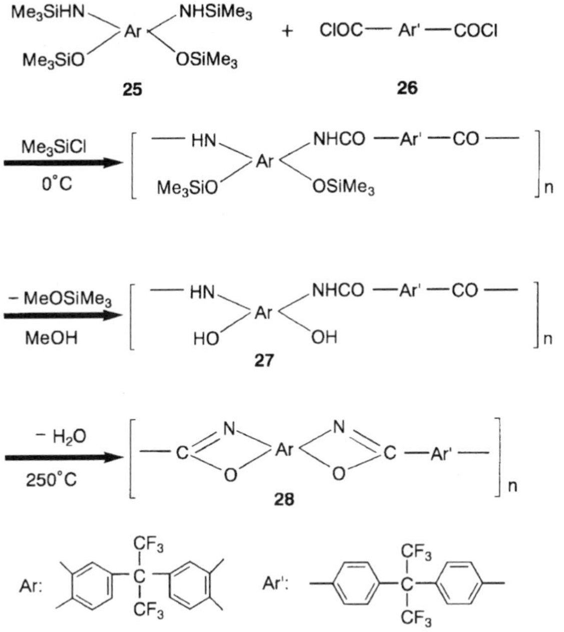

Scheme 15

In the second stage, the poly(o-hydroxy amides) (27) are subjected to thermal cyclodehydration to convert to poly(benzoxazole) (28). The conversion requires 15 to 20 h at 250°C *in vacuo* for its completion. The resulting poly(benzoxazole) (28) has an inherent viscosity of 0.49 dl/g.

Fluorine-containing aromatic poly(benzothiazole)s are synthesized by direct polycondensation of 2,2-bis(4-carboxyphenyl)-1,1,1,3,3,3-hexafluoropropane (15) with 2,5-diamino-1,4,-benzenedithiol dihydrochloride (29) using PPMA or PPA as both condensing agent and solvent (Scheme 16).[26]

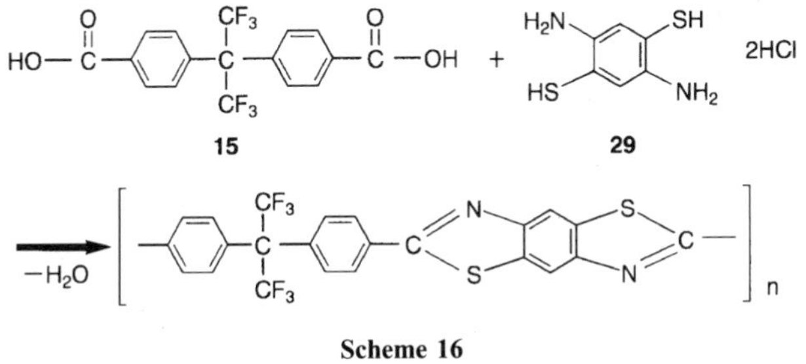

Scheme 16

The direct polycondensation of **15** with **29** in PPMA at 140°C for 24 h does not yield polymers having high enough reduced viscosity to produce a tough film. The prolonged reaction is not successful. The low reactivity of **15** is associated with a strongly electron-withdrawing hexafluoroisopropylidene group at the *para* position to the carboxyl groups. The conventional PPA method conducted at 200°C affords only a low-molecular-weight polymer in low yield owing to the low solubility of **15** in PPA and its sublimation. High-molecular-weight polymer is successfully obtained by a two-step reaction, first at 140°C for 48 h and then at 200°C for 24 h.

High-molecular-weight aromatic poly(benzothiazole)s can be obtained from dichloride (**26**) of **15** and **29** in PPA under similar reaction conditions (Scheme 17).

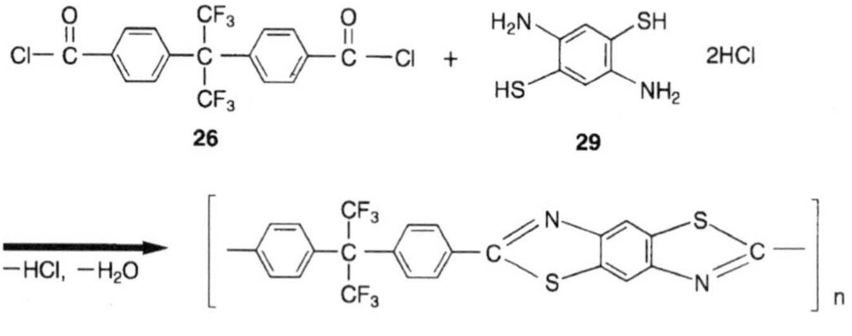

Scheme 17

9.6.2. Properties

Poly(benzimidazole)s possess excellent thermal stability, flame resistance, and outstanding chemical resistance. The solubility of hexafluoroisopropylidene-unit-containing poly(benzimidazole)s is remarkably improved.[24] They are readily soluble in strong acids such as formic acid, concentrated sulfuric acid, and methanesulfonic acid and in aprotic polar solvents such as DMAc and NMP. The polymer from tetramine (**23**) is soluble even in *m*-cresol and pyridine.

These poly(benzimidizole)s are amorphous.[24] The T_g of the polymer from tetramine (**23**) is 330°C, and no glass transition is observed for the polymer from tetramine (**24**). The former T_g value is higher (280 to 325°C) than those of the hexafluoroisopropylidene-unit containing poly(benzoxazole)s.[25]

The thermal stability of poly(benzimidazole) is further improved by the introduction of hexafluoroisopropylidene units in the main chain. The DT_{10} is 520°C in air and greater than 520°C in nitrogen for the polymer from tetramine **(23)** and 506°C in air and greater than 520°C in nitrogen for that from tetramine **(24)**.

The increased solubility and amorphous nature of this polymer is ascribed to reduced intermolecular forces between the polymer chains owing to the introduction of fluorine atoms and to looser packing owing to the highly distorted diphenylhexafluoroisopropylidene units in the polymer backbone.

Aromatic poly(benzoxazole)s exhibit excellent thermal stability. Rigid-rod poly(benzoxazole)s are fabricated into high-strength and high-modulus fibers. Fluorine-containing aromatic poly(benzoxarole)s are expected to have unique properties.

The precursor of poly(penzoxazole) **(28)**, poly(o-hydroxy amide) **(27)** is amorphous and readily soluble in DMF, NMP, DMSO, pyridine, THF, and acetone.[25] Transparent, flexible, and tough film of **27** can be obtained by casting. However, fluorine-containing poly(benzoxazole) **(28)** from **25** dissolves only in concentrated sulfuric acid and o-chlorophenol.

The poly(benzoxazole) **(28)** has a T_g of 295°C. The DT_{10} is 525°C in air and 530°C in nitrogen, respectively.[25]

Aromatic poly(benzothiazole)s are thermally and thermooxidatively stable and have outstanding chemical resistance and third-order nonlinear optical susceptibility. Aromatic poly(benzothiazole)s can be spun into highly-oriented ultrahigh strength and ultrahigh modulus fibers. However, this type of polymer is insoluble in most organic solvents. Therefore, hexafluoroisopropylidene units are introduced in the polymer backbone to obtain soluble or processable aromatic poly(benzothiazole)s.

The aromatic poly(benzothiazole) from **15** and **29** is almost amorphous and easily soluble in strong acids such as concentrated sulfuric acid and methanesulfonic acid.[26] It also dissolves in organic solvents such as HMPA and o-chlorophenol. The increased solubility and amorphous nature of this polymer is also ascribed to reduced intermolecular forces and to looser packing owing to the presence of highly distorted diphenylhexafluoroisopropylidene units in the polymer backbone.

Transparent, flexible, and tough films with deep orange color are cast from HMPA solution of poly(benzothiazole) from **15** and **29**.[26] The tensile strength of 69 Mpa and the tensile modulus of 2.7 GPa are obtained for aromatic poly(benzoxazole) **(28)** from **15** and **29**. These values are higher than those of hexafluoroisopropylidene-unit-containing poly(benzimidazole).

The T_g of hexafluoroisopropylidene-unit-containing poly(benzthiazole) is 327°C, which is almost the same as that of hexafluoroisopropylidene-unit-containing poly(benzimidazole). The hexafluoroisopropylidene-unit-containing poly(benzthiazole) is stable up to 470°C in both air and nitrogen. The thermal stability of the hexafluoroisopropylidene-unit-containing poly(benzimidazole) is almost comparable to that of the hexafluoroisopropylidene-unit-containing poly(benzimidazole). The DT_{10} is 527°C in air and 537°C in nitrogen.

9.7. POLY(SILOXANE)

9.7.1. Synthesis

Novel poly(aryloxydiphenylsilane) is prepared from Bisphenol AF (2,2-bis(4-hydroxyphenyl)-1,1,1,3,3,3-hexafluoropropane) (1) and dianilinodiphenylsilane (30) by melt polycondensation at elevated temperatures under reduced pressure of 1 to 2 Torr (Scheme 18).[27] The molecular weight of the poly-(aryloxydiphenylsilane) derived from anilinosilane and bisphenols have been reported to be highly dependent on the reaction temperature in the melt polycondensation.[28]

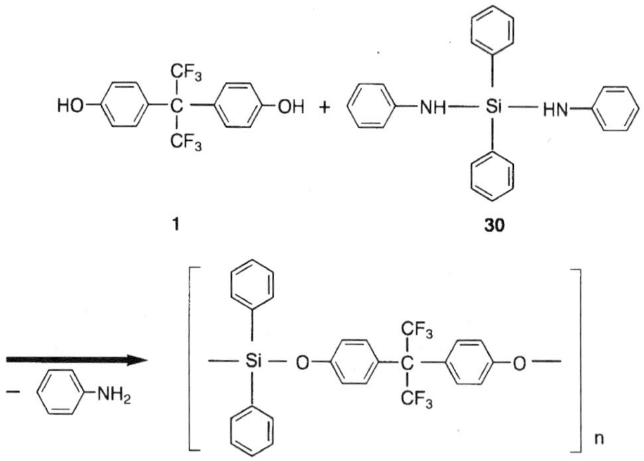

Scheme 18

The most suitable reaction temperature for obtaining a high-molecular-weight polymer is 320°C, and the chloroform-soluble polymer having the reduced visibility of 0.31 dl/g is obtained in a reaction time of 2 h.[27] At higher reaction temperatures, the insoluble fraction occurs owing to cross-linking of polymers. The reduced viscosity of Bisphenol-AF-derived poly(aryloxydiphenylsilane) is not high, which is attributed to the decrease in nucleophilicity of the bisphenol induced by strong electron attraction of fluorine atoms.

9.7.2. Properties

Poly(organosiloxane)s have a number of outstanding properties such as thermal and oxidative stability, water and chemical resistance, electric insulating capacity, selective permeability to gases, and biocompatibility.

Usually, silicon polymers have extremely low T_g's owing to the highly flexible Si—O linkages in the main chain. Their low T_g's are very advantageous for low-temperature applications, whereas some other applications are limited. One approach to raise the T_g of silicon polymers is the introduction of aromatic rings into the polymer backbone.

X-ray diffraction measurements have shown that Bisphenol-AF-derived poly(aryloxydiphenylsilane) is amorphous.[28] The T_g value of 106°C is higher than that of the poly(aryloxydiphenylsilane)s derived from dianilinodiphenylsilane and bisphenols such as 4,4'-biphenol, 2,7-dihydroxynaphthalene, and hydroquinone.[28] The aromatic units in poly(aryloxy-diphenylsilane(s) have a remarkable effect on the T_g. The thermal stability of this polymer is somewhat lower than those of poly(aryloxydiphenylsilane)s derived from dianilinodiphenylsilane and bisphenols such as 4,4'-biphenol, 2,7-dihydroxynaphthalene, and hydroquinone. The DT$_{10}$ is 362°C and the residual weight at 500°C in air is 54%.

Bisphenol-AF-derived poly(aryloxydiphenylsilane) dissolves easily in a wide variety of organic solvents, including chlorinated and aromatic hydrocarbons, cyclic ethers, and aprotic polar solvents.

The UV transmission of Bisphenol-AF-derived poly(aryloxydiphenylsilane) shows a very sharp cut-off at about 285 nm (Figure 9.2).

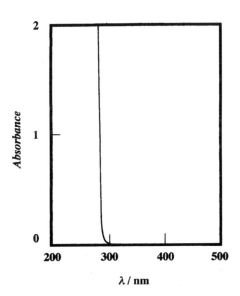

Figure 9.2. UV transmission of poly(aryloxydiphenylsilane) film obtained from Bisphenol AF and dianilinodiphenylsilane.

9.8. CONCLUSIONS

The solubility, water and oil repellency, thermal and thermooxidative stability, and T_g are enhanced and crystallinity and water absorption are decreased by introducing hexafluoroisopropylidene units, rather than units of isopropylidene, into polymer backbone of aromatic condensation polymers.

9.9. REFERENCES

1. P. E. Cassidy, T. M. Aminibhavi, and J. M. Farley, *J. Macromol. Sci.—Rev. Macromol. Chem. Phys.* C29, 365–429 (1989).
2. I. L. Knunyants, T.-Y. Chen, N. P. Gambaryan, and D. I. Rokhlin, *Zh. Vses, Khim. O-va* 5, 114–116 (1960) [in Russian].
3. D. W. Fox, in *Macromolecular Synthesis, Coll. Vol. 1* (J. A. Moore, ed.), Wiley-Interscience, New York (1977), pp. 9–12.
4. Y. Wada, T. Ito, and S. Suzuki, *Jpn. Kokai Tokkyo Koho* 53, 526 (1982).
5. Y, Wada, M. Wada, S. Suzuki, and T. Ito, *Jpn. Kokai Tokkyo Koho* 57, 121, 031 (1982).
6. Y. Saegusa, M. Kuriki, A. Kawai, and S. Nakamura, *J. Polym. Sci. Pt. A: Polym. Chem.* 28, 3327–3335 (1990).
7. S. Nakamura, unpublished data.
8. R. E. Baier, *Surface Properties of Materials for Prosthetic Implants*, Calspan Corporation, CAL Report VH-2801-p-2 (February 1970).
9. J. C. Jeferis, *Polymer Handbook, 3rd ed.* (J. Brandrup and E. H. Immergut, eds.), Wiley-Interscience, New York (1989), VI 451–461.
10. A. S. Hay, F. J. Williams, H. M. Relles, B. M. Boulette, P. E. Donahue, and D. S. Johnson, *J. Polym. Sci.: Polym. Lett. Ed.* 21, 449–457 (1983).
11. A. S. Hay, F. J. Williams, H. M. Relles, and B. M. Boulette, *J. Macromol. Sci.—Chem.* A21, 1065–1080 (1984).
12. Y. Saegusa, M. Kuriki, A. Kawai, and S. Nakamura, *J. Polym. Sci. Pt. A: Polym. Chem.* 32, 57–63 (1993).
13. S. Nakamura, Y. Suzuki, T. Kojima, and K. Tago, *Thermochim. Acta* 267, 231–237 (1995).
14. R. N. Johnson, A. G. Farnham, R. A. Clendinning, W. F. Hale, and C. N. Merriam, *J. Polym. Sci.: A-1* 5, 2375–2398 (1967).
15. G. L. Tullos, P. E. Cassidy, and A. K. St. Clair, *Macromolecules* 24, 6059–6064 (1991).
16. M. Ueda and K. Kato, *Makromol. Chem.: Rapid Commun.* 5, 833–836 (1984).
17. Y. Iwakura, K. Uno, and T. Takiguchi, *J. Polym. Sci.: A-1* 6, 3345–3355 (1968).
18. Y. Saegusa, A. Kojima, and S. Nakamura, *Macromol. Chem.* 194, 777–789 (1993).
19. H. R. Kricheldorf and. G. Schwarz, in *Handbook of Polymer Synthesis, Part B* (H. R. Kricheldorf, ed.), Dekker, New York (1992), pp. 1673–1682.
20. Y. Saegusa, M. Kuriki and S. Nakamura, *Macromol. Chem. Phys.* 195, 1877–1889 (1994).
21. H. Vogel and C. C. Marvel, *J. Polym. Sci.* 50, 511–539 (1961).
22. Y. Iwakura, K. Uno, and Y. Imai, *J. Polym. Sci. Pt. A* 2, 2605–2615 (1964).
23. M. Ueda, M. Sato, and A. Mochizuki, *Macromolecules* 18, 2723–2726 (1985).
24. Y. Saegusa, M. Horikiri, and S. Nakamura, *Macromol. Chem. Phys.* 198, 619–625 (1997).
25. Y. Muruyama, Y. Oishi, M. Kakimoto, and Y. Imai, *Macromolecules* 21, 2305–2309 (1994).
26. Y. Saegusa, M. Horikiri, D. Sakai, and S. Nakamura, *J. Polym. Sci. Pt. A: Polym. Chem.* 36, 429–435 (1998).
27. Y. Saegusa, T. Kato, H. Oshiumi, and S. Nakamura, *J. Polym. Sci. Pt. A: Polym. Chem.* 30, 1401–1406 (1992).
28. W. R. Dunnavant, R. A. Markle, P. B. Stickney, J. E. Curry, and J. D. Byad, *J. Polym. Sci. A-1* 5,

10

Novel Fluorinated Block Copolymers

Synthesis and Application

M. ANTONIETTI and S. OESTREICH

10.1. INTRODUCTION

Block copolymers possess unique and novel properties for industrial applications. During the past 20 years, they have sparked much interest, and several of them have been commercialized and are available on the market. The most common uses of block copolymers are as thermoplastic elastomers, toughened thermoplastic resins, membranes, polymer blends, and surfactants. From a chemist's point of view, the most important advantage of block copolymers is the wide variability of their chemical structure. By choice of the repeating unit and the length and structure of both polymer blocks, a whole range of properties can be adjusted.

A special class of block copolymers with blocks of very different polarity is known as "amphiphilic" (Figure 10.1). In general, the word "amphiphile" is used to describe molecules that stabilize the oil–water interface (e.g., surfactants). To a certain extent, amphiphilic block copolymers allow the "generalization of amphiphilicity." This means that molecules can be designed that stabilize not only the oil–water interface but any interface between different materials with different cohesion energies or surface tensions (e.g., water–gas, oil–gas, polymer–metal, or polymer–polymer interfaces). This approach is straightforward, since the wide variability of the chemical structure of polymers allows fine and specific adjustment of both polymer parts to any particular stabilization problem.

As useful as block copolymers are, their synthesis is as difficult. In general, high purity during the reactions, tedious isolation procedures, or the use of

M. ANTONIETTI and S. OESTREICH · Max-Planck Institute of Colloids and Interfaces, D-14513 Teltow-Seehof, Germany.

Fluoropolymers 1: Synthesis, edited by Hougham *et al.*, Plenum Press, New York, 1999.

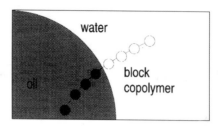

Figure 10.1. Schematic drawing of an amphiphilic block copolymer stabilizing the oil–water interface.

protecting group chemistry is required. This makes block copolymers very expensive and precludes many technological applications. In addition, adjustment of block copolymers to a special interface problem is made difficult by the necessity of synthesizing new polymerizable monomers to fit the problem and adapting the polymerization process to these monomers.

In light of these difficulties we have developed a "modular" synthesis route to transform preformed commercially available block copolymers into many different block copolymers with very different polarities. Figure 10.2 shows the advantage of decoupling the adjustment of amphiphilicity from the polymerization process and the control of the polymer architecture. A complicated block copolymer synthesis is avoided, the molecular parameters of the prepolymer are maintained, and a whole range of different polarities is accessible from one polymer by simply varying the side groups. For this synthesis route, mild and quantitative reactions have to be used to avoid cross-linking between the single polymer chains by side reactions.

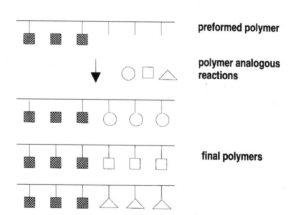

Figure 10.2. Schematic drawing of the modular concept. By simple attachment of different side groups to one block, block copolymers with very different properties are obtained.

Here, we focus on one class of block copolymers synthesized by this method: polystyrene-*b*-poly(vinylperfluorooctanic acid ester) block copolymers (Figure 10.33). After describing the synthesis and characterization, we will treat some properties and the potential applications of this new class of block copolymers. The amphiphilicity of the polymers is visualized by the ability to form micelles in diverse solvents that are characterized by dynamic light scattering (DLS). Then the use of these macromolecules for dispersion polymerization in very unpolar media is demonstrated by the polymerization of styrene in 1,1,2-trichlorotrifluoroethane (Freon 113).

In the third part of the chapter the solid state properties of our block copolymer are examined. The surface energies of these materials are characterized by contact angle measurements. The organization of the polymer chains in the solid state phase is investigated by small-angle X-ray scattering (SAXS) and the gas selectivity of porous membranes coated with these block copolymers is characterized by some preliminary permeation measurements.

10.2. POLYMER SYNTHESIS

We use polystyrene-*b*-polybutadiene block copolymers as the starting material with preformed polymer architecture. These polymers are comparatively cheap and easily accessible.[1] For the present problems a series of narrowly distributed polystyrene-*b*-polybutadiene block copolymers with rather different molecular weights were synthesized via anionic polymerization (Figure 10.4, Table 10.1). As a test for the modification of technological products, a commercial triblock copolymer was also used.

We chose the π-bonds of the polybutadiene block as the functional groups to connect different side groups to the polymer block. A direct connection of side groups to the π-bonds is difficult and leads to unstable products; therefore, a two-step synthesis was employed. The first step is a conversion of the π-bonds into a more easily accessible species. This can be done by either hydroxylation (hydro-

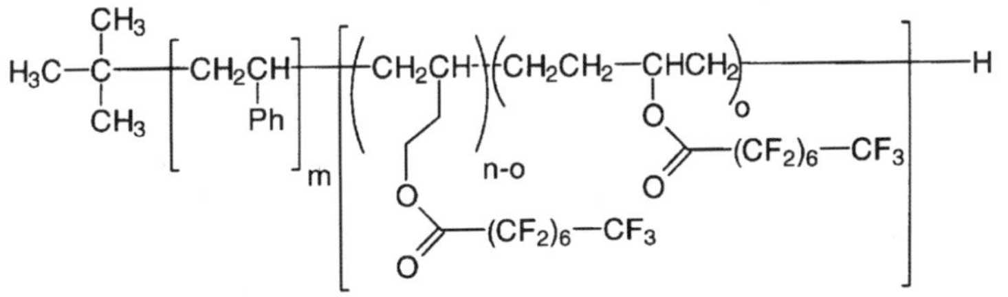

Figure 10.3. Drawing of the final product, a polystyrene-*b*-poly(vinylperfluorooctanic ester) ($o \approx n \cdot 10\%$)

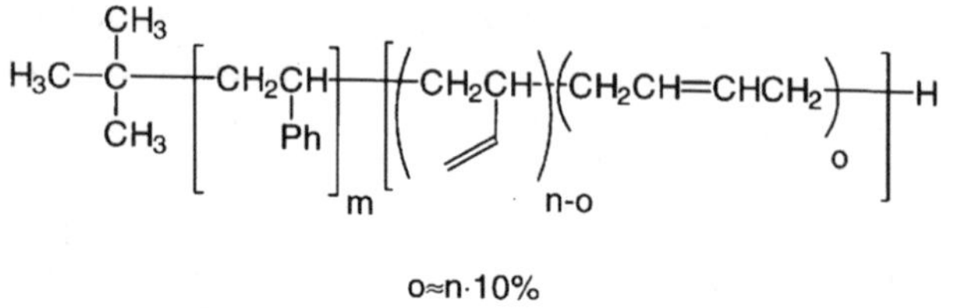

$o \approx n \cdot 10\%$

Figure 10.4. Drawing of the polystyrene-*b*-polybutadiene block copolymers used for amphiphilic modification.

Table 10.1. Polystyrene-*b*-Polybutadiene Block Copolymers Used as Starting Materials[a]

Copolymer	m	n	$D = M_W/M_N$
PSB-II	423	390	1.023
PSB-III	205	74	1.025
PSB-IV	106	34	1.031
PSB-V	435	26	1.020
PSB-VI	49	4	1.042
PSB-VIII	58	19	1.124
PSB-IX	45	10	1.066
PSB-XII	83	102	1.167
PSB-XIII	83	114	1.079
PSB-XIV	33	10	1.048
Triblock	—	—	10.0

[a] The absolute number of monomer units of the polystyrene (m) and polybutadiene (n) blocks were determined by combined ^{1}H-NMR and GPC measurements. The triblock copolymer is a commercially available polystyrene-*b*-polybutadiene-*b*-polystyrene polymer and was purchased from Aldrich ($M_W = 100,000$; 28 wt% polystyrene).

boration/oxidation) or epoxidation. Both functionalization procedures are very effective and lead to a "close-to-complete conversion" without any cross-linking by side reactions. We found that the hydroboration/oxidation sequence, in particular, leads to a quantitative reaction.[2-4] In the second step the functionalized block is esterified by perfluorinated carboxylic acid chlorides. For this chapter we used perfluorooctanic acid chloride (C_8). The whole reaction scheme is sketched in Figure 10.5. The final product, a polystyrene-*b*-poly(vinylperfluorooctanic acid ester), is shown in Figure 10.3.

The hydroboration/oxidation sequence does not change the molecular-weight distribution. Gel permeation chromatography (GPC) measurements in dimethylformamide (DMF) with the resulting polystyrene-*b*-polyalcohol polymers show very similar polydispersity indexes (Table 10.2). Here, the hydroboration/oxidation sequence is clearly superior to the epoxidation reaction, which leads to a

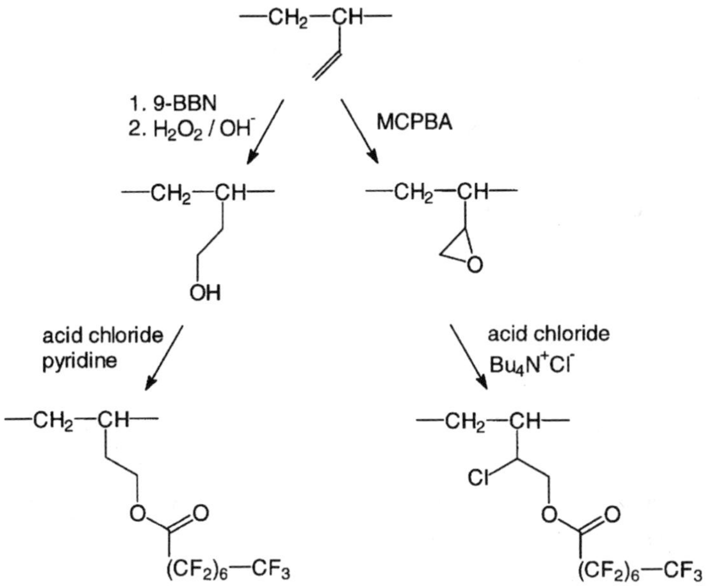

Figure 10.5. Drawing of the whole synthesis route. Hydroxylation followed by esterification on the left side and epoxidation followed by addition of a carboxylic acid chloride on the right side.

small amount of cross-linked polymer. In each case, ^{1}H-NMR and IR data ensures the completeness of the hydroxylation or epoxidation reactions, respectively.

As the final products—polystyrene-*b*-poly(vinylperfluorooctanoic ester)— form micelles in tetrahydrofuran (THF) as well as in DMF, there are not direct GPC data to characterize molecular parameter. For this reason, we employed esterification of the hydroxylated block copolymers with benzoylchloride as a model reaction to obtain a comparable product with molecular solubility that can easily be characterized by DMF–GPC. The GPC data from PSB-II—our largest and therefore most sensitive block copolymer—are summarized in Table 10.2. Results for all the other polymers are similar.

Table 10.2. Molecular-Weight Distribution Throughout the Synthesis Process

Process step	$D = M_W/M_N{}^a$
Starting material (PSB-II)	1.023
After hydroxylation (Alc-PSB-II)	1.050
After esterification with benzoylchloride (BC-PSB-II)	1.073

^{a} Polydispersity of the unmodified, hydroxylated, and esterified block copolymer. Esterification was performed with benzoylchloride as a model reaction.

Completeness of esterification is confirmed by ^{1}H-NMR and IR spectroscopy. Only a small number of hydroxyl groups can be detected by IR spectroscopy. ^{1}H-NMR spectroscopy does not show any remaining hydroxyl group signal.

The above data prove that the polystyrene-b-polybutadiene prepolymer is quantitatively transformed into block copolymers with perfluorinated side chains. The narrow molecular-weight distribution $(D = M_W/M_N)$ of the prepolymers is maintained by the described reaction sequence.

10.3. PROPERTIES

10.3.1. Micelle Formation

Block copolymers in a selective solvent, i.e., a good solvent for one block but a precipitant for the other, behave like typical amphiphiles. The copolymer molecules aggregate reversibly to form micelles in a manner analogous to the aggregation of classical surfactants. Our block copolymers are very amphiphilic in the sense described above and form well-defined micelles in a wide range of selective solvents. In solvents for polystyrene, the polystyrene block is located in the micelle corona, while the modified block is "hidden" in the micelle core.

The micelle formation is not restricted to solvents for polystyrene but also occurs in very unpolar solvents, where the fluorinated block is expected to dissolve. Comparing the data, we have to consider that the micelle structure is inverted in these cases, i.e., the unpolar polystyrene chain in the core and the very unpolar fluorinated block forming the corona. The micelle size distribution is in the range we regard as typical for block copolymer micelles in the superstrong segregation limit.[2,5,6] The size and polydispersity of some of these micelles, measured by DLS, are summarized in Table 10.3.

Micelle formation in standard organic solvents such as toluene or THF is very useful, since fluorinated polymers are usually not soluble in standard solvents; micelle formation therefore enables processing of fluorinated products with classical technologies, e.g., for coating applications.

10.3.2. Dispersion Polymerization of Styrene in Fluorinated Solvents

Micelle formation of our block copolymers in fluorinated solvents indicates that these polymers might act as stabilizers or surfactants in a number of stabilization problems with high technological impact, e.g., the surface between standard polymers and media with very low cohesion energy such as short-chain hydrocarbons (isopentane, butane, propane), fluorinated solvents (hexafluorobenzene, perfluoro(methylcyclohexane), perfluorohexane) and supercritical CO_2. As

Table 10.3. Micelle Sizes and Size Distribution in Various Solvents[a]

Solvent	Fluoro-PSB-II		Fluoro-PSB-III		Fluoro-PSB-IV		Fluoro-PSB-VIII		Fluoro-triblock	
	d_h (nm)	σ	d_h (nm)	σ	d_h (nm)	σ	d_h (nm)	σ	d_h (nm)	σ
Toluene	176.1	0.321	89.2	0.251	Soluble	None	Soluble	None	Insoluble	Swelling
Freon 113	102.5	0.346	36.8	0.174	26.9	0.264	Soluble	None	Soluble	None
Perfluoro(methylcyclohexane)	111.8	0.315	Insoluble		32.1	0.367	43.9	0.537	Insoluble	Swelling

[a] Hydrodynamic diameters d_h and corresponding Gaussian width σ of block copolymer micelles in solvents of decreasing polarity. "None" means that no micelles could be detected by DLS.

the importance of this stabilization problem is usually underestimated, a brief introduction follows.

Heterophase polymerization in general is a current trend in polymer science as it allows solvent-free polymer synthesis; polymer powders are obtained when the dispersion agent is removed. DeSimone improved this procedure significantly by performing the heterophase polymerization in supercritical CO_2, which simplifies the synthesis of powders with excellent handling of the polymerization and evaporation process.[7–9]

To keep the precipitating polymers in the dispersed state throughout the polymerization, requires steric stabilizers. This problem is classically tackled via copolymerization with fluoroalkylmethacrylates or the addition of fluorinated surfactants, both being only weak steric stabilizers. DeSimone et al. also applied a fluorinated block copolymer,[9] proving the superb stabilization efficiency of such systems via a rather small particle size. One goal of the present chapter is therefore an investigation of our fluorinated block copolymers as steric stabilizers in low-cohesion-energy solvents.

Instead of performing the polymerization reactions in supercritical CO_2 (we do not yet have access to this technology), we chose the dispersion polymerization in 1,1,2-trichlorotrifluoroethane (Freon 113), to serve as a model for supercritical CO_2. We only used the short diblock copolymers for the polymerization of styrene in Freon 113, since it is known that such diblock copolymers are the most efficient steric stabilizers.

Table 10.4 summarizes the compositions of some experiments as well as the colloid-analytical data of the final polystyrene lattices. A particle diameter of about 100 nm (including the shell of the adsorbed block copolymers in an extended conformation) is rather low for the product of a dispersion polymerization in unpolar solvents. In addition, a mean deviation (σ) of about 20% of the particle size indicates a well-controlled and stable latex.

Table 10.4. Results of Dispersion Polymerization of Styrene in Freon 113

Sample	Applied stabilizer[a]	S	d_h (nm)	σ
1	Fluoro-PSB-IV	0.05	344.7	0.141
2	Fluoro-PSB-IV	0.10	189.9	0.186
3	Fluoro-PSB-IV	0.20	107.1	0.212
4	Fluoro-PSB-VIII	0.05	203.1	0.279
5	Fluoro-PSB-VIII	0.10	134.2	0.331
6	Fluoro-PSB-VIII	0.20	110.7	0.287

[a] Fluoro-PSB-IV and Fluoro-PSB-VIII are steric stabilizers. The resulting particle diameter d_h and corresponding Gaussian width σ depend on the relative amount of block copolymer (S), which is varied between 5–20 wt% with respect to the monomer.

Some typical transmission electron micrographs of these polystyrene lattices are shown (Sample 2 and Sample 3) in Figure 10.6. The effects of the amount of stabilizer (S is the relative amount of stabilizer) on the particle size is strong: the more stabilizer applied, the smaller the particles are. It must be emphasized that this effective stabilization of nanopowders by our fluorinated block copolymers is not restricted to polymerization processes, but can be generalized to the fabrication of all organic nanopowders in media with low cohesion energy density, e.g., to the dispersion of dyes, explosives, or drugs.

10.3.3. Surface Energies of the Block Copolymers

Another field where low-cohesion-energy block copolymers will certainly find some applications is in reduction of the surface energy (γ), i.e., interfaces with gases (which generally have a very low cohesion energy). A minimized γ-value is also important for the stability of foams, where a low surface energy ensures a fine foam structure. Surface energies smaller than 20 mN/m are called ultralow as most standard solvents, such as oil and water, cannot wet those surfaces; their importance for protective and nonpolluting coatings, water repelling fabrics, or self-lubricating machine parts is currently being discussed in the literature[10–12] and is well known in fluorine chemistry. Typical γ-values for such materials are in the range of 11 mN/m $< \gamma <$ 16 mN/m; the well-known polytetrafluoroethylene (PTFE) exhibits a γ-value of 18.6 N/m, but has the disadvantages of being difficult to process and having a porous surface. With polystyrene-b-polydimethylsiloxane (PDMS) block copolymers it was shown that even doping of a bulk material (e.g., polystyrene) with such block copolymers enables signficant lowering of the surface energies of the compound since the PDMS blocks get enriched in the interface.[13] Ober *et al.* also found a significant lowering of the surface energies when they blended homopolystyrene with partially fluorinated block copolymers.[14]

It is well known that the polymer/air interface heavily influences, orients, and modifies the phase structure of block copolymers. In close proximity to the surface, the microphase with the lower-cohesion energy density is energetically more favorable. This is most easily investigated through contact angle measurements of thin polymer films toward hexadecane/air. The resulting values are summarized in Table 10.5. We found rather large contact angles for our polymers toward hexadecane/air. This indicates a ultralow to low energy surface. The equations from Li and Neumann[15] allow us to estimate γ_{SV} (also shown in Table 10.5). Owing to the inhomogeneous surface of our polymer films, exact γ_{SV} values are very difficult to obtain.

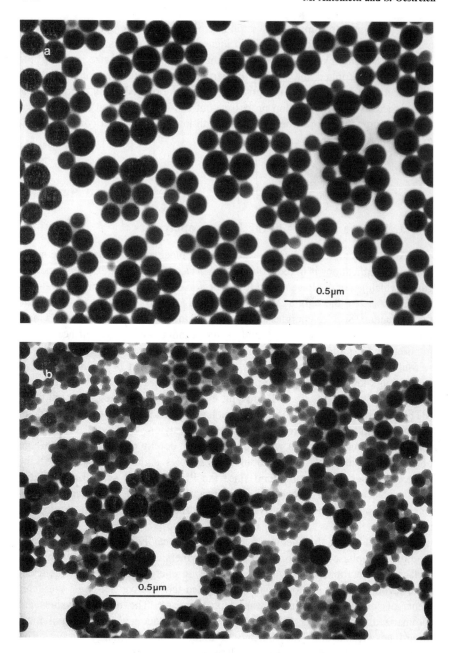

Figure 10.6. Transmission electron micrographs of polystyrene particles prepared by dispersion polymerization in Freon 113 and stabilized by Fluoro-PSB-IV: (a) Sample 2; (b) Sample 3.

Table 10.5. Results of Contact Angle Measurements toward Hexadecane/Air[a]

Film	θ_{mean} (deg)	γ_{sv}^{HD} (m N/m)
Fluoro-PSB-II	66.5	14.2
Fluoro-PSB-III	60.5	16.0
Fluoro-PSB-IV	61.0	15.8
Fluoro-PSB-VIII	61.4	15.7
Fluoro-triblock	65.7	14.4

[a] Average contact angles θ and the corresponding surface energies γ_{sv} are given [$\gamma_{lv} = 27.76$ mN/m; $\beta = 0.0001247$ (m^2/mJ)^2].[15]

10.3.4. Mesophase Formation

Solid films of our block copolymers and their surface behavior have been examined using a variety of techniques. Block copolymers composed of incompatible polymer blocks are known for mesophase formation as a consequence of the microphase separation of the chains. Our fluorinated block copolymers form a microphase-separated structure with a high degree of order. This can easily be visualized by polarization microscopy and SAXS.

For all films of fluorinated block copolymers with noncubic symmetry, we observe a spontaneous birefringence, which is a measure of the high anisotropy and degree of order of these films. The degree of order increases remarkably with orientation, which is easy to do by simple mechanical elongation of the thermoplastic elastomer at appropriate temperatures. In case of high elongation, we obtain monooriented phases, as can be seen by a monochromatic appearance of polarization micrographs.

SAXS measurements on solid films of the block copolymers underscore the high degree of order inside these films. Figure 10.7 shows some SAXS diffractograms and the corresponding microstructures that we found in our block copolymers. Here, we are only able to present some preliminary data. A detailed discussion of all the investigated block copolymers will follow in a later publication.[16]

The phase morphology of block copolymers can also be visualized by transmission electron microscopy. Figure 10.8 shows the lamellar structure of Fluoro-PSB-IX. From diblock copolymers it is well known that the resulting microphase morphology depends on the volume fraction (ϕ) of the two phases. By simple adjustment of the relative block lengths we are able to synthesize block copolymers with specific structures.[17,18]

The mechanical properties also depend on the relative block lengths. For instance, Fluoro-PSB-II is a soft, elastic material. Differential scanning calorimetry (DSC) measurements and wide-angle X-ray scattering (WAXS)

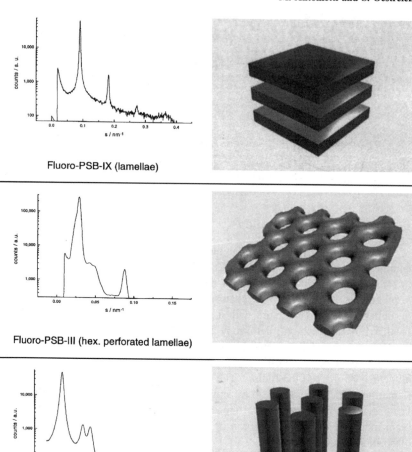

Fluoro-PSB-IX (lamellae)

Fluoro-PSB-III (hex. perforated lamellae)

Fluoro-PSB-VIII (hex. packed cylinders)

Figure 10.7. Small-angle X-ray scattering curves of microphase-separated samples and drawings of the corresponding structure morphologies.

diffractograms reveal the absence of any side chain crystallinity of the fluorinated tails. With DSC we obtain a second heating curve characterized by two glass transition temperatures, typical for microphase-separated block copolymers. The glass transition temperatures of the fluorinated phase and the polystyrene phase are T_g^1 is $T_g^1 = -25°C$ and T_g^2 is $T_g^2 = 103°C$, respectively. These values have to be regarded as limiting for high-molecular-weight samples; smaller chains exhibit the typical depression of the glass transition temperature.

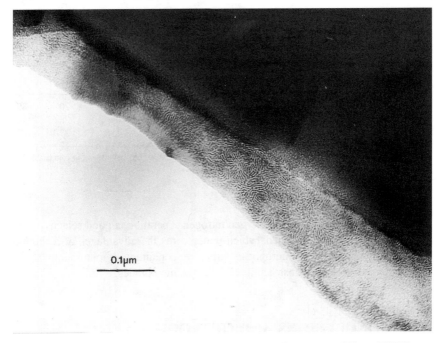

Figure 10.8. Transmission electron micrograph of the lamellar structure of Fluoro-PSB-IX.

Therefore, at room temperature Fluoro-PSB-II a thermoplastic elastomer with a soft polymer phase (fluorinated block) and a hard phase (PS-block), similar to the parental polystyrene-*b*-polybutadiene block copolymer. Depending on the relative volume fraction of both components and the continuity of the phases, the resulting bulk material is rubbery or a high-impact solid.

10.3.5. Gas Permeation Measurements

It is known that fluorinated compounds dissolve considerable amounts of oxygen[19] and that membranes containing ultrahydrophobic moieties show a high selectivity in gas-separation processes, e.g., in O_2/N_2 separation.[20]

Encouraged by the X-ray and contact angle results, we performed some preliminary gas-permeation measurements. Here, a self-supporting film is required and only the longer block copolymers were used. Fluoro-PSB-II and Fluoro-triblock were coated on porous Celgard 2400 membranes; the measurements were taken at room temperature at a driving pressure of 5 bars. Since no absolute polymer layer thickness has been determined, only relative values of the permeability are given (Table 10.6). In the case of the separation of CO_2 from

Table 10.6. Data for Gas Permeation Measurements on Porous Celgard 2400 Membranes Coated with Thin Films of Fluoro-PSB-II and Fluoro-Triblock[a]

	Flow (cm^3/min)	CO_2 (%)	CH_4 (%)	C_2H_4 (%)	Flow (cm^3/min)	O_2 (%)	N_2 (%)
Feed	10	20	40	40	10	20	80
Fluoro-PSB-II	1.2–2.0	43.6	25.1	31.3	0.6	28.5	71.5
Fluoro-triblock	2.8–5.6	37.5	29.2	33.2	1.4	27.1	72.9

[a] It can be seen that both systems exhibit a remarkable enrichment of CO_2 or O_2, respectively, at comparably high permeabilities.

hydrocarbons as well as for the oxygen/nitrogen separation, a good selectivity at a very high flow rate is obtained, which underscores the advantages of a thermoplastic polymer with a microphase structure containing a continuous liquid transport phase for the construction of effective membranes.

10.4. CONCLUSIONS AND OUTLOOK

The synthesis of block copolymers with blocks of ultralow cohesion energy densities on the basis of polystyrene-*b*-polybutadiene via two highly efficient polymer analogous reactions has been presented.

Owing to their amphiphilicity and a balanced molecular architecture these molecules from micelles in all solvents for polystyrene as well as in solvents for the fluorinated block. The structure parameters of these micelles have to be regarded as typical for other block copolymers in the superstrong segregation limit.[5,6]

These block copolymers can act as effective steric stabilizers for the dispersion polymerization in solvents with ultralow cohesion energy density. This was shown with some polymerization experiments in Freon 113 as a model solvent. The dispersion particles are effectively stabilized by our amphiphiles. However, these experiments can only model the technically relevant case of polymerization or precipitation processes in supercritical CO_2 and further experiments related to stabilization behavior in this sytem are certainly required.

Solid polymer films of the block copolymers exhibit ultralow surface energies, thus proving the enrichment of perfluorinated side chains at the surface. This enables the construction of stable polymer coatings using the technologically interesting procedures of solvent-casting and melt-processing. In addition, only the polymer/air interface exhibits ultralow surface energies. Films of our block copolymers adhere strongly to the underlying glass plates. This is again easily explained by the pronounced amphiphilicity of these polymers where the polystyrene blocks are oriented to the glass plates. This effect leads to interesting possibilities for the synthesis of good-adherence coatings with ultralow surface energies.

The solid state phase behavior of our block copolymers is similar to that of other diblock copolymers. SAXS measurements show that highly ordered nanostructures are formed. The size and morphology of these structures can be changed by tuning the length of both polymer blocks. It is thus possible to adjust the mechanical properties of the bulk material. Gas permeation measurements reveal a remarkable gas selectivity at a very high flow, which, in connection with the low water sensitivity of such materials, offers some interesting perspectives for membrane technology.

Further work will be related to the development of curable systems as well as to a broader examination of the phase morphologies in the solid state. In addition, the dependence of the technologically relevant selectivity in gas permeation on the type and size of the resulting mesomorphous phases is currently under investigation.

ACKNOWLEDGMENTS: Many thanks to H. D. Lehmann for his gas-diffusion measurements. We thank C. Burger for help during X-ray analysis, H. Kamusewitz for contact angle measurements, and S. Förster for the synthesis of the PS-*b*-PBd block copolymers. S. Oestreich wants to thank the Fonds der Chemischen Industrie for a Kekulé fellowship. Financial support by the Max-Planck Society is gratefully acknowledged.

10.5. REFERENCES

1. H. F. Mark, N. M. Bikales, C. G. Overberger, G. Menges, Block Copolymers, in *Encyclopedia of Polymer Science and Engineering, 2nd Ed.*, Vol. 2, John Wiley and Sons, New York (1985), pp. 324–434.
2. M. Antonietti, S. Förster, J. Hartmann, S. Oestreich, *Macromolecules 29*, 3800–3806 (1996).
3. S. Ramakrishnan, *Macromolecules 24*, 3753–3759 (1991).
4. T. Nishikubo and A. Kameyama, *Prog. Polym. Sci. 18*, 963–995 (1993).
5. M. Antonietti, S. Heinz, M. Schmidt, and C. Rosenauer, *Macromolecules 27*, 3276–3281 (1994).
6. S. Förster, E. Wenz, M. Zisenis, and M. Antonietti, *J. Chem. Phys. 104* (24), 9956–9970 (1996).
7. J. M. DeSimone and Z. Guan, *Macromolecules 27*, 5527–5532 (1994).
8. J. M. DeSimone, J. B. McCain, D. E. Betts, D. A. Canelas, E. T. Samulski, J. D. Londono, H. D. Cochran, G. D. Wignall, D. Chillura-Martino, and R. Triolo, *Science 274*, 2049–2052 (1996).
9. J. M. DeSimone, D. A. Canelas, and D. E. Betts, *Macromolecules 29*, 2818–2821 (1996).
10. M. Antonietti, S. Henke, and A. Thünemann, *Adv. Mater. 8*, 41 (1996).
11. D. L. Schmidt, C. E. Coburn, B. M. DeKoven, G. E. Potter, G. F. Meyers, and D. A. Fischer, *Langmuir 12*, 518–529 (1996).
12. D. L. Schmidt, C. E. Coburn, M. M. DeKoven, G. E. Potter, G. F. Meyers, and D. A. Fischer, *Nature 368*, 39–41 (1994).

13. X. Chen and J. A. Gardella, *Macromolecules 27*, 3363–3369 (1994).

14. D. R. Iyengar, S. M. Perutz, C. A. Dai, C. K. Ober, and E. J. Krmaer, *Macromolecules 29*, 1229–1234 (1996).

15. D. Li, A. W. Neumann, *J. Coll. Interf. Sci. 148* (1), 190 (1992).

16. M. Antonietti, C. Burger, M. A. Micha, S. Oestreich, and S. Förster, *Europhys. Lett. 42* (4), 452–429 (1998).

17. F. S. Bates, J. H. Rosedale, H. E. Bair, and T. P. Russell, *Macromolecules 22*, 2557–2564 (1989).

18. A. K. Khandpur, S. Förster, F. S. Bates, I. W. Hamley, A. J. Ryan, W. Bras, K. Almdal, and K. Mortensen, *Macromolecules 28* (26), 8796–8806 (1995).

19. J. G. Riess, *New J. Chem. 19* (8–9), 891–909 (1995).

20. Y. Yamashita (ed.), *Chemistry and Industry of Macromonomers*, Hüthig and Wepf, Basel (1993).

11

Synthesis and Structure–Property Relationships of Low-Dielectric-Constant Fluorinated Polyacrylates*

HENRY S.-W. HU and JAMES R. GRIFFITH

11.1. INTRODUCTION

Dielectric constant is directly proportional to the capacitance of a material. Present computer operations are limited by the coupling capacitance between circuit paths and integrated circuits on multilayer boards since the computing speed between integrated circuits is reduced by this capacitance and the power required to operate is increased.[1†] If the dielectric constant is reduced a thinner dielectric provides equivalent capacitance, and the ground plane can be moved closer to the line, so that additional lines can be accommodated for the same cross-talk. Thus, the effect of a low dielectric constant will be to increase the speed of the signal and improve the density of the packaging, and this will result in improved system performance.[2]

With recent trends toward microminiaturization and utilization of very thin conductor lines, close spacings, and very thin insulation, greater demands are being placed on the insulating layer. Reductions in such parasitic capacitance can

---

* This chapter is the fifth in a series of articles by these authors, entitled "Processable Fluoropolymers with Low Dielectric Constants."
† The relationship of capacitance C with dielectric constant K_s can be expressed as $C = AK_s\epsilon_0/d$, where A is area, d is distance, and $\epsilon_0 = 8.85418 \times 10^{-14}$ F/cm.

HENRY S.-W. HU · Geo-Centers, Inc., Fort Washington, Maryland 20744. Present address: Sensors for Medicine and Science, Inc., Germantown, Maryland 20874. JAMES R. GRIFFITH · Naval Research Laboratory, Washington, D.C. 20375.

Fluoropolymers 1: Synthesis, edited by Hougham *et al.*, Plenum Press, New York, 1999.

be achieved in a number of ways through the proper selection of materials and the design of circuit geometry. In 1988 St. Clair et al.[3] reported a reduction of dielectric constant to 2.39 by chemically altering the composition of a polyimide backbone to reduce the interactions between linear polyimide chains and by the incorporation of fluorine atoms. In 1991 Cassidy et al.[4] reported a reduction of dielectric constant to 2.32 for the hexafluoroisopropylidene-containing polyarylates and copolyarylates. In 1992 Snow et al.[5] reported that the thermally induced trimerization of a perfluorohexamethylene-linked aromatic cyanate resin to a cyanurate-linked network giave a dielectric constant between 2.3 and 2.4. In 1993 Babb et al.[6] stated that the thermally induced cyclodimerization of a trifluorovinyl aryl ether to a perfluorocyclobutane aromatic ether polymer gave a dielectric constant of 2.40. What is common to all four cases is that they involve incorporating fluorine into polymers as a means of reducing the dielectric constant.

Fluorine is different from the other halogens in the elemental state or when bonded in chemical compounds. Fluorine is the most electronegative element, and when bonded to other atoms, it polarizes the bond, drawing electrons to it. As fluorine replaces hydrogen, the C—F bond shortens and simultaneously the bond strength increases. This shortening and strengthening of the C—F bond with increasing substitution is in striking contrast to the chlorinated or brominated methanes, in which essentially constant bond lengths and bond strengths are observed for all substituted methanes. These are general characteristics of the C—F bond in almost all fluorinated organic compounds.

The highly polar nature of the C—F bond has been used to provide some high-performance characteristics in comparison with their hydrogen-containing or other halogen-containing analogues. Fluorine-containing epoxies or acrylics generally exhibit resistance to water penetration, chemical reaction, and environmental degradation; they also show a combination of unusual properties including low surface tension, low friction coefficient, high optical clarity, low refractive index, low vapor transmission rate, and exceptional electromagnetic radiation resistance.

Poly(tetrafluoroethylene) (PTFE), which is also known by DuPont's tradename Teflon, is a solid at room temperature and has a dielectric constant in the range of 2.00–2.08,* while its monomer, tetrafluoroethylene, is a gas at room temperature. PTFE is exceptionally chemically inert, has excellent electrical properties and outstanding stability, and retains its mechanical properties at high temperatures. The problem with PTFE is that it is not processable. A family of commercial polymeric materials known as Teflon AF is believed to be a

---

* See Licari and Hughes[1] pp. 378–379, Table A-13: Dielectric Constants for Polymer Coatings (at 25°C).

terpolymer of tetrafluoroethylene, perfluoropropylene, and 2,2-bis(trifluoromethyl)-4,5-difluoro-1,3-dioxole (a derivative of hexafluoroacetone). It is reported to have a dielectric constant in the range of 1.89–1.93 and to be more processable than PTFE.[7]

Teflon
$$-\left[CF_2{-}CF_2\right]_n-$$

Teflon AF
$$-\left[CF_2{-}CF_2\right]_x-\left[CF_2{-}\underset{CF_3}{CF}\right]_y-\left[\underset{\underset{\underset{CF_3}{}\diagdown\underset{CF_3}{}}{\underset{C}{}}}{\underset{O\diagup\quad\diagdown O}{CF{-}\quad\quad{-}CF}}\right]_z-$$

Very recently we[8] reported on a class of processable heavily fluorinated acrylic resins that exhibit dielectric constants as low as 2.10, very close to the minimum known values. In this chapter we report on the preparation of a series of processable heavily fluorinated acrylic and methacrylic homo- and copolymers that exhibit dielectric constants as low as 2.06, and the factors that affect the reduction of dielectric constant from structure–property relationships is elucidated.[9]

11.2. EXPERIMENTAL

Only two typical examples in monomer preparation are described here.

11.2.1. Materials

Triethylamine was fractionally distilled from lithium aluminum hydride under nitrogen and acryloyl chloride was distilled under nitrogen. 1,1,2-Trichlorotrifluoroethane (Freon 113) was distilled from phosphorus pentoxide under nitrogen. Azobisisobutyronitrile (AIBN) was used as received. Methyl ethyl ketone peroxide (MEKP) (9% Organic Peroxide VN 2550) was obtained from Witco. Alumina (neutral, Brockman activity 1; 80–200 mesh) was obtained from Fisher Scientific Co. All reagents were obtained from Aldrich unless otherwise specified.

11.2.2. Techniques

Infrared spectra were obtained with a Perkin-Elmer 1800 and a Nicolet Magna-IR 750 FTIR spectrophotometer, and the absorption frequencies are reported in wave numbers (cm^{-1}). NMR spectra were obtained with BZH-300 and CA-F-300 Bruker FTNMR 300 MHz spectrometers. Chloroform-d was used as solvent, and all chemical shifts are reported in parts per million downfield (positive) of the standard. ^{1}H-NMR and ^{13}C-NMR chemical shifts are reported relative to internal tetramethylsilane, while ^{19}F-NMR chemical shifts are reported relative to internal fluorotrichloromethane. R_f values were obtained from silica gel thin-layer chromatography developed with a mixture of 1.5 mL methylene chloride and three drops of acetone. The number of hydrate water molecules was calculated from the integration of ^{1}H-NMR spectra.

11.2.2.1. Preparation of Etherdiacrylate 6 from Etherdiol

1,3-Bis-(2-hydroxyhexafluoro-2-propyl)-5-(4-heptafluoroisopropoxy-1,1,2,2,3,3,4,4-octafluoro-1-butyl)benzene, b.p. 95°C/1.0 mm, was prepared by the multistep route reported by Griffith and O'Rear.[10] Elemental analysis showed it to be anhydrous, but after long storage it had 0.5 H_2O as water of hydration. R_f 0.36; ^{1}H-NMR δ 8.35 (s, 1H, 2-ArH), 8.09 (s, 2H, 4, 6-ArH), 4.14 (s, 2H, −OH), 2.13 (s, 1H, H_2O); IR (neat) 3620 and 3515 br (OH), 3125, 1618, 1468, 1365, 1350–1100 (CF), 1005, 992, 980, 948, 938, 900, 872, 841, 800, 775, 768, 758, 738, 728, 707, 690, 668, 621 cm^{-1}.

The etherdiacrylate **6** was prepared by a modification of the procedure of Griffith and O'Rear.[11] To a solution of etherdiol (3 g, 3.78 mmol) in Freon 113 (10 ml) in an ice-water bath under nitrogen, triethylamine (0.791 g, 7.81 mmol) in Freon 113 (5 ml) was added dropwise in 10 min. After 10 min more, acryloyl chloride (0.707 g, 7.81 mmol) in Freon 113 (5 ml) was added dropwise in 20 min; a precipitate formed immediately. After stirring for 2 h at room temperature, filtration through Celite to remove the solid, followed by evaporation at room temperature *in vacuo*, 3.02 g of a viscous liquid was obtained.

The liquid was dissolved in a mixture of methylene chloride (20 ml) and Freon 113 (10 ml); filtration through neutral alumina (1 g) gave a clear filtrate. It was cooled in an ice bath, washed twice with 1.3 N sodium hydroxide (10 ml), washed with water (10 ml), dried over anhydrous sodium sulfate, percolated and washed through a column of neutral alumina (3 g) twice, and evaporated *in vacuo* at room temperature for 3 h to give a colorless viscous liquid **6** (1.38 g, yield 40%). R_f 0.92; IR (neat) 3118, 1772, 1637, 1468, 1410, 1365, 1350–1100 (CF), 1055, 1040, 992, 900, 878, 848, 800, 770, 759, 740, 728, 718, 670 cm^{-1}; ^{1}H-NMR δ 7.68 (s, 2H, 4,6-ArH), 7.63 (s, 1H, 2-ArH), 6.61 (d, $J = 17$ Hz, 2H), 6.29 (d, d, $J = 17$, 10 Hz, 2H), 6.13 (d, $J = 10$ Hz, 2H), 1.57 (s, 2H, H_2O) indicated 1.0 H_2O as water of hydration; ^{19}F-NMR −70.86 [−c$(CF_3)_2$−], −81.10

[$-CF(CF_3)_2$], -81.35 ($-O-CF_2-$), -112.28 (Ph$-CF_2-$), -123.09 ($-OCF_2-CF_2-$), -124.99 (Ph$-CF_2-CF_2-$), -145.75 [$-CF(CF_3)_2$]; ^{13}C-NMR 82.42 [hept, $J = 30.6$ Hz, $-O-C(CF_3)_2-$], 121.26 [q, $J = 289.7$ Hz, $-OC(CF_3)_2-$], 125.80 (s, 4,6-aromatic C—H), 127.07 (br. s, 1,3-aromatic C), 128.96 (s, 2-aromatic C—H), 130.32 t, $J = 25.0$ Hz, 5-aromatic C), 129.42 (s, CH_2=CH—), 135.29 (s, CH_2=CH—), 160.74 (s, $-COO-$).

11.2.2.2. Preparation of Triacrylate 3 from Triol

1,3,5-Tris (2-hydroxy-hexafluoro-2-propyl)benzene was prepared by a multi-step route according to the procedure of Soulen and Griffith.[12] This compound was hygroscopic as also observed by Griffith and O'Rear.[10] R_f 0.15; IR (KBr pellet) 3660, 3618, 3590 and 3513 (OH), 3125, 1465, 1300–1100 (CF), 1026, 1013, 980, 891, 775, 730, 709, 693 cm^{-1}; ^{1}H-NMR δ 8.25 (s, 1H, ArH), 4.44 (s, 1H, —OH), 2.11 (s, 1.5H, H_2O) indicated 2.25 H_2O as water of hydration.

The triacrylate 3 was prepared by a procedure similar to that described for the synthesis of etherdiacrylate 6. Triol (2.176 g, 3.78 mmol), triethylamine (1.517 g, 15.0 mmol) and acryloyl chloride (1.358 g, 15.0 mmol) were used. After filtration and evaporation, 1.82 g of a viscous liquid was obtained. It was dissolved in a mixture of methylene chloride (20 ml) and Freon 113 (20 ml), and a gelatinous solid was removed by filtration through neutral alumina (1 g). The filtrate was purified by percolation and washing through a column of neutral alumina (3 g) twice. Evaporation at room temperature *in vacuo* for 3 h gave 1.12 g (40% yield) of 3 as a very viscous colorless liquid, which upon standing at -13°C became a semisolid. ^{1}H-NMR δ 7.53 (s, 3H), 6.57 (d, $J = 17$ Hz, 3H), 6.24 (d, d, $J = 17$, 10 Hz, 3H), 6.06 (d, $J = 10$ Hz, 3H) indicated that no water of hydration was present. This semisolid is, however, hygroscopic. R_f 0.85; IR (KBr pellet) 3515, 3130, 2920, 2850, 1770, 1635, 1630, 1412, 1350–1100 (CF), 1090, 1055, 995, 940, 900, 802, 750, 740, 725, 715, 689, 669 cm^{-1}; ^{1}H-NMR δ 7.51 (s, 3H), 6.60 (d, $J = 17$ Hz, 3H), 6.28 (d, d, $J = 17$, 10 Hz, 3H), 6.11 (d, $J = 10$ Hz, 3H), 1.56 (s, 3H, H_2O) indicated 1.5 H_2O as water of hydration. ^{19}F-NMR -70.86; ^{13}C-NMR 80.53 [hept. $J = 30.5$ Hz, $-C(CF_3)_2-$], 121.30 [q, $J = 289.4$ Hz, $-C(CF_3)_2-$], 125.82 (s, 2,4,6-aromatic C—H), 127.11 (br. s, 1,3,5-aromatic C), 128.80 (s, CH_2=CH—), 135.10 (s, CH_2=CH—), 160.61 (s, $-COO-$).

11.2.2.3. Preparation of Polymer "Cylindrical Donuts" from Monomers

In order to prepare the samples for dielectric-constant measurements, etherdiacrylate 6 was mixed with a trace amount of AIBN at room temperature in a cylindrical donut mold made from General Electric RTV 11 silicon molding compound. The donuts had an outer diameter of 7.0 mm, an inner diameter of 3.0 mm, and a thickness of 3.0 mm; the semisolid triacrylate 3 was mixed with a trace amount of liquid MEKP with some heating to obtain a clear liquid; equal masses of **6** and **3** were also mixed with MEKP with some heating.

For polymerization the filled donut molds were kept under an inert atmosphere, the temperature was raised to 85°C over 2 h, and then kept at 85–100°C for 20 h. Homopolymers and 50/50 (w/w) copolymers were obtained.

11.2.2.4. Dielectric Constant Measurements

Dielectric constant (DE) values are reported as "permittivity" with the symbol ϵ or K_s. The polymer cylindrical donuts were used for the measurement of DE on a Hewlett-Packard 8510 automated network analyzer. The analyzer is capable of measuring 401 data points over a frequency band of 500 MHz to 18.5 GHz. Typically S11 and S21 values, which correspond to reflection and transmission, respectively, are measured and then these values are used to calculate the permittivity and permeability.

Samples stood at room temperature in air prior to testing; measurements were run at room temperature and approximately 25% relative humidity (RH). A virgin PTFE sheet (MMS-636-2) was obtained from Gilbert Plastics & Supply Co. and cut into the same cylindrical donut size. The result for PTFE reported here is the average of the DE values measured for three samples.

11.3. RESULTS AND DISCUSSION

11.3.1. Monomers and Polymers

The preparation of fluorinated alcohols was carried out in multistep routes according to the reported procedures.[10,12] The synthesis of acrylic and methacrylic esters as shown in Table 11.1 was carried out in a fluorocarbon solvent such as Freon 113 by the reaction of the respective fluorinated alcohol with acryloyl chloride or methacryloyl chloride and an amine acid acceptor such as triethylamine with examples shown in Scheme 1. Other attempts to esterify the fluoroalcohols directly with acrylic acid or acrylic anhydride were not successful.[11] Product purification by distillation was not feasible because of the temperature required, but purification by percolation of fluorocarbon solutions through neutral alumina resulted in products of good purity identified by TLC, FTIR, and ^{1}H-, ^{13}C-, and ^{19}F- FTNMRs.

Owing to their liquid or semisolid nature, monomers are easy to process into polymers. For radical polymerization the use of solid AIBN for liquid monomers at room temperature and liquid MEKP for semisolid monomers or a mixture of liquid and semisolid monomers with some heating is convenient. During the course of curing at 85–100°C for 22 h the problem of surface inhibition of free radicals by oxygen from the air can be avoided by inert-gas blanketing.

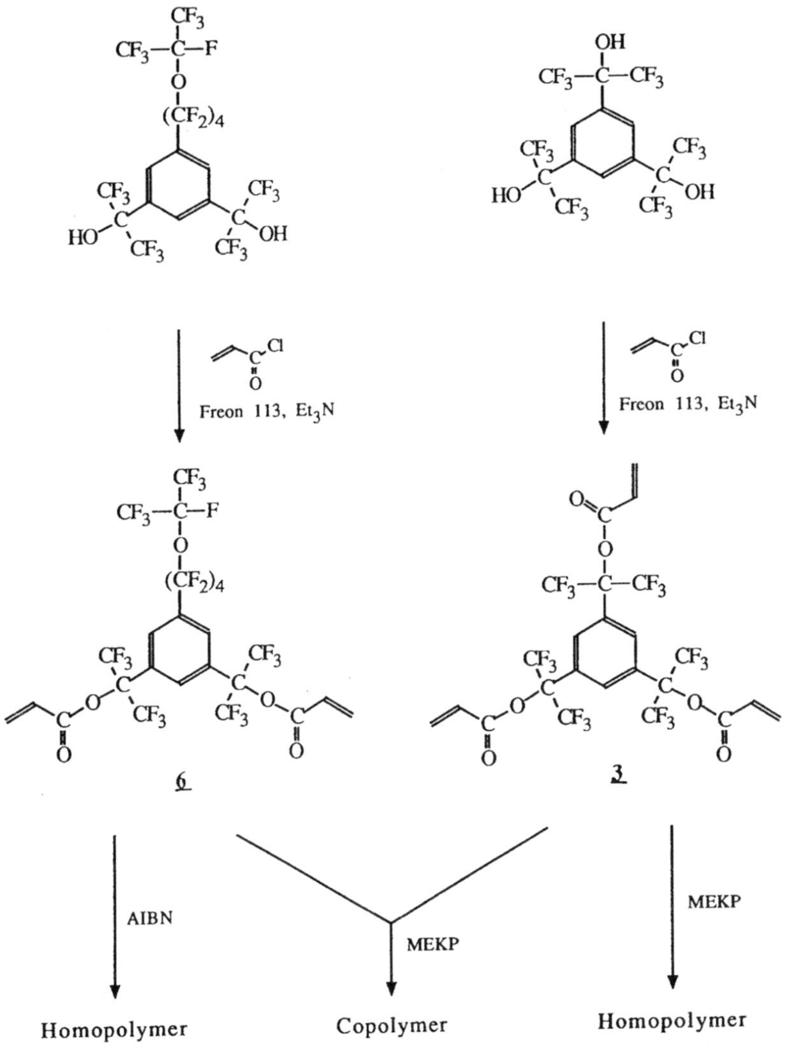

Scheme 1

In ^{1}H-NMR spectra, the acrylates showed a characteristic ABX pattern in the region of δ 6.8–6.0 with a pair of doublet couplings for each vinyl proton, while the methacrylates showed a characteristic AB pattern in the region of δ 6.5–5.8 with an equivalent singlet peak for each vinyl proton. The monomers purified by percolation over alumina contained no detectable hydrate water or polymerized impurities.

Henry S.-W. Hu and James R. Griffith

Table 11.1. Summary of Data for Dielectric Constant Measurements[a]

Polymer		Dielectric constant				
Monomer used	Fluorine content (%)[d]	0.0 GHz	3.0 GHz	9.0 GHz	15.0 GHz
1	41.50	2.37	2.41	2.36	2.36
2	43.99	2.26	2.25	2.28	2.27
3	46.32	2.23	2.23	2.23	2.22
(3 + 6)[b]	51.58	2.22	2.23	2.24	2.22
(4 + 5)[b]	52.27	2.21	2.22	2.22	2.19
5	52.60	2.18	2.21	2.16	2.15
6	56.85	2.11	2.10	2.12	2.13
7	55.14	2.07	2.06	2.08	2.10
PTFE[c]	75.98	1.96	1.98	1.99	—

[a] See Experimental Section for details.
[b] 50/50 (w/w).
[c] Polytetrafluoroethylene (MMS-636-2) from Gilbert Plastics & Supply Co.
[d] Calculated.

As expected, the homopolymers and the 50/50 (w/w) copolymers are semitransparent, hard solids, and some shrinkage in volume is observed during curing. In an undercured state they are frangible, but when totally cured they acquire a more resilient character. The degree of polymerization can be easily monitored with an FTIR spectrophotometer as shown in Figure 11.1, e.g., by examining the intensity of the absorption frequencies at 1635 and 1410 cm^{-1} that

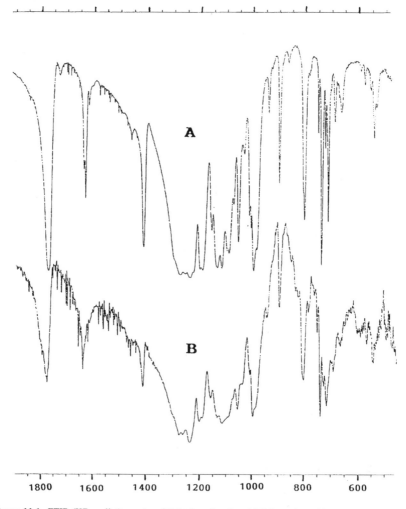

Figure 11.1. FTIR (KBr pellet) spectra of (A) triacrylate **3** and (B) its curing with AIBN at 100°C for 18 h.

are assigned to the acrylate functional groups. In all cases, the filled donut molds were degassed and kept under inert gas for curing at up to 100°C for 20 h.

Since each acrylate group is difunctional, the diacrylates are tetrafunctional while the triacrylates are hexafunctional. For polymerization of the polyfunctional monomers at sufficiently high degrees of conversion, the branching must result in the formation of cross-links to give a three-dimensional network.

The fluoroacrylic polymers are high-modulus, low-elongation plastics, which are brittle in the sense that all thermosetting polymers are brittle. However, they are tough, rugged materials not easily damaged by impact or mechanical abuse.

11.3.2. Dielectric Constant and Structure–Property Relationships

In order to validate the accuracy of our measurements, three samples of virgin PTFE in the same cylindrical donut size were measured and found to have average DE values around 1.96–1.99, which are close to the reported values 2.0–2.08.* Summarized results of the DE measurements on the polymer donuts are shown in Table 11.1. All the new polymer exhibit low DEs, around 1.06–2.41, over a wide frequency region of 500 MHz to 18.5 GHz; the variation of DE values over the measured frequency region is within 0.03, which indicates a wide-range frequency independence.

Basically, the effect of a higher fluorine content on the polymer backbone or side chain will be the reduction of polymer chain–chain electronic interactions, which further results in a reduction of DE. This effect was observed by St. Clair *et al.*,[3] but only on the polymer backbone.

In a comparison of **5** and **6** (see Table 11.1), the existence of a perfluoroalkyl ether linkage seems to play an important role in further reducing the DE values. Teflon AF and **6** both have low dielectric constants and contain ether linkages. Since the dielectric constant of Teflon AF is lower than that of Teflon, which has a higher fluorine content, it appears that the ether linkage has an effect in lowering the DE. The Teflon AF has been reported by Resnick[7] as the lowest dielectric constant (1.89–1.93) of any polymer thus far.

It has been reported[13] that in the preparation of polyimides the more flexible *meta*-linked diamine systematically gave lower DE values than the corresponding *para*-linked system and that this may be related to free volume in the polymer since the *meta*-substituted systems should have a higher degree of entropy. Therefore, the symmetrical *meta*-substitution in our aromatic structure may result in a reduction of DE.

In a comparison of **6** and **7**, a reduction of DE was observed when acrylate was replaced by methacrylate and this may be related to the free volume in the

* See Licari and Hughes[1] pp. 378–379, Table A-13: Dielectric Constants for Polymer Coatings (at 25°C).

polymer since the bulkier methacrylate should have more steric hindrance than the acrylate.

The wide-range frequency independence of these low DEs and the processability of these monomers suggest many potential applications. In particular, it is known that the addition of a fluorine-containing group to the polymer backbone will reduce the polymer chain–chain electronic interactions, which will result in a reduction of DE as reported by St. Clair *et al.*[13] However, the DE for the polymer from **6** (fluorine content 57%) is at most 0.13 below that for the polymer from **3** (fluorine content 46%), which may indicate that a minimum value has nearly been reached.

11.3.3. Processability and Applications

The liquid or low-melting solid monomers can be cured to the solid state by incorporating a curing catalyst and heating the mixtures below the decomposition temperature. Moreover, the cured solids are transparent and hard polymers formed of three-dimensional networks with moderate thermostability.

In the course of curing, the viscosity varies from thin to syrupy liquids. This is convenient for impregnating reinforcing materials, such as fiberglass scrim, used in making wiring or circuit boards for electronics applications.

A compound containing two or three unsaturated groups would be expected to polymerize to a cross-linked, solid, and nonlinear thermosetting polymer. Fluorinated compounds that have only one unsaturated group can be used in a mixture with our monomers to polymerize to a solid polymer of a thermoset nature and this is a convenient way to enhance the ease of processing.

11.3.4. Lower Dielectric Constants?

Dielectric constants of these materials can be further lowered by known means such as by incorporating air bubbles into the materials or by inhibiting crystallization. A difference of a couple of hundredths in the DE value may be important when one is at the low extremes. Recently Singh *et al.* calculated the DEs of polyimide films from the measured free volume fraction and found that the calculated values, are close to the experimental results.[14,15]*

In 1991 Groh and Zimmerman[16] estimated the theoretical lower limit of the refractive index of amorphous organic polymers by using the Lorentz–Lorentz equation and reported the lower limit to be very close to 1.29, while in 1979 Dislich[17] proposed a limit of about 1.33 from a screening of published polymer data.

---

* The calculation was based on the relation: $1/\epsilon = (1-f)/\epsilon_R + f/\epsilon_{Air}$, where ϵ_R is the value of ϵ for zero free volume fraction, which can be obtained from the plot of ϵ vs. f.

The amorphous Resnick Teflon AF terpolymer has the lowest refractive index in the range of 1.29–1.31 and also the lowest DE in the range of 1.89–1.93. Furthermore, Groh and Zimmermann[16] reported that functional groups with a high fluorine content, such as CF_3 and CF_2, have the lowest refractive index contribution. The value for the ether group is also remarkably low, while the values for the carbonyl and carboxyl groups are high. In view of the good agreement between the refractive index and the DE on amorphous organic polymers, the opportunity to obtain the former in the range of 1.89–2.06 by modifying our synthesis is high.

11.4. CONCLUSIONS

In this work we have demonstrated that a new class of heavily fluorinated acrylic and methacrylic resins can be efficiently synthesized and then cured to solid form with radical initiator at elevated temperatures. These cured resins were found to have low dielectric constants, which are close to the minimum known values for Teflon and Teflon AF. In contrast to tetrafluoroethylene, our monomers are processable owing to the fact that they are liquids or low-melting solids, and moreover are soluble in common organic solvents. Lower dielectric constants are obtained as fluorine contents on the polymer backbone or side chain increase, when acrylate is replaced by methacrylate, when ether linkages are present in the fluorocarbon, and when aromatic structure is symmetrically *meta*-substituted.

ACKNOWLEDGMENT: We are indebted to Mr. Jones K. Lodge of SFA, Inc. for the dielectric constant measurements. Partial funding support from the Office of Naval Research is gratefully acknowledged.

11.5. REFERENCES

1. J. J. Licari and L. A. Hughes, *Handbook of Polymer Coating for Electronics*, Noyes Publications, Park Ridge (1990), p. 114.
2. R. R. Tummala, R. W. Keyes, W. D. Grobman, and S. Kapur, in *Microelectronics Packaging Handbook* (R. R. Tummala and E. J. Rymaszewski, eds.), Van Nostrand–Reinhold, New York (1989), p. 691.
3. A. K. St. Clair, T. L. St. Clair, and W. P. Winfree, *Polym. Mater. Sci. Eng.* **59**, 28 (1988).
4. K. M. Kane, L. A. Wells, and P. E. Cassidy, *High Perform. Polym.* **3**(3), 191 (1991).
5. A. W. Snow, J. R. Griffith, R. L. Soulen, J. A. Greathouse, and J. K. Lodge, *Polym. Mater. Sci. Eng.* **66**, 466 (1992).
6. D. A. Babb, B. R. Ezzell, K. S. Clement, W. R. Richey, and A. P. Kennedy, *Polym. Preprints 34* (1), 413 (1993) and *J. Polym. Sci.: Polym. Chem. Ed. 31*, 3465 (1993).
7. P. R. Resnick, *Polym. Preprints 31* (1), 312 (1990).

8. H. S.-W. Hu and J. R. Griffith, *Polym. Mater. Sci. Eng.* 66, 261 (1992) [No. 1]. H. S.-W. Hu and J. R. Griffith, in *Polymers for Micorelectronics*, G. Willson, L. F. Thompson, and S. Tagawa, ed., ACS Symposium Series, 537, 507 (1994) [No. 2]. J. R. Griffith and H. S.-W. Hu, U. S. Patent 5, 292,927 (1994) [No. 3]*

9. H. S.-W. Hu and J. R. Griffith, *Polym. Preprints 34*, 401 (1993) [No. 4]*; also presented at the 205th National Meeting of the American Chemical Society, Denver, Colo. March 28–April 2, 1993.

10. J. R. Griffith and J. G. O'Rear, *Polym. Mater. Sci. Eng.* 53, 766 (1985).

11. J. R. Griffith and J. G. O'Rear, in *Biomedical and Dental Applications of Polymers* (C. G. Gebelein and F. F. Koblitz, eds.), Plenum Press, New York (1981), pp. 373–377.

12. R. J. Soulen and J. R. Griffith, *J. Fluorine Chem.* 44, 210 (1989).

13. T. L. St. Clair, in *Polyimides* (D. Wilson, H. D. Stenzenberger, and P. M. Hergenrother, eds.), Blackie, Glasgow (1990), p. 74.

14. J. J. Singh, A. Eftekhari, and T. L. St. Clare, NASA Memorandum 102625 (1990).

15. A. Eftekhari, A. K. St. Clare, D. M. Stoakley, S. Kuppa, and J. J. Singh, *Polm. Mater. Sci. Eng. 66*, 279 (1992).

16. W. Groh and A. Zimmermann, *Macromolecules 24*, 6660 (1991).

17. H. Dislich, *Angew. Chem.—Int. Ed. Engl. 18*, 49 (1979).

---

* These references are part of a series of articles, entitled "Processable Fluoropolymers with Low Dielectric Constants."

12

Epoxy Networks from a Fluorodiimidediol*

HENRY S.-W. HU and JAMES R. GRIFFITH

12.1. INTRODUCTION

The highly polar nature of the C—F bond has been exploited to provide some high-performance characteristics of fluorine-containing epoxies in comparison with their hydrogen-containing or other halogen-containing analogues. Fluorine-containing epoxies or acrylics generally exhibit resistance to water penetration, chemical reaction, and environmental degradation; they also show different degrees of surface tension, friction coefficient, optical clarity, refractive index, vapor transmission rate, and electromagnetic radiation resistance.

Several classes of polymers including epoxies have been developed from 2-phenyl-1,1,1,3,3,3-hexafluoropropan-2-ol and its derivatives.[1] Fluorinated epoxy resins are the key intermediates for the new types of practical organic coatings and plastics, which have fluorocarbon properties and convenient characteristics such as hydrophobicity, oleophobicity, light stability, low friction, and, in some cases, possibly high thermal stability.

A series of studies has been carried out in our laboratory to design and synthesize new fluorinated epoxy and acrylic resins for the many applications that are possible because of their unique properties.[2-5]

Polyimides offer outstanding properties, such as glass transition temperature, oxidative stability, toughness, adhesion, permeability, and the capability of being fabricated into useful products. It is this versatility that has established the reputation of polyimides for many applications.[6]

---

*This chapter is the seventh in a series of articles by these authors, entitled Processable Fluoropolymers with Low Dielectric Constants.

HENRY S.-W. HU · Geo-Centers, Inc., Fort Washington, Maryland 20744. Present address: Sensors for Medicine and Science, Inc., Germantown, Maryland 20874.   JAMES R. GRIFFITH · Naval Research Laboratory, Washington, D.C. 20375.

Fluoropolymers 1: Synthesis, edited by Hougham *et al.*, Plenum Press, New York, 1999.

In this chapter, which is a follow-up to an earlier discussion,[7] we report on two specific synthetic methodologies using a fluorodiimidediol **3** to prepare the heavily fluorinated epoxy networks as shown in Schemes 1 and 2.

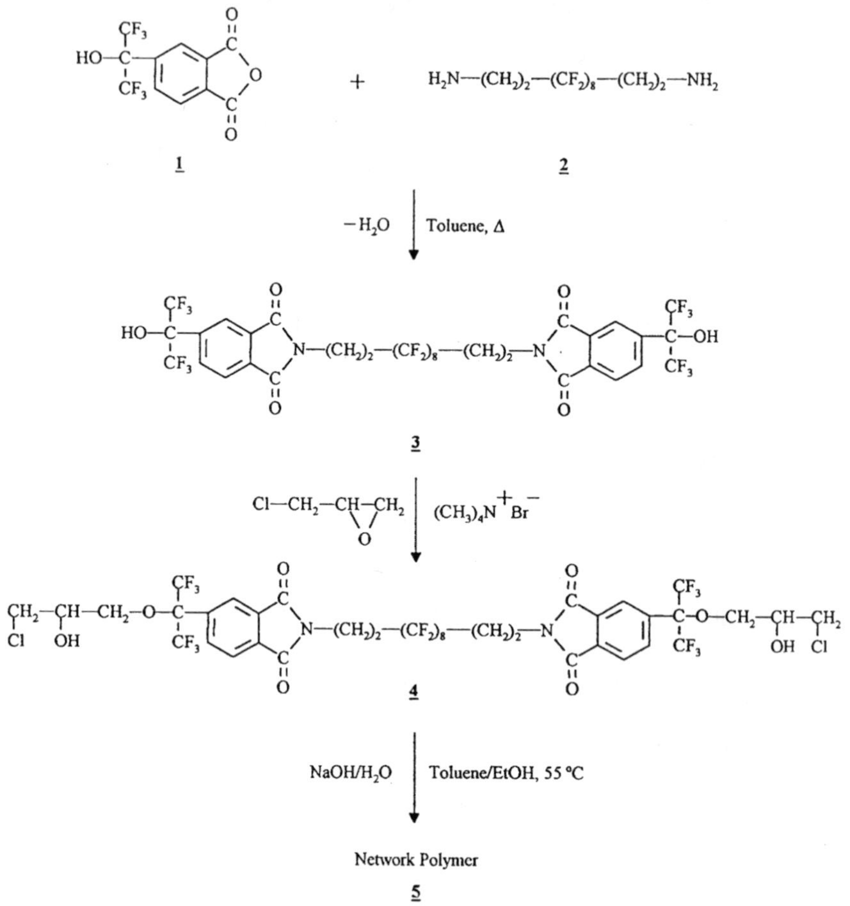

$$+ \quad H_2N-(CH_2)_2-(CF_2)_8-(CH_2)_2-NH_2$$

1   **2**

$-H_2O$ | Toluene, Δ

3

$Cl-CH_2-CH-CH_2$ (epoxide) $(CH_3)_4N^+Br^-$

4

NaOH/H_2O | Toluene/EtOH, 55 °C

Network Polymer

5

Scheme 1

This task represents a continuation of efforts to maximize the hydrophobicity of acrylic, epoxy, and other polymeric systems for resistance to water penetration and environmental degradation, and to minimize the dielectric constant and improve the processability for adhesives and coatings, without compromising the necessary structural characteristics for materials used for, e.g., structural elements, liners, paints, and microelectronic devices.

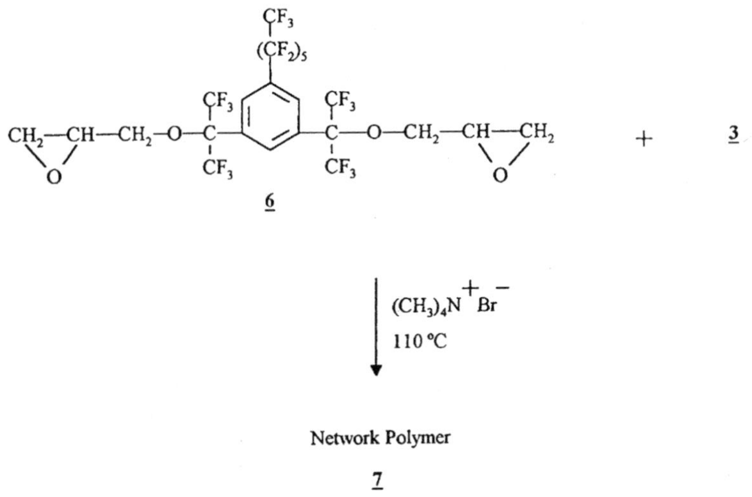

Scheme 2

12.2. EXPERIMENTAL

12.2.1. Methods

Infrared spectra were obtained with a Perkin-Elmer 1800 and a Nicolet Magna-IR 750 FTIR spectrophotometer, and the absorption frequencies are reported in wave numbers (cm^{-1}). NMR spectra were obtained with a BZH-300 and a CA-F-300 Bruker FT NMR 300 MHz spectrometer, and a Varian EM390 90-MHz spectrometer. All chemical shifts are reported in parts per million downfield (positive) of the standard. ^{1}H-NMR and ^{13}C-NMR chemical shifts are reported relative to internal tetramethylsilane, while ^{19}F-NMR chemical shifts are reported relative to internal fluorotrichloromethane. Unless specified, R_f values were obtained from silica gel thin-layer chromatography developed with a mixture of 1.5 ml of methylene chloride and three drops of acetone. Melting points were determined with an Electrothermal melting-point apparatus without correction. Elemental analyses were performed by Galbraith Laboratories, Inc., Knoxville, Tenn.

12.2.2. Materials

Hexadecafluoro-1,12-dodecanediamine 2 was obtained from Fluorochem, Inc. (680 Ayon Ave., Azusa, CA 91702) and has a melting point of 109–111°C. All other reagents were used as received or purified by standard procedures.

12.2.3. Techniques

12.2.3.1. Preparation of Diimidediol 3

4-(2-Hydroxyhexafluoro-2-propyl)phthalic anhydride 1 was prepared by a multistep route from o-xylene and hexafluoroacetone according to the procedure of Griffith et al.[8] with mp, IR, and elemental analysis reported. R_f 0.48 and 0.85 (acetone). ^{1}H-NMR (acetone-d$_6$/CDCl$_3$) 8.48 (s, 1H), 8.37 (d, $J = 8$ Hz, 1H), 8.14 (d, $J = 8$ Hz, 1H), 7.5 (br. s, —OH, 1H); ^{19}F-NMR −75.18; ^{13}C-NMR 161.90 and 161.77 (—CO—), 139.55, 134.7, 132.43, 131.55, 125.60, 124.7 (s, aromatic C), 122.25 (q, $J = 288$ Hz, −CF$_3$). This anhydride is hygroscopic and its phthalic acid derivative is readily formed upon treatment with water. R_f 0.01 and 0.48 (acetone). ^{1}H-NMR (acetone-d$_6$/CDCl$_3$) 8.28 (s, 1H), 8.1 (br. s, −OH), 7.99 (d, $J = 8$ Hz, 1H), 7.88 (d, $J = 8$ Hz, 1H); ^{19}F-NMR −75.32; ^{13}C-NMR 169.08 and 168.55 (−COOH), 134.32, 134.07, 132.12, 130.06, 129.40, and 128.41 (s, aromatic C), 122.83 (q, $J = 288$ Hz, −CF$_3$).

Anhydride 1 (41 g, 13- mmol) and diamine 2 (25 g, 51 mmol) in 300 ml of toluene were heated to reflux for 20 h under a Dean–Stark trap. Upon cooling in an ice bath, a red oil settled to the bottom. It was separated and solidified. This crude product was stirred as a suspension in water at 50°C until the water became neutral. It was filtered and dried $in\ vacuo$ to give a white powder 18.8 g, yield 34%, recrystallization from hexanes/THF (10 : 1) with 90% recovery, m.p. 145–148°C. R_f 0.27. IR (KBr pellet) 3415 (OH), 1780 and 1720 (imide), 1450, 1388, 1300–1100 (F), 980, 970, 750, 722 cm^{-1}. ^{1}H-NMR (acetone-d$_6$) 7.68 (d, $J = 8$ Hz, 2H), 7.53 (s, 2H), 7.45 (d, $J = 8$ Hz, 2H), 4.08 (t, $J - 7$ Hz, 4H, N—CH$_2$—), 2.86 (s, 2H, −OH), 2.76 (m, 4H, −CH$_2$—CF$_2$—); ^{19}F-NMR −74.29 (s, CF$_3$, 12F), −114.17 (t, $J = 12$ Hz, 4f), −121.14 (br. s, 4F), −121.34 (br. s, 4F), −123.12 (br. s, 4F); ^{13}C-NMR 167.43 (−CO, imide), 137.87, 134.71, 133.90, 135.64, 125.58, 124.34 (aromatic C), 123.68 (q, $J = 287$ Hz, CF$_3$). For partially hydrolyzed imide, 167.91, 167.54 and 167.43 for −CO are observed. Analysis: Calcd. For $C_{34}H_{16}F_{28}N_2O_6$: C, 37.79; H, 1.49; F, 49.24; N, 2.59. Found: C, 38.71; H, 1.79; F, 49.59; N, 2.49.

12.2.3.2. Preparation of 5 from 3 via 4

A mixture of 3 (0.20 g, 0.186 mmol), epichlorohydrin (1.74 g, 18.80 mmol) and tetramethylammonium bromide (0.030 g, 0.195 mmol) and stirred and maintained at 90°C for 70 h. Evaporation at 90°C/20 mm left a viscous liquid 0.36 g of dichlorohydrin 4. IR (neat film) 3480, and 3300 (OH), 3070, 3006, 2960, 1781 and 1720 (imide), 1485, 1450, 1385, 1300–1050 (F), 1015, 982, 945, 915, 858, 750, 726, 708 cm^{-1}. NMR (CDCl$_3$/TMS) δ showed only 8.3–7.9 in the aromatic region indicated that no 3 was present, but additional integration in 4.3–2.2 indicated that there was some epichlorohydrin left.

Dehydrochlorination of **4** was accomplished by first mixing it with toluene (5 g) and ethanol (1.7 g) under nitrogen and stirring in an ice bath; an 18% sodium hydroxide solution (4 g) was then added dropwise over 10 min. After stirring at room temperature overnight and then at 55°C for 1 h, a yellow oil stayed between the top organic layer and bottom aqueous layer. Tetrahydrofuran (2 ml) was added and the mixture was stirred at room temperature for 10 min during which the yellow oil dropped to the bottom. It was separated and dried *in vacuo* at room temperature for 2 h to a semisolid that became a white solid **5** on standing: 0.161 g, yield 78%, no m.p. observed but there was softening and color change. IR (KBr pellet) 3400 (OH), 3080, 2980, 2890, 1655, 1625, 1600, 1570, 1390, 1300–1100 (F), 1020, 995, 970, 840, 735, 715, 700, 543 cm^{-1}.

12.2.3.3. Preparation of **7** from **3** and **6**

The diglycidyl ether **6** of 1,3-bis-(2-hydroxyhexafluoro-2-propyl)-5-trideca-fluorohexyl-benzene was prepared by a multistep route according to the general procedure of Griffith and O'Rear.[9,10] A colorless viscous liquid, R_f 0.66. IR (neat) 3090, 3007, 2945, 1460, 1400–975 (F), 915 (epoxide), 862, 843, 811, 780 cm^{-1}. ^{1}H-NMR (CDCl$_3$) 6.39 (s, 1H, 2-ArH), 6.27 (s, 2H, 4,6-ArH), 3.95 (d, $J = 10.3$ Hz, 2H), 3.57 (t, $J = 7.5$ Hz, 2H), 3.31 (m, 2H), 2.88 (t, $J = 4.5$ Hz, 2H), 2.71 (m, 2H); ^{19}F-NMR −71.07 and −71.71 (—OC(CF$_3$)$_2$—, 6F each), −81.48 (—CF$_2$—CF$_3$, 3F), −111.63, −121.66, −122.13, −123.09, −126.56 (—CF$_2$—CF$_2$—CF$_2$—CF$_2$—CF$_2$—CF$_3$, 2F each); ^{13}C-NMR 131.82 (s), 131.15 (t, $J = 25$ Hz), 130.48 (s), and 129.16 (s) (aromatic-C), 121.90 [q, $J = 288$ Hz, —OC(CF$_3$)$_2$—], 120–100 (m, —CF$_2$—CF$_2$—CF$_2$—CF$_2$—CF$_2$—CF$_3$), 82.27 (hept, $J = 29$ Hz, —OC(CF$_3$)$_2$—), 67.76, 49.31, and 43.80 (glycidyl-C).

Diepoxy **6** (0.84 g, 1 mmol) and diimidediol **3** (1.07 g, 1 mmol) were mixed well and the mixture heated at 110°C overnight, during which time no change took place. Tetramethylammonium bromide in a catalytic amount was added, mixed well, and heated at 110°C. After one day a brittle transparent solid was obtained. During two more days it became a very hard solid **7**. IR (KBr pellet) 3480 (OH), 1785 and 1730 (imide), 1452, 1388, 1300–1100 (F), 1050, 1018, 983, 750, 730, 716, 708 cm^{-1}.

12.3. RESULTS AND DISCUSSION

The absence of an IR absorption peak around 913 cm^{-1} (10.95 μm) for both **5** and **7** indicates that the glycidyl ether rings are mostly converted into resins through cross-linking.[11] Heavily fluorinated compounds, **5** and **7**, have very strong IR absorption in the region of 1300–1100 cm^{-1}.

Upon the formation of **5**, the absence of 1780 and 1720 cm^{-1} peaks for an imide ring and the appearance of an amide peak at 1655 cm^{-1} as well as a carboxylate salt peak at 1570 cm^{-1} strongly indicated that the imide ring reacted with the glycidyl ether ring. Since amides are curing reagents for epoxies[12] and imides are in equilibrium with amic acids in the presence of water,[13] the imide ring may be partially hydrolyzed into secondary amide and carboxylate salt under the basic conditions in the presence of tetramethylammonium bromide as catalyst; it may have further reacted with epoxy to form tertiary amide as shown in Scheme 3.

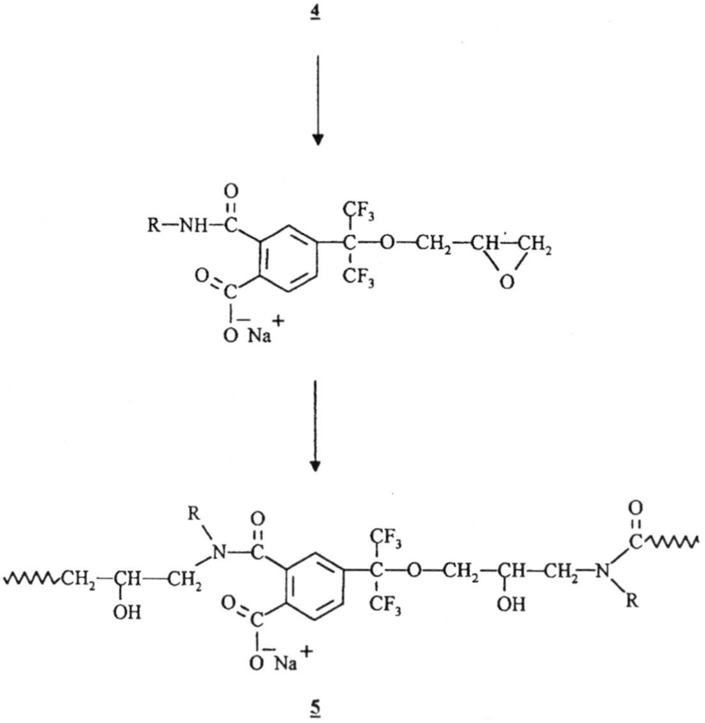

Scheme 3

Similar to our results, it has been reported[14] that epoxy compounds bearing azo-methine linkages with —COOH-, —OH-, and —NH$_2$-substituted aromatic amines can be self-cured with thermal polymerization. The epoxy group first reacts with the —COOH, —OH, and —NH$_2$ functional groups of the other molecule and the resulting pendant —OH group may then react with another epoxy group to produce an ether linkage. It has been reported that hydroxy-substituted diacrylates were produced from diglycidyl ethers by reaction with acrylic acid[15] a diacrylate was also prepared from m-12F-diol by reaction with propylene oxide and subsequently with acryloyl chloride.[16]

Since each epoxy group is difunctional, the diepoxy compounds are tetra-functional. For polymerization of the polyfunctional monomers at sufficiently high degrees of conversion, the branching must result in the formation of cross-links to give a three-dimensional thermosetting network. As expected, the resulting epoxy networks **5** and **7** are semitransparent, hard solids, and some shrinkage in volume is observed during curing. Fluorinated compounds that have only one unsaturated group can be used in a mixture with our monomers to polymerize to a solid polymer of a thermoset nature and to enhance processing ease.

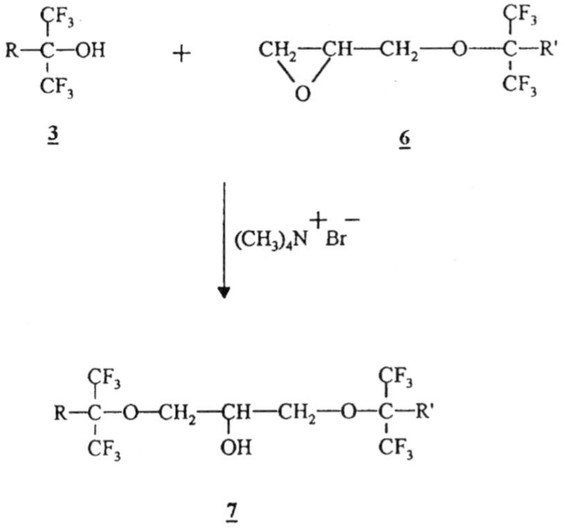

$$
\begin{array}{ccc}
& \text{CF}_3 & \\
& | & \\
\text{R}-\text{C}-\text{OH} & + & \text{CH}_2-\text{CH}-\text{CH}_2-\text{O}---\text{C}-\text{R}' \\
& | & \\
& \text{CF}_3 & \\
\underline{\mathbf{3}} & & \underline{\mathbf{6}}
\end{array}
$$

$$(\text{CH}_3)_4\text{N}^+\ \text{Br}^-$$

$$
\begin{array}{c}
\text{CF}_3 \qquad\qquad\qquad \text{CF}_3 \\
| \qquad\qquad\qquad\quad | \\
\text{R}-\text{C}-\text{O}-\text{CH}_2-\text{CH}-\text{CH}_2-\text{O}-\text{C}-\text{R}' \\
| \qquad\qquad | \qquad\qquad | \\
\text{CF}_3 \qquad\quad \text{OH} \qquad\quad \text{CF}_3 \\
\underline{\mathbf{7}}
\end{array}
$$

Scheme 4

For the formation of **7** from equivalent amounts of **3** and **6**, reaction took place only in the catalytic presence of tetramethylammonium bromide. In contrast to the behavior of **5**, the imide ring structure remained but with a shift in IR absorption from 1780 and $1720\,\text{cm}^{-1}$ to 1785 and $1730\,\text{cm}^{-1}$. One might expect the product generated from **3** and **6** to have less intermolecular hydrogen bonding. Therefore, the classical reaction[12] between epoxy and alcohol is suggested as shown in Scheme 4 to produce less acidic alcohol. A fluorodiepoxide (C8) similar to **6** has been reported to cure with a silicone diamine hardener (1SA) to form a network polymer as a hydrophobic model for investigations of postcuring and transition behavior,[17] while in our study a catalyst such as tetramethylammonium bromide is required for the reaction of **6** with **3**. The chain-extension of nonfluorinated liquid epoxy resins by reacting with bisphenol-A and catalyst to a high-molecular-weight solid resin has been reported for use in water-borne can coating.[18] Since the fluorinated alcohol is a stronger acid in comparison with bisphenol-A, which is a phenol, the reactivity of our fluorinated system is higher.

In examining the ^{1}H-NMR spectra, the five protons of the glycidyl ether group in **6** appear as a typical complex series of five bands in the chemical-shift range from 4.0 to 2.7 ppm, the opening of the epoxide ring to form a polymer **5** and **7** results in the concurrent decrease of peaks in the range from 2.7 to 3.5 ppm and an attendant increase in the glycerol resonance centered around 4.0 ppm, similar to the case of bisphenol-A diepoxide.[19]

12.4. CONCLUSIONS

A novel fluorodiimidediol has been synthesized utilizing diamine, anhydride, and existing solution imidization techniques. Two synthetic methodologies leading to heavily fluorinated epoxy networks are reported. One approach to convert it into diepoxy via dichlorohydrin in basic solution permits the opening of the imide rings and subsequently reacting with the epoxy rings to an amide, carboxylate-containing epoxy network. The other approach, to cure it with a fluorodiepoxide in the catalytic presence of tetramethylammonium bromide, results in an imide-containing epoxy network with intact imide rings.

ACKNOWLEDGMENT: Partial funding support from the Office of Naval Research is gratefully acknowledged.

12.5. REFERENCES

1. P. E. Cassidy, T. M. Aminabhavi, V. Screenivasulu Reddy, and J. W. Fitch, *Eur. Polym. J. 31* (4), 353 (1995).
2. J. R. Griffith and R. F. Brady, Jr., *Chemtech 19* (6), 370 (1988).
3. J. R. Griffith, *Chemtech 12* (5), 290 1982).
4. J. R. Griffith, J. G. O'Rear, and S. A. Reines, *Chemtech 2* (5), 311 (1972).
5. H. S.-W. Hu and J. R. Griffith, *Polym. Mater. Sci. Eng. 66*, 261 (1992) [No. 1]. H. S.-W. Hu. and J. R. Griffith, in *Polymers for Microelectronics* (G. Willson, L. F. Thompson, and S. Tagawa, ed.), ACS Symposium Series *537*, 507 (1994) [No. 2]. J. R. Friffith and H. S.-W. Hu, U.S. Patent 5,292,927 (1994) [No. 3]. H. S.-W. Hu and J. R. Griffith, *Polym. Preprints 34*, 401 (1993) [No. 4]. H. S. -W. Hu and J. R. Griffith, this volume Ch. 11 [No. 5].*
6. D. Wilson, H. Stenzenberger, and P. M. Hergenrother (eds.), in *Polyimides*, Blakie, Glasgow (1990).
7. H. S.-W. Hu and J. R. Griffith, *Polym. Preprints 32* (3), 216 (1991); also presented at the 202nd National Meeting of the American Chemical Society, New York, August 25–30, 1991.

---

*These references are part of a series of articles, entitled "Processable Fluoropolymer with Low Dielectric Constants."

8. J. R. Griffith, J. G. O'Rear, and J. P. Reardon, in *Adhesion Science and Technology, Vol. 9A* (L. H. Lee, ed.), Plenum Press, New York (1975), pp. 429–435.
9. J. G. O'Rear and J. R. Griffith, *Org. Coatings and Plastics Chem 33* (1), 657 (1973).
10. J. R. Griffith and J. G. O'Rear, *Synthesis 1974*, 493.
11. H. Lee and K. Neville, *Handbook of Epoxy Resins*, McGraw-Hill, New York (1967), Appendix 4-1.
12. H. Lee and K. Neville, *Handbook of Epoxy Resins*, McGraw-Hill, New York (1967), Appendix 5-1.
13. C. D. Smith, R. Mercier, H. Watson, and B. Sillion, *Polymer 34* (23), 4852 (1993).
14. J. A. Mikroyannidis, *J. Appl. Polym. Sci. 41*, 2613 (1990).
15. A. M. Zeig, U.S. Patent 4,914,171 (1990).
16. A. Washimi, M. Yoshida, and K. Kimura, Japanese Kokai Tokyo Koho, JP01,199,937 (1989).
17. T. E. Twardowski and P. H. Geil, *J. Appl. Polym. Sci. 41*, 1047 (1990).
18. J. Gannon, *Polym. News 15*, 138 (1990).
19. *Guide to the NMR Spectra of Polymers*, Sadtler Research Laboratories, Inc., Philadelphia (1973), pp. 137–146.

13

Synthesis of Fluoropolymers in Liquid and Supercritical Carbon Dioxide Solvent Systems

JAMES P. DeYOUNG, TIMOTHY J. ROMACK, and JOSEPH M. DeSIMONE

13.1. INTRODUCTION

13.1.1. Alternative Solvent Technologies

Prompted by public and regulatory demand to reduce the emission of toxic compounds into the environment, solvent-intensive industries are actively seeking alternatives to traditionally employed organic and chlorinated solvents. Many sectors of industry have turned to aqueous systems, only to discover that the drawbacks associated with mixed aqueous waste disposal and energy-intensive drying offer little improvement over solvent-based systems. Carbon dioxide in liquid or supercritical form is a potential alternative for many of the industrial segments under fire. It is not regulated, it is nontoxic, inexpensive, readily available, and has the potential to be easily separated from other components in a waste stream, reducing the volume of mixed waste produced. As a result, carbon dioxide is actively being pursued as an alternative technology by many solvent- and waste-intensive industries—from precision and textile cleaning to chemical manufacture and polymer synthesis and processing. A detailed discussion of many of the important areas of interest is beyond the scope of this manuscript, but several reviews are available.[1–4]

JAMES P. DeYOUNG and TIMOTHY J. ROMACK · MICELL Technologies Inc., Raleigh, North Carolina 27613.   JOSEPH M. DeSIMONE · Department of Chemistry, University of North Carolina at Chapel Hill, Chapel Hill, North Carolina 27599-3290.

Fluoropolymers 1: Synthesis, edited by Hougham *et al.*, Plenum Press, New York, 1999.

13.1.2. Solvent Properties of Carbon Dioxide

A complete understanding of the solvent properties of dense CO_2 has been essential as an increasing emphasis has been placed on its use in a variety of manufacturing and processing applications, but to date these parameters are still somewhat undefined. A traditional view of solubility parameters ascribed to CO_2 based on widely held thermodynamic data would suggest that dense CO_2 has a solvent strength similar to that of toluene or hexane. However, a large contribution to the CO_2 solubility parameter emanates from its large quadrupole moment, similar to that of benzene.[5,6] Although the large quadrupole moment plays an important role in the physical properties of CO_2 and perhaps in the solubility of many small molecules, its role in the solvation of nonvolatile materials such as siloxanes and amorphous fluoropolymers is not completely clear. Work is ongoing to clarify the role of the strong quadrupole moment in the dissolution of polymeric materials.[7] It can generally be said that the solvent strength of dense CO_2 is a debatable issue and perhaps best determined through experimentation.[8]

Carbon dioxide in both its liquid and supercritical states is of interest in the synthesis and processing of polymeric materials. Historically, CO_2 has received much attention as a supercritical fluid, owing to its readily accessible critical point ($T_c = 31.8°C$; $P_c = 73.8$ bars, ca. 1100 psi). By adjusting the pressure above the critical temperature, a range of physiochemical properties (e.g., density, dielectric constant, and viscosity) are achievable (Figure 13.1). Liquid CO_2 can be accessed below the critical temperature and is of great interest for many applications, including the manufacture of polymers, because it has most of the advantages of supercritical CO_2 (e.g., low viscosity and ease of separation from waste components) and can be accessed at greatly reduced pressures. The vapor

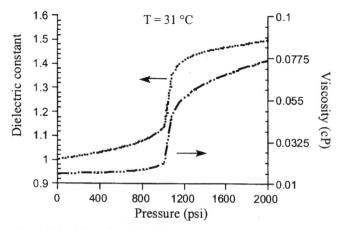

Figure 13.1. Solvent "tunability," viscosity/dielectric versus CO_2 pressure.

pressure of CO_2 at 0°C is less than 40 bars ($\approx$ 600 psi), only five to ten times the pressure in an ordinary bicycle tire. To put this in context, low-density polyethylene is commonly manufactured on a commercial scale at pressures of over 2000 bars (ca. 30,000 psi).[9] For liquid and supercritical CO_2 the dielectric constant is from 1.2 to 1.5, consistent with CO_2's extremely nonpolar nature, and its viscosity ranges from gaslike to approaching that of a liquid solvent.[10] Liquid and supercritical CO_2, under reasonable conditions (pressure ≤ 500 bars, temperature ≤ 100°C), are good media for low-molecular-weight, nonionic compounds, but poor solvents for most macromolecules.[11] The only high-molecular-weight materials that dissolve to a significant degree in CO_2 under readily accessible conditions are siloxanes and some amorphous fluoropolymers, as discussed below.

13.2. FLUOROALKYL ACRYLATE POLYMERIZATION IN CARBON DIOXIDE

13.2.1. Homopolymers

Historically the widespread use of liquid and supercritical CO_2 has been limited by the low solubility of most nonvolatile materials at reasonably low pressures. However, studies carried out over a decade ago established the solubility of oligometric perfluoropolyethers and poly(chlorotrifluoroethylene) in liquid CO_2,[12] and a few years later it was discovered that other highly fluorinated polymers such as fluorinated acrylates were also quite soluble in liquefied CO_2 at relatively low pressures.[13]

The first synthesis of amorphous fluoroacrylate polymers using supercritical CO_2 as an inert reaction medium was reported by DeSimone et al. in 1992 and involved the free-radical polymerization of 1,1-dihydroperfluorooctyl acrylate using AIBN initiator[14] (Scheme 1). These studies showed that conditions commensurate with conventional solvent-based systems can be used to produce high conversions of high-molecular-weight materials in CO_2. It has been noted that for traditional types of homogeneous and heterogeneous polymerizations the elevation of pressure can affect free-radical polymerizations by: (1) increasing the concentration of gaseous monomers; (2) modificying rate constants for initiation, propagation, termination, and chain transfer; (3) changing the equilibrium constant for polymerization. The cumulative effect of higher pressure is generally increased reaction rates and larger molar masses.[15] However, owing to the highly compressible nature of supercritical fluids, and the variation in solvent strength that occurs, particularly in the case of supercritical CO_2, the effect of pressure on reaction rate in these media is more complicated than in traditional liquid solvents.

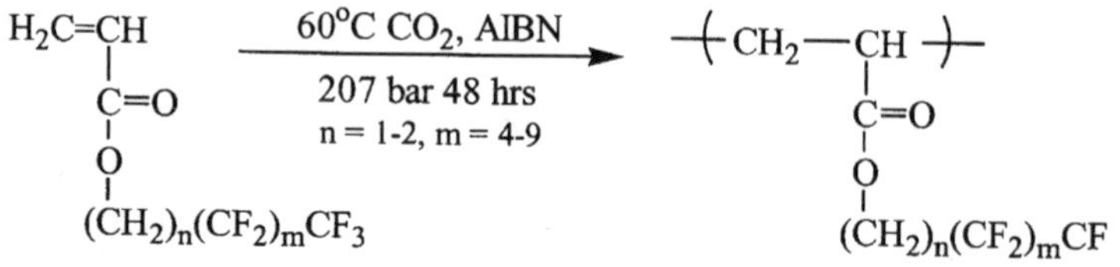

$$H_2C=CH \qquad \xrightarrow[\substack{207 \text{ bar } 48 \text{ hrs} \\ n = 1\text{-}2, \, m = 4\text{-}9}]{60^\circ C \; CO_2, \; AIBN} \qquad \left(CH_2 - CH \right)$$

(structure on left)
C=O
|
O
|
$(CH_2)_n(CF_2)_mCF_3$

(structure on right)
C=O
|
O
|
$(CH_2)_n(CF_2)_mCF$

Scheme 1. Free radical polymerization of fluorocarbon acrylates in supercritical CO_2.

Initiator decomposition studies of AIBN in supercritical CO_2 carried out by DeSimone *et al.* showed that there is kinetic deviation from the traditionally studied solvent systems.[16] These studies indicated a measurable decrease in the thermal decomposition of AIBN in supercritical CO_2 over decomposition rates measured in benzene. Kirkwood correlation plots indicate that the slower rates in supercritical CO_2 emanate from the overall lower dielectric constant (ϵ) of CO_2 relative to that of benzene. Similar studies have shown an analogous trend in the decomposition kinetics of perfluoroalkyl acyl peroxides in liquid and supercritical CO_2.[17] Rate decreases of as much as 30% have been seen compared to decomposition measured in 1,1,2-trichlorotrifluoroethane. These studies also served to show that while initiator decomposition is in general slower in supercritical CO_2, overall initiation is more efficient. Uv–visual studies incorporating radical scavengers concluded that primary geminate radicals formed during thermal decomposition in supercritical CO_2 are not hindered to the same extent by "cage effects" as are those in traditional solvents such as benzene. This effect noted in AIBN decomposition in CO_2 is ascribed to the substantially lower viscosity of supercritical CO_2 compared to that of benzene.[18]

13.2.2. Random Copolymers

Statistical copolymers of fluorinated acrylic monomers with conventional hydrocarbon monomers such as butyl acrylate, methyl methacrylate, styrene, and ethylene, among others, have also been synthesized under homogeneous conditions in supercritical CO_2.[19] The hydrocarbon homopolymers from these monomers, being completely insoluble in supercritical CO_2, differ from these copolymers as homogeneous conditions can be maintained even at high hydrocarbon incorporation levels and relatively low pressures (330 bars at 65°C). In some cases, depending on the hydrocarbon monomer, these conditions can be maintained at molar incorporations greater than 50%. The synthesis of amorphous fluoropolymers in CO_2 also offers the unique benefit associated with an inherently low-viscosity solvent, and thus avoids autoacceleration at high polymer conversions. Moreover, typical polydispersities of 1.5 indicate that termination by radical combination is predominant in these systems, as may be expected with a completely inert solvent. Integrated reaction and separation processes resulting from the variable density and solvent strength of supercritical CO_2, not available with traditional solvent systems, provide an added benefit in these sytems.

13.2.3. Applications of Amphiphilic Copolymers

The insolubility of hydrocarbon polymers in CO_2 has been a limiting factor for the application of CO_2 technologies in the production and processing of many polymeric materials. Heterogeneous polymerizations utilizing CO_2 without the use of efficient dispersants generally result in low yields of unsatisfactory materials having low molecular weight and high polydispersities. However, the development of amphiphilic copolymers, designed for CO_2 applications and predicated on the discovery that CO_2 is a good solvent for fluoroacrylate synthesis, has opened numerous doors for the use of this alternative technology in the production of a broad range of hydrocarbon polymers.[20-22] These materials, generally containing a "CO_2-philic" segment and a lyophobic segment, self-assemble in CO_2 into well-defined structures providing stable polymeric colloidal dispersions. A detailed discussion of this work is beyond the scope of this paper.

13.3. FLUOROOLEFIN POLYMERIZATION IN CARBON DIOXIDE

13.3.1. Overview

Tetrafluoroethylene (TFE)-based copolymers have become premium high-performance materials for a broad range of applications requiring superior chemical and thermal resistance, and melt-processing capability.[23] Materials embodied by copolymers of TFE with hexafluoropropylene (HFP), perfluoro-(propyl vinyl ether) (PPVE), or sulfonyl fluoride functionalized perfluoro(alkyl vinyl ether) monomer (PSEVPE) have often been synthesized in nonaqueous solvent systems to avoid prcblems associated with traditional aqueous dispersion or emulsion polymerizations. Aqueous suspension or dispersion methods result in increased occurrence of carboxylic acid end groups that are deleterious to the products during melt-processing steps and to the function of the polymer. This is particularly true of the copolymers with perfluorovinyl ether comonomers that are generally used to modify the crystallinity of the fluoroplastic. Fluoropolymer materials containing significant levels of these acid end groups may require finishing steps such as high-temperature hydrolysis or fluorination. Processing without these steps can result in product decomposition, discoloration erratic changes in molecular weight, and the emission of hazardous and damaging hydrogen fluoride. To minimize the occurrence of these acid end groups as much as possible, copolymerizations employing perfluorinated vinyl ether monomers have generally been carried out in chlorofluorocarbons (CFCs). Owing to the highly electrophilic nature of fluoroolefin radicals,[24] hydrocarbon solvents must be avoided in the polymerization process to obtain high-molar-mass materials. Otherwise, chain-transfer to the solvent becomes prevalent, limiting the molecular weight of the final polymer. Perfluorocarbon and hydrofluorocarbon solvents have been proposed as suitable alternatives for these polymerizations[25-27]; however, these solvents are generally quite expensive.

At the forefront of the development of this technology has been the need to replace detrimental halogenated solvents with viable alternative technologies. Aside from the many advantages afforded by CO_2 over many traditional solvent systems, further advantages may exist in the storage of TFE with CO_2. Many associated dangers of working with TFE, such as disproportionation and auto-polymerization, can be avoided by storage with CO_2.[28] Currently to avoid these hazards, TFE is often stored as an azeotrope with hydrogen chloride.

13.3.2. Melt-Processable Fluoropolymers

The first reactions of fluorinated olefins in CO_2 reported by DeSimone *et al.* involved the free-radical telomerization of 1,1-difluoroethylene[29] and tetrafluor-oethylene.[30] This work demonstrated the feasibility of carrying out free-radical reactions of highly electrophilic species in solvents other than expensive fluor-ocarbons and environmentally detrimental chlorofluorocarbons. The work has since been more broadly applied to the synthesis of tetrafluoroethylene-based, nonaqueous grades of fluoropolymers,[31,32] such as poly(tetrafluoroethylene-co-perfluoropropylvinyl ether) (Scheme 2). These reactions were typically carried out at between 20 and 40% solids in CO_2 at initial pressures of between 100 and 150 bars, and 30–35°C (Table 10.1).

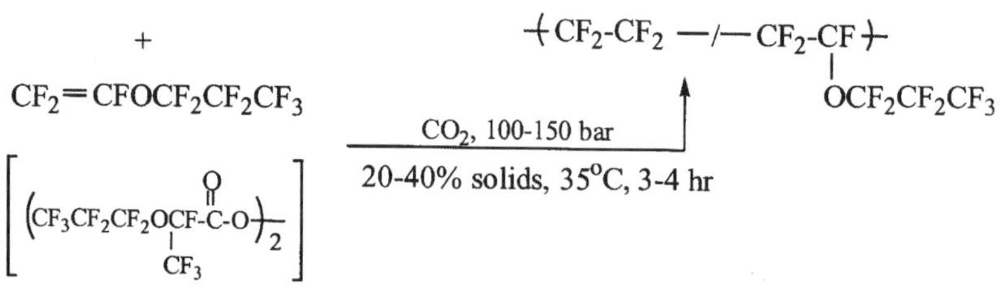

Scheme 2. Copolymerization of TFE with PPVE in CO_2.

Several observations could be made from these copolymerizations. High yields of high-molecular-weight copolymer can readily be formed at incorporation levels well in excess of the 2–4 wt% PPVE desired in commercial products. For reactions done in the CO_2 system, even at high vinyl ether incorporations, melt-

Table 13.1. Selected Data for TFE/PPVE Copolymerization in CO_2

TFE	PPVE	Yield (%)	PPVE (incorporation)	MV^a (poise)	T_m
1.9	0.18	99	2.9%	*	321.5
2.2	0.55	100	8.9%	*	318.6
2.0	0.55	100	5.2%	*	312.7
2.2	0.92	100	5.8%	*	313.7
1.2	0.35	70	7.8%	$**7.3(10^4)$	297.1

^{a} * too high to measure; ** methanol added as chain-transfer agent.

flow properties are suppressed, suggesting that chain-transfer reactions and β-scission of the propagating vinyl ether-containing chains are substantially minimized in CO_2, leading to very high-molecular-weight materials. In fact, the addition of small amounts of chain-transfer agents was needed to produce lower-molecular-weight materials having the desired melt-flow properties. Quantitative IR analysis confirmed the substantially reduced occurrence of acid end groups. The extremely low number of unstable acid end groups observed in most samples prepared in CO_2 is unprecedented, detecting only zero to three end groups per 10^6 carbon atoms. This an order of magnitude or so lower than for materials prepared in traditional solvents and comparable to materials that have been prepared in conventional nonaqueous systems and then subjected to high-temperature aggressive fluorination with F_2 or an alternative posttreatment process to remove unstable end groups. Scanning electron micrographs (SEM) (Figure 13.2) show little differentiation at 30,000× magnification between resin synthesized in CO_2 versus that made in chlorofluorocarbon solvent.

13.3.3. Ion-Exchange Resins

Perfluorinated ion-exchange resins such as Nafion[TM], used in chloro alkali cells[23] and as solid acid catalysts[33,34] are often manufactured in nonaqueous media to avoid loss of the expensive vinyl ether monomers from partitioning into the water phase and the susceptibility of these monomers to hydrolysis, as well as to aid in the avoidance of acid end group formation. The resulting copolymers are then fluorinated with F_2 and fabricated into membranes using melt-extrusion techniques. Membrane production is accomplished by lamination with polytetrafluoroethylene cloth for increased strength followed by conversion of the polymer to the acid or ionic form. The limiting synthetic parameter in the preparation of high-molecular-weight copolymers at high vinyl ether incorporation is chain-transfer stemming from β-scission of the vinyl ether macroradical. If undesirable side reactions are somehow suppressed through the implementation of a CO_2-

Figure 13.2. SEM of TFE/PPVE copolymer synthesized in (A) CO_2, and (B) chlorofluorocarbon.

based polymerization process, this process could open the door to a new generation of high-performance perfluorinated ion exchange membranes.

Carbon dioxide has also proven to be an exemplary medium for the polymerization of TFE with perfluorinated alkylvinyl ether monomers containing sulfonyl fluoride such as $CF_2{=}CFOCF_2CF(CF_3)OCF_2CF_2SO_2F$ (PSEVPE). As seen in Table 13.2, the dramatic difference in the number of acid end groups between the commercial sample and those made in CO$_2$ indicates that chain-transfer processes stemming from vinyl ether radical arrangement are not nearly as prevalent in CO$_2$ as in conventional systems.

Another series of PSEPVE/TFE copolymerizations was carried out on a larger scale to allow assessment of relative molecular weights through melt-flow analysis and to allow screening of properties important in commercial applications. Reagent concentrations for these reactions are given in Table 13.3 along with equivalent weights, wt% comonomer incorporation, glass transition temperature (T_g), melt index, and end group analysis for each sample and for a commercially prepared example. A melt-flow index corresponds to how quickly the polymer flows under a constant pressure at high temperature (270°C). Typical commercial products exhibit a melt-flow index of 10–15. PSEPVE/TFE copolymers that were prepared in CO$_2$ had melt-flow indexes near the range of commercial products, as well as values corresponding to substantially higher molecular weight than commercial varieties (melt-flow index of 2.4). Samples 1, 2, and 3 in Table 13.4 were prepared targeting increasing molecular weight at similar levels of PSEPVE incorporation and conversion. This was accomplished by moving to higher percent solids (monomer) and lower initiator concentrations. The fact that a significant number of acid end groups are evidenced for two of the copolymers made in CO$_2$ (Samples 2 and 3) may be due to some oxygen contamination introduced while the reagents were being added. Reactions such as these run on a larger scale required the use of more than one TFE/CO$_2$ cylinder, necessitating the changing of cylinders during the monomer-charging step. The increasing numbers of acid end groups in the later series could also be

Table 13.2. Copolymerization of (SO$_2$F)-Functional Perfluorinated Alkyl Vinyl Ether Monomer with TFE

Equivalent weight	Incorporation (wt%)	Acid end groups[a]
2280 g/eq	21%	1
1120 g/eq	40%	0
1070 g/eq	42%	0
975 g/eq	46%	0
975 g/eq[b]	46%	365

[a] Determined by high-resolution IR, parts per 10^6 carbon atoms.
[b] Representative high-incorporation commercial sample.

Table 13.3. Results and Product Analysis for TFE/PSEVPE Copolymerizations in CO_2

	Sample			
	1	2	3	Control
TFE	0.72	0.85	1.0	—
PSEPVE	1.3	1.5	1.8	—
Initiator	4.3×10^{-4}	3.0×10^{-4}	1.5×10^{-4}	—
EN	975	930	999	975
T_g (°C)[a]	0.90	3.6	5.7	−2.8
wt% PSEPVE	46	48	45	46
Melt flow	20.7	16.8	2.4	10–15
End groups/10^6 Carbons	0	177	145	365

[a] As determined by dynamic mechanical analysis (DMA).

due to the change in solvent characteristics brought on by the inclusion of larger amounts of PSEPVE and TFE monomers in the higher-solids runs. PSEPVE monomer comprised the bulk of the monomer charge and is a nonpolar fluorocarbon liquid that could act in a way similar to conventional solvents.

The plasticizing and dissolution characteristics[5,6] of CO_2 toward fluoropolymers is further demonstrated by the swelling effect that it has on sulfonic acid derivatives of these perfluoroalkyl vinyl ether-containing polymers. The highly insoluble nature of these materials in most common organic solvents substantially inhibits the efficacy of these materials as solid acid catalysts. With small surface area, most potentially active catalytic sites remain buried in the resin, incapable of catalyzing reactions. However, the extremely acidic nature of these materials and their efficacy as solid acid catalysts has been demonstrated by Sun *et al.* in the application of the resins as sol gel products.[34] More recent work has demonstrated that the catalytic efficacy of perfluorinated resins containing sulfonic acid can also be improved through the use of supercritical CO_2 in catalytic reactions,[37] Kinetic

Table 13.4. Results for the Photooxidation of Hexafluoropropylene in Liquid CO_2[a]

HFP (g)	Solvent (volume ml)	PFPE yield (%)	C_3F_6O/CF_2O[b]	M_n[c] (g/mol)
10	CO_2 (8.9)	31	1.4	590
11	CO_2 (6.9)	34	2	800
20	CO_2 (6.8)	16	5.3	1300
17	None	29	10	2700

[a] Reactions in CO_2 carried out at −40°C.
[b] Molar ratio of hexafluoropropylene oxide repeat units to difluoromethylene repeat units determined from ^{19}F-NMR.
[c] Estimated from ^{19}F-NMR

rate constants for the dimerization of α-methyl styrene have proven to be as much as an order of magnitude higher than rates measured in traditional solvents.

13.3.4. β-Scission and Acid End Groups

During polymerization, a polymeric radical with a perfluoro(alkyl vinyl ether)-derived active center can have one of two fates: it can cross-propagate to tetrafluoroethylene or it can undergo β-scission to yield an acid-fluoride-terminated polymer chain and generate a perfluoroalkyl radical capable of initiating further polymerization (i.e., chain transfer to monomer). These scenarios are illustrated in Scheme 3.

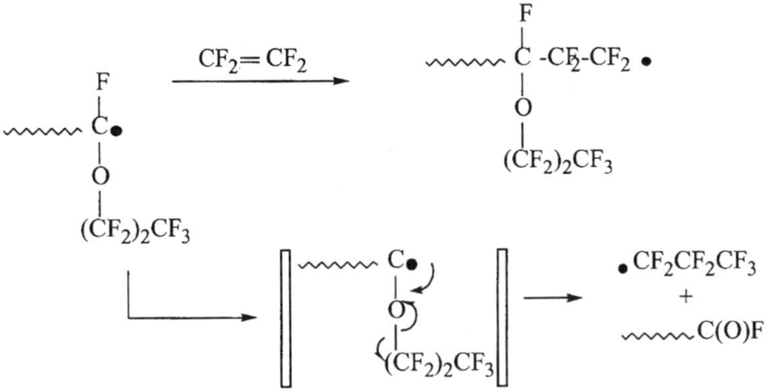

Scheme 3. Fate of the vinyl ether macroradical during copolymerizations.

The large reduction in the level of acid end groups occurring in copolymers of TFE and perfluoroalkyl vinyl ether monomers synthesized in CO_2 implies that propagation of terminal vinyl ether-containing radicals is overwhelming favored over β-scission, more so than in traditional solvent systems. There are a couple of possible explanations that could account for the difference between the two systems. The ability of CO_2 to plasticize polymeric materials and its excellent transport properties make it a very efficient tool for the conveyance of small molecules into even highly crystalline materials.[38,39] It is therefore quite likely that the presence of CO_2 facilitates diffusion of TFE monomer into the precipitated polymer phase, increasing the effective bimolecular rates for cross-propagation. It is also plausible to assume that a completely nonpolar solvent such as CO_2, with a dielectric constant between 1.3 and 1.5, would be less able to stabilize the intermediate transition state leading to β-scission than would 1,1,2-trichlorotri-fluoroethane. The fact that samples of PPVE/TFE copolymer made in perfluoro-N-methyl morpholine, a very nonpolar liquid solvent ($\epsilon \approx 2$), do not exhibit a substantially reduced number of acid end groups makes this explanation less likely.

13.3.5. Other Fluoropolymers

The polymerization of other fluoroolefins such as TFE with hexafluoropropylene (HFP), TFE with ethylene, and vinylidine difluoride[31,40] further demonstrates the broad applicability of liquid and supercritical CO_2 in the production and processing of fluorinated polymers. Many of the aforementioned advantages associated with CO_2, including tunable solvent properties, integrated synthesis, separation and purification processes, negligible chain transfer in the presence of highly electrophilic species, and relative ease of recycling, make it an ideal solvent for fluoroolefin polymerization.

13.4. OTHER SYSTEMS OF INTEREST

13.4.1. Photooxidation of Fluoroolefins in Liquid Carbon Dioxide

Perfluoropolyethers (PFPEs) like many other fluoropolymers exhibit exceptional thermal and oxidative stability. However, unlike fluoropolymers synthesized from most fluoroolefins, their properties are modified by a backbone of ether linkages, providing a high degree of chain flexibility. The cumulative result of these properties is manifested in materials that can be functional in a broad range of chemically demanding environments, as well as at very high and very low temperatures.[41] This versatility has presented opportunities for the use of PFPEs in applications ranging from high-performance lubrication for magnetic recording media, to aerospace and electronic devices and equipment, to cosmetics and monument protection.[42]

Photooxidation reactions of fluoroolefins in the presence of oxygen is one commercial method used in the production of PFPEs, generally employing either TFE or HFP. Fluorolefin concentration, oxygen level, light intensity, and temperature are all variables that have substantial impact on reaction rates, product distributions, polymer microstructure, peroxide content, and molecular weight. While HFP photooxidations are often carried out in bulk at low temperatures, TFE photooxidation must be carried out in an inert solvent, historically chlorofluorocarbons.

Carbon dioxide, shown to be an ideal medium for reactions involving very electrophilic radicals,[31,43] has also been shown to be useful in the photooxidation of fluorinated olefins in the synthesis of PFPEs.[4] Table 13.4 shows some representative results of the photooxidation of HFP in liquid CO_2. Overall, the resulting products of photooxidations in liquid CO_2 are similar to those emanating from reactions carried out in bulk, with lower C_3F_6O/CF_2O ratios and lower molecular weights seen in the former. This trend is likely a result of the lower HFP concentration in reactions in CO_2 leading to an increased occurrence of β-scission, which competes with propagation in these free-radical reactions involving perfluoroalkoxy radicals. Since TFE is substantially more reactive than HFP and photooxidations must be carried out diluted in an inert medium, CO_2 represents an inexpensive and environmentally benign alternative to chlorofluorocarbons. The easily varied solvent strength and density of CO_2 may also provide unique separation and purification opportunities that are not possible in commercial photooxidation processes.

13.4.2. Hybrid Carbon Dioxide/Aqueous Systems

Poly(tetrafluoroethylene) (PTFE) is manufactured primarily by free-radical methods in aqueous media. These heterogeneous processes, which may or may not involve the use of dispersing agents, result in either coagulated granular resins or fine PTFE resin particles.[41] Recently, DeSimone has shown that a CO_2/aqueous hybrid system is a useful medium for the production of granular and spherical high-molecular-weight PTFE resins.[45] This system represents a substantial deviation from traditional systems as CO_2 and water exhibit low mutual solubilities allowing for the compartmentalization of monomer, polymer, and initiator based on their solubility characteristics. Storage of TFE with CO_2 forming a pseudo-azeotrope, reducing potential disproportionation hazards,[46] clearly lends itself well to the use of this technology.

13.5. CONCLUSIONS

Unlike a large portion of the many billions of pounds of organic and halogenated solvents used in industry every year, CO_2 is inexpensive, of low toxicity, and environmentally and chemically benign. The prime factors inhibiting the widespread use of this attractive solvent replacement have been the disappointingly low solubility of most materials in CO_2 in both liquid and supercritical states, and a less than complete understanding of its physical properties. Discoveries made over the past decade identifying the solubility of siloxanes and many amorphous fluoropolymers in CO_2 have facilitated a greater understanding of the solvency properties of this uniquely "tunable" solvent. The subsequent development of amphiphilic surfactants incorporating lyophilic fluorocarbon or siloxane segments with lipophilic or hydrophilic moieties opens the door for the synthesis of a broad range of hydrocarbons. Carbon dioxide technology is of particular significance with respect to the manufacture of many fluoropolymers that require nonaqueous synthesis. Reduction of unwanted side

James P. DeYoung *et al.*

reactions such as β-scission and opportunities for unique separation and purification processes demonstrate that CO_2 is not only an adequate replacement for chlorofluorocarbons, but in many cases a superior one.

13.6. REFERENCES

1. M. McHugh and V. J. Krukonis, *Supercritical Fluid Extraction: Principles and Practice*, Butterworths, Boston (1986).
2. K. D. Bartle, A. A. Clifford, S. A. Jafar, and G. F. Shilstone, *J. Phys. Chem. Ref. Data 20*, 713–778 (1991).
3. K. P. Johnson and J. M. L. Penninger (eds), *Supercritical Fluid Science and Technology*, American Chemical Society, Washington D.C. (1989).
4. C. L. Phelphs, N. G. Smart, and C. M. Wai, *J. Chem. Ed. 73*, 1163–1168 (1996).
5. J. M. Prausnitz, R. N. Lichtenthaler, and E. G. deAzevedo, *Thermodynamics of Fluid Phase Equilibria*, Prentice-Hall, Englewood Cliffs, N.J. (1986).
6. C. S. Murthy, K. Singer, and I. R. McDonald, *Mol. Phys. 44*, 135 (1981).
7. F. Rindfleisch, T. P. DiNoia, and M. A. McHugh, *J. Phys. Chem. 28*, 15581 (1996).
8. M. A. McHugh, V. J. Krukonis, and J. A. Pratt, *TRIP 2*, 301–307 (1994).
9. G. Odian, *Principles of Polymerization, 3rd Ed.*, John Wiley and Sons, New York (1993), pp. 303–306.
10. V. Vesovic, W. A. Wakeham, G. A. Olchowy, J. V. Sangers, J. T. R. Watson, and J. Millat, *J. Phys. Chem. Ref. Data 19*, 763 (1990).
11. J. A. Hyatt, *J. Org. Chem. 49*, 5097 (1984).
12. M. McHugh and V. J. Krukonis, *Supercritical Fluid Extraction: Principles and Practice*, Butterworths, Boston (1986), Ch. 9.
13. Z. Guan, Homogeneous free radical polymerization in supercritical carbon dioxide, University of North Carolina, Dissertation, Ch. 2.
14. J. M. DeSimone, Z. Guan, and C. S. Eisbernd, *Science 257*, 945 (1992).
15. P. W. Moore, F. W. Ayscough, and J. G. Clouston, *J. Polym. Sci: Polym. Chem. Ed. 15*, 1291 (1977).
16. Z. Guan, J. R. Combes, Y. Z. Menceloglu, and J. M. DeSimone, *Macromolecules 26*, 2663 (1993).
17. J. P. DeYoung, J. Kalda, and J. M. DeSimone, Decomposition kinetics of perfluoroalkyl acyl peroxides in liquid and supercritical carbon dioxide, in preparation.
18. K. E. O'Shea, J. R. Combes, M. A. Fox, and K. P. Johnson, *Photochem. Photobiol. 54*, 571 (1991).
19. J. M. DeSimone, and K. A. Shaffer, *Trends Polym. Sci. 3*, 146 (1995).
20. J. B. McClain, D. E. Betts, D. A. Canalas, E. T. Samulski, J. M. DeSimone, J. D. Londono, H. D. Cochran, G. D. Wignall, D. Chillura-Martino, and R. Trillo, *Science 274*, 2049 (1996).
21. Y. Hsiao, E. E. Maury, J. M. DeSimone, S. Mawson, and K. P. Johnson, *Macromolecules 28*, 8159 (1995).
22. J. M. DeSimone, E. E. Maury, Y. Z. Menceloglu, J. B. McClain, T. R. Romack and J. R. Combes, *Science 265*, 356–359 (1994).
23. A. E. Fiering, in *Organofluorine Chemistry: Principles and Commercial Applications* (R. E. Banks, B. E. Smart, and J. C. Tatlow, eds.), Plenum Press, New York (1994).
24. D. V. Avila, K. U. Ingold, J. Lusztyk, W. R. Dolbier, H. Q. Pan, and M. Muir, *J. Am. Chem. Soc. 116*, 99–104 (1994).
25. A. E. Feiring, C. G. Krespan, P. R. Resnick, B. E. Smart, T. A. Treat, and R. C. Wheland, U.S. Patent 5182342 (1993).
26. C. G. Krespan and R. C. Wheland, U.S. Patent 5,286,822 (1994).
27. A. Funaki, K. Kato, and T. Takakura, Jpn. Patent 06,157,675 (1994).

28. D. J. Van Bramer, M. B. Shiflett, and A. Yokozeki, U.S. Patent 5,345,013 (1994).
29. J. R. Combes, Z. Guan, and J. M. DeSimone, *Macromolecules 27*, 865 (1994).
30. T. J. Romack, J. R. Combes, and J. M. DeSimone, *Macromolecules 28*, 1724 (1995).
31. T. J. Romack and J. M. DeSimone, *Macromolecules 28*, 8429 (1995).
32. J. P. DeYoung, T. J. Romack, and J. M. DeSimone, *Polym. Preprints 38* (2), 424, (1997).
33. M. A. Harmer, W. E. Farneth, and Q. Sun, *J. Am. Chem. Soc. 118* (33), 7708 (1996).
34. G. A. Olah, P. S. Iyer, and S. Prakash, *Synthesis 1986*, 513.
35. M. F. Vincent, S. G. Kazarian, and C. A. Eckert, *AiChE J. 43* (7), 1838 (1997).
36. C. A. Mertdogan, T. P. DiNoia, and m. A. McHugh, *Macromolecules 30*, 7511 (1997).
37. J. P. DeYoung, B. E. Kipp, and J. M. DiSimone, Solid acid catalysis using Nafion[®] in supercritical carbon dioxide: A quantitative kinetics study of α-methyl styrene dimerization, unpublished results.
38. K. P. Johnson, P. D. Condo, and D. R. Paul, *Macromolecules 27*, 365–371 (1994).
39. J. J. Watkins and T. J. McCarthy, *Macromolecules 27*, 4845–4847 (1994).
40. T. J. Romack, Dissertation, University of North Carolina at Chapel Hill (1996).
41. R. E. Banks, B. E. Smart, and J. C. Tatlow, in *Organofluorine Chemistry: Principles and Commercial Applications* (R. E. Banks, B. E. Smart, and J. C. Tatlow, eds), Plenum Press, New York (1994).
42. G. Marchionni and P. Srinivasan, *Proceedings of Fluoropolymer Conference* (1992).
43. J. M. DeSimone, E. E. Maury, J. R. Combes, and Y. Z. Menceloglu, U.S. Patent 5,382,623 (1995).
44. W. C. Bunyard, T. J. Romack, and J. M. DeSimone, *Polym. Preprints 38* (2), 424, (1997).
45. T. J. Romack, B. E. Kipp, and J. M. DeSimone, *Macromolecules 28*, 8432 (1995).
46. D. J. Van Bramer, M. B. Shiflett, and A. Yokozeki, U.S. Patent 5,345,013, (1994).

II

Direct Fluorination

14

Direct Fluorination of Polymers

RICHARD J. LAGOW and HAN-CHAO WEI

14.1. INTRODUCTION

The direct fluorination of inorganic,[1,2] organometallic,[3-5] and organic compounds,[6-8] employing the LaMar[9,10] and Exfluor–Lagow[11] methods, has impacted the synthesis of fluorinated compounds over the past 25 years. Among the most important applications of direct fluorination are the synthesis of fluoropolymers from hydrocarbon polymers and the conversion of the surface of the hydrocarbon polymers to fluoropolymer surfaces.[12,13] The direct fluorination process is an excellent approach to the synthesis of fluoropolymers.

Idealized structures and reaction schemes for these fluorination studies of hydrocarbon polymers are shown in Figure 14.1 and other structures have been discussed in the literature.[10] The new fluoropolymers differ from the idealized structures primarily because, in such high-molecular-weight species, carbon–carbon cross-linking occurs to a significant extent during fluorination. This cross-linking may be controlled, but in general the fluoropolymer is of higher molecular weight than the hydrocarbon precursor. Polymer chain fission can be almost entirely eliminated under the proper conditions. Most of these fluoropolymers are white solids with high thermal stability. The elemental analyses indicate complete conversion to fluoropolymers and correspond closely to the idealized compositions. The infrared spectral analyses and other spectroscopic evidence are consistent with perfluorocarbon materials. A number of very interesting X-ray photoelectron spectroscopy studies by D. T. Clark et al.[14,15] have confirmed that hydrocarbon polymers are converted to fluoropolymers by direct fluorination with very few or no structural rearrangements. These studies, in addition to further

RICHARD J. LAGOW and HAN-CHAO WEI · Department of Chemistry and Biochemistry, University of Texas at Austin, Austin, Texas 78712-1167

Fluoropolymers 1: Synthesis, edited by Hougham et al., Plenum Press, New York, 1999.

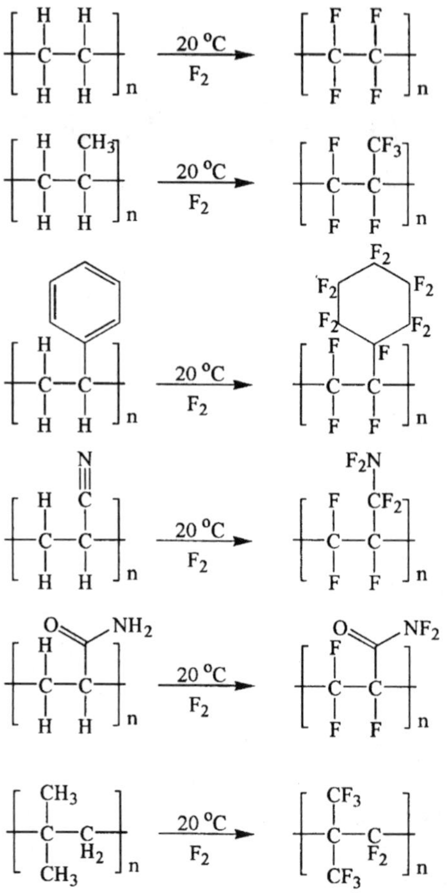

Figure 14.1. Direction fluorination of polymers.

elucidating the structure of the fluoropolymers, provide information on subtle surface effects and comment on the rate and depth of the fluorination.

The primary advantage of direct fluorination for synthesizing fluoropolymers is that many monomers are difficult to polymerize because of the steric repulsion of fluorine, which leads to relatively low-molecular-weight species; also, it is difficult or impossible to synthesize many of the corresponding monomers.

If solid polymer objects are fluorinated or polymer particles much larger than 100 mesh are used, only surface conversion to fluoropolymer results. Penetration of fluorine and conversion of the hydrocarbon to fluoropolymers to depths of at least 0.1 mm is a result routinely obtained and this assures nearly complete conversion of finely powdered polymers. These fluorocarbon coatings appear to have a number of potentially useful applications ranging from increasing the thermal stability of the polymer surfaces and increasing their resistance to solvents and corrosive chemicals, to improving their friction and wear properties. It is also possible to fluorinate polymers and polymer surfaces partially to produce a number of unusual surface effects.

We are going to discuss the syntheses of fluoropolymers, poly(carbon monofluoride), perfluoropolyethers, perfluorinated nitrogen-containing ladder polymers, and surface fluorination of polymers by direct fluorination.

14.2. POLY(CARBON MONOFLUORIDE)

Poly(carbon monofluoride), $(CF_x)_n$ (Figure 14.2) has been known since 1934 when Ruff and co-workers[16] prepared a gray compound of composition $CF_{0.92}$. In 1947 W. and G. Rudorff[17,18] reported a series of compositions of $CF_{0.68}$ to $CF_{0.99}$, varying in color from black in the case of $CF_{0.68}$ through gray to white in the case of $CF_{0.99}$.

Poly(carbon monofluoride) is a white compound, often reported to be explosive and unstable but found in our research results to be stable in air at temperatures up to at least 600°C. In fact, poly(carbon monofluoride) is the most thermally stable fluoropolymer known. It decomposes upon heating at 800°C or under a high vacuum at 580°C to form a series of polyolefinic fluorocarbons.[19] The compounds with compositions in the range of $CF_{0.68}$ to $CF_{0.8}$ are nearly black. The $CF_{0.8}$ to $CF_{0.95}$ compounds become gray and the $CF_{0.95}$ to $CF_{1.12}$

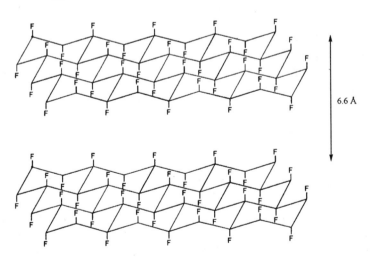

Figure 14.2. Structure of poly(carbon monofluoride).

develop into a snow-white solid. In earlier syntheses[16-18] poly(carbon mono-fluoride) was prepared by passing fluorine over graphite at 450–600°C for several hours. In some cases hydrogen fluoride is used as a catalyst. With these methods, a gray material of stoichiometry $CF_{0.68}$ to $CF_{0.8}$ is normally obtained.

Careful control of the reaction temperature at higher temperatures $(627° \pm 3°C)$ was necessary to reproducibly obtain a completely fluorinated pure white material. A snow-white material was obtained by direct fluorination of graphite at this temperature. Elemental analysis showed the calculated empirical formula to be $CF_{1.12\pm0.03}$.[20]

To produce $(CF_x)_n$ of the highest fluorine content, it was necessary to maintain a temperature of $627° \pm 3°C$. This 6°C range was very critical, which was not recognized in earlier work and consequently the process for reproduction of $(CF_x)_n$ gained a reputation for being irreproducible. $(CF_x)_n$ is formed from about 540 to 630°C at atmospheric fluorine pressure. It has been found that at higher pressures the critical temperature decreases.[21] At 540°C, the empirical formula is $CF_{0.68}$ and the product is black. If the furnace has a hot zone within this 6°C range smaller than the length of the nickel boat, only the center region of the boat produces white $(CF_x)_n$ and on either side the color ranges from gray to bladck. At temperatures over 630°C, the compound is unstable in fluorine and burns to produce CF_4 and black soot. $(CF_x)_{1.1}$ is therefore a solid subfluoride of carbon that should be considered metastable with respect to CF_4. These observations of reaction variations with temperature can be explained by consequent changes in fluorine atom concentrations and subsequent kinetics.

Studies at the Lewis Research Center of the National Aeronautic and Space Administration[22] and at the Frankford Arsenal of the U.S. Army[23] have shown that poly(carbon monofluoride) is a superior solid lubricant under heavy loads, in high temperatures, in oxidizing atmospheres, and under other extreme conditions. Researchers at the U.S. Army Electronics Command at Fort Monmouth, N.J.[24,25] and industrial scientists in Japan have recently demonstrated a high potential for the use of poly(carbon monofluoride) as a cathode material in high-energy batteries.

14.3. PERFLUOROPOLYETHERS

Perfluoropolyethers are the most extraordinary and exciting high-performance lubricants known today and represent perhaps the most important discovery in lubrication in the last 20 years.[26-30] For example, perfluoropolyether greases formulated from perfluoropolyether oils are used as the lubricants of choice for almost all civilian and military space activities (these greases often have fillers of either finely powdered Teflon or molybdenum disulfide) for lubricating bearings and other moving parts in satellites. The satellite grease applications are very important because the satellites must often operate for 10 years in both high-temperatures and in unusually cold environments, and failure makes for severe losses. These lubricants are also used on the space shuttle and will soon find their way into use as jet engine lubricants in aircraft.

Perfluoropolyethers have extraordinary properties and the materials are stable in some cases to 430°C, i.e., stable at a temperature 60°C higher than poly(tetra-fluoroethylene), thus giving very high thermal performance, because carbon–oxygen bonds are thermochemically stronger than carbon–carbon bonds. In addition, the fluoropolymer backbones are protected by a sheath of fluorine and the cloud of projecting nonbonding electrons associated with protruding fluorine atoms covering the chains provide lubrication properties.

On the other hand, perfluoropolyethers are the most extraordinary low-temperature lubricants known because at very low temperatures the carbon–oxygen–carbon ether linkage acts as hinge for storing vibrational energy. At even lower temperatures, often less than -100°C, these materials still have free rotation around the ether linkages giving a lower operating temperature.

Perfluoropolyethers emerged on the market in the early 1970s. The first perfluoropolyether was the homopolymer of hexafluoropropylene oxide produced by DuPont, which has the structure $[-CF_2CF(CF_3)O-]_n$ and this new lubricant material was called Krytox.[31,32] Krytox was and is used in most of the vacuum pumps and diffusion oil pumps for the microelectronics industry because it does not produce any hydrocarbon or fluorocarbon vapor contamination. It also has important applications in the lubrication of computer tapes and in other data processing as well as military and space applications.

The second material to come onto the market was the Montecatini Edison fluoropolymer called Fomblin Z[33,34] with the structure $(-CF_2O-)_m$ $(-CF_2CF_2O-)_n$. This fluoropolymer has better low-temperature properties than Krytox, but is more expensive. Fomblin Z is made by photo-chemical polymerization of a mixture of oxygen and tetrafluoroethylene to prepare the random copolymer. The methylene oxide unit $(-CF_2O-)$ imparts even more extraordinary low-temperature properties than those derived from vibration and free rotation of other perfluoroether linkages.

The third material, Demnum,[35] was only introduced in 1984 with the structure of $(-CF_2CF_2CF_2O-)_n$ and no new materials had been commercialized since. It was clear that the full potential of this new area had not yet been fulfilled.

Work from our research group at the University of Texas on the synthesis of perfluoroethers and perfluoropolyethers by direct fluorination is revolutionizing and, in fact, has revolutionized this field. Our laboratory has been responsible for putting at least 30 new perfluoropolyethers into the literature and in some cases they are now on the commercial chemical market. Exfluor Research Corporation prepared several hundred new perfluoropolyethers expanding upon the work from

$$(CH_2CH_2O)_n \xrightarrow{\text{F}_2/\text{He}} R_f(OCF_2CF_2)_nOR_f$$

Figure 14.3. Direct fluorination of poly(ethylene oxide).

our laboratory. This represents great flexibility in the direct fluorination technique and substantially lower costs as well.

The Lagow group first entered the perfluoropolyethers field in 1977, by reacting fluorine with inexpensive hydrocarbon polyethers to prepare perfluoropolyethers. In the simplest case (Figure 14.3) poly(ethylene oxide) is converted to perfluoroethylene oxide polymer, a simple reaction chemistry that we first reported in the literature.[27] As will be seen later, this direct fluorination technology as well as many new patents from Exfluor Research Corporation have been non-exclusively licensed to the 3M Corporation by the Lagow research group.[36–39]

New perfluoropolyethers were important because the only differences in the physical properties among different Krytox and Fomblin Z materials were produced by varying the degree of polymerization. Direct fluorination technology offers great flexibility in controlling molecular weights as well as in providing as new structures.

Subsequently, we were able to make perfluorinated analogues of Krytox from the hydrocarbon poly(propylene oxide)[40] (Figure 14.4). In 1985, we published three interesting perfluoropolyethers. First we copolymerized hexafluoroacetone with ethylene oxide, propylene oxide, and trimethylene oxide. Subsequent fluorination yielded the new perfluoropolyethers (Figure 14.5).

We discovered another synthetic technique that involves the conversion by direct fluorination of hydrocarbon polyesters to perfluoropolyesters followed by treatment with sulfur tetrafluoride to produce new perfluoropolyethers.[42] The first paper in this area of reasearch reported that conversion of poly(2,2-dimethyl-1,3-propylene succinate) and poly(1,4-butylene adipate) by using the direct fluorination to produce novel branched and linear perfluoropolyethers, respectively. The structures are shown in Figure 14.6. The second paper concerns the application of the direct fluorination technology base directed toward oligomers, diacids, diesters, and surfactants.[43]

Two of the fluid structures are very interesting. The first structure is a strictly alternating copolymer of ethylene oxide and methylene oxide,

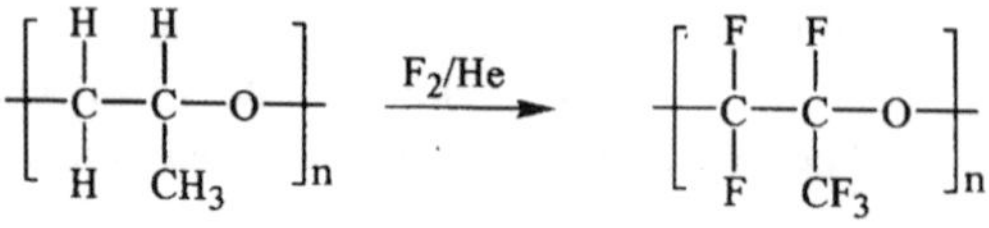

Figure 14.4. Direct fluorination of poly(propylene oxide).

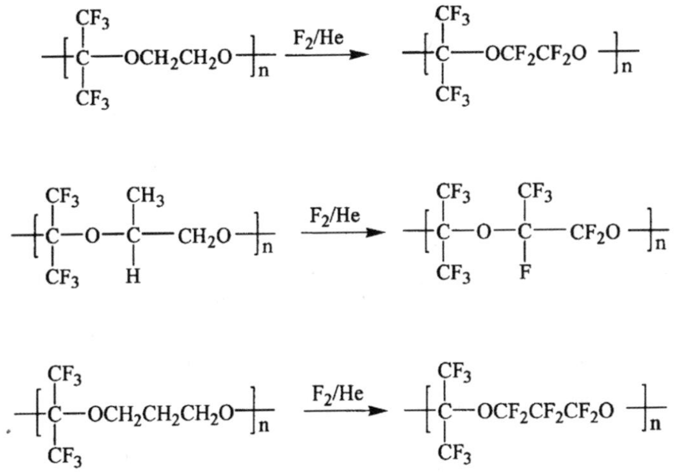

$$\begin{array}{c}
\overset{CF_3}{\underset{CF_3}{\big[\text{C}}}\text{—OCH}_2\text{CH}_2\text{O}\big]_n \xrightarrow{\ F_2/\text{He}\ } \overset{CF_3}{\underset{CF_3}{\big[\text{C}}}\text{—OCF}_2\text{CF}_2\text{O}\big]_n
\end{array}$$

$$\begin{array}{c}
\overset{CF_3}{\underset{CF_3}{\big[\text{C}}}\text{—O}\overset{CH_3}{\underset{H}{\text{—C}}}\text{—CH}_2\text{O}\big]_n \xrightarrow{\ F_2/\text{He}\ } \overset{CF_3}{\underset{CF_3}{\big[\text{C}}}\text{—O}\overset{CF_3}{\underset{F}{\text{—C}}}\text{—CF}_2\text{O}\big]_n
\end{array}$$

$$\begin{array}{c}
\overset{CF_3}{\underset{CF_3}{\big[\text{C}}}\text{—OCH}_2\text{CH}_2\text{CH}_2\text{O}\big]_n \xrightarrow{\ F_2/\text{He}\ } \overset{CF_3}{\underset{CF_3}{\big[\text{C}}}\text{—OCF}_2\text{CF}_2\text{CF}_2\text{O}\big]_n
\end{array}$$

Figure 14.5. Direct fluorination of polyethers (copolymerization of hexafluoroacetone with ethylene oxide, propylene, and trimethylene oxide).

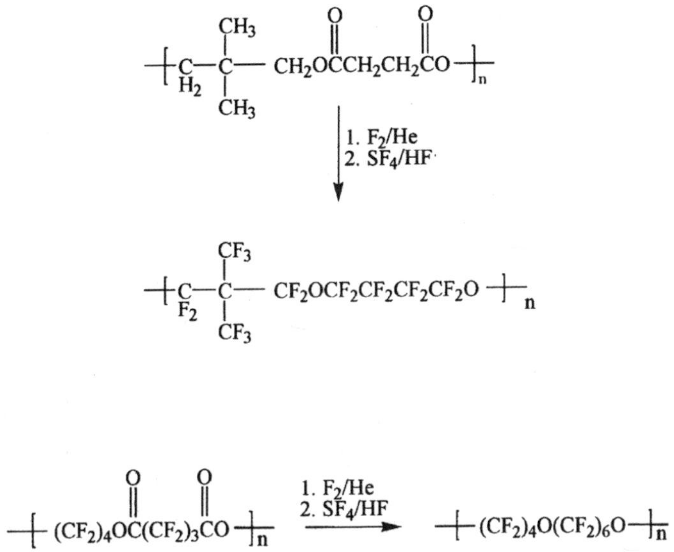

$$\begin{array}{c}
\overset{CH_3}{\underset{CH_3}{\big[\underset{H_2}{\text{C}}\text{—C}}}\text{— CH}_2\text{O}\overset{O}{\overset{\|}{\text{C}}}\text{CH}_2\text{CH}_2\overset{O}{\overset{\|}{\text{C}}}\text{O}\big]_n
\end{array}$$

$$\begin{array}{c} 1.\ F_2/\text{He} \\ 2.\ SF_4/\text{HF} \end{array}\Big\downarrow$$

$$\begin{array}{c}
\overset{CF_3}{\underset{CF_3}{\big[\underset{F_2}{\text{C}}\text{—C}}}\text{— CF}_2\text{OCF}_2\text{CF}_2\text{CF}_2\text{CF}_2\text{O}\big]_n
\end{array}$$

$$\big[(CF_2)_4\text{O}\overset{O}{\overset{\|}{\text{C}}}(CF_2)_3\overset{O}{\overset{\|}{\text{C}}}\text{O}\big]_n \xrightarrow[2.\ SF_4/\text{HF}]{1.\ F_2/\text{He}} \big[(CF_2)_4\text{O}(CF_2)_6\text{O}\big]_n$$

Figure 14.6. Syntheses of perfluoropolyethers from perfluoropolyesters.

$(-CF_2O-CF_2CF_2O-)_n$, which currently is the longest liquid range of any carbon-containing molecule.[44] While it is a liquid down to as low as $-100°C$, its decomposition temperature is about $350°C$. The second structure, $(-CF_2O-)_n$, was long sought by a number of laboratories. Other groups had attempted to polymerize carbonyl fluoride to synthesize material and had used a number of other approaches as well. We succeeded in capping this structure with a perfluoro-ethyl group to provide stability.[45] This is one of the lowest-temperature high-molecular-weight liquids known. It remains a liquid lubricant down to $-112°C$ but unfortunately when it is in contact with metals or in acidic or basic environments it is only stable to about $250°C$. Our new perfluoropolyether lubricants where the subject of a paper published jointly with Bill Jones of the Tribology Branch at NASA's Lewis Research Center.[46]

The fluorination of chlorine-containing polyethers has given rise to a new class of materials that exhibit outstanding properties. For example, telomers of epichlorohydrin can be fluorinated to give perfluoropolyepichlorohydrin, $(-CF_2C(F)(CF_2Cl)O-)_n$.[47]

Functional perfluoropolyethers[11] (Figure 14.7) can also be prepared by direct fluorination in high yields. Difunctional perfluoropolyethers based on fluorinated poly(ethylene glycol) are of particular interest as possible precursors for elasto-mers, which should have outstanding high-temperature and low-temperature properties.

A breakthrough in our laboratory has involved the synthesis of perfluorinated crown ethers and cryptands.[48-51] Several manuscripts in collaboration with J. S. Brodbelt[52-55] describe syntheses in which both perfluoro crown ethers and perfluoro cryptands tenaciously encapsulate O^{2-} and F^-. It will be extremely interesting to make perfluoro crown ether polymers and investigate their binding abilities toward oxygen and other molecules. Recently, we have used the direct fluorination technique to make the first perfluorinated crown ether polymer (Figure 14.8).[56] The hydrocarbon polymer, poly[(dibenzo-18-crown-6)-coformaldehyde] (1), was reacted with fluorine in carbon tetrachloride with excess sodium fluoride to make the perfluoro crown ether polymer (2). The excess sodium fluoride was removed by washing the fluorinated product with excess water. The yield of perfluoro crown either polymer is 94%. The results of elemental analysis of the fluoropolymer $(C_{21}F_{36}O_6)_n$ is consistent with the proposed structure (Calcd: C,

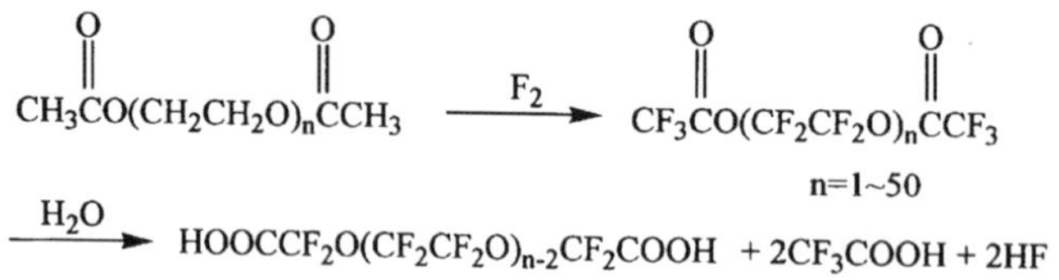

$$CH_3\overset{\displaystyle O}{\overset{\displaystyle \|}{C}}O(CH_2CH_2O)_n\overset{\displaystyle O}{\overset{\displaystyle \|}{C}}CH_3 \xrightarrow{F_2} CF_3\overset{\displaystyle O}{\overset{\displaystyle \|}{C}}O(CF_2CF_2O)_n\overset{\displaystyle O}{\overset{\displaystyle \|}{C}}CF_3$$

$$n=1\sim50$$

$$\xrightarrow{H_2O} HOOCCF_2O(CF_2CF_2O)_{n-2}CF_2COOH + 2CF_3COOH + 2HF$$

Figure 14.7. Synthesis of functional perfluoropolyethers.

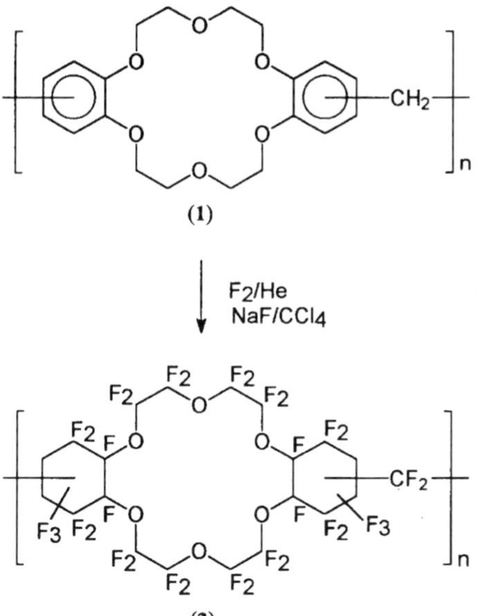

(1)

F₂/He
NaF/CCl₄

(2)

Figure 14.8. Synthesis of perfluoro crown ether polymer.

24.42; F, 66.28; H, 0. Found: C, 24.20; F, 66.44; H, 0.25). Trace hydrogen was found because of contributions from carboxylic acid end groups or hidden unfluorinated hydrogen. The results of thermal gravimetric anlaysis (TGA) showed that approximately 50% of perfluoro crown ether polymer's original weight is lost by 390°C and only 10% of fluoropolymer's original weight is lost at 261°C. XPS spectroscopy showed carbon, fluorine, and oxygen peaks at 289, 695, and 540 eV, respectively. No sodium peak was found. Solid state ^{19}F-NMR (external CFCl₃) δ −40.51, −57.47, −62.26, −74.81, −82.03, −88.84, −123.71, −165.35, and −207.00 ppm.

The new perfluoro crown ether polymer is currently under investigation in regard to its binding abilities of several gas molecules.

14.4. PERFLUORINATED NITROGEN-CONTAINING LADDER POLYMERS

Two nitrogen-containing polymeric materials with extended aromatic ladder structures have been chosen for direct fluorination studies (Figure 14.9).[57] Pyrolyzed polyacrylonitrile (**3**) and paracyanogen (**4**) [poly(pyrazinopryazine)] have been subjected to direct fluorination to produce perfluorinated analogues.

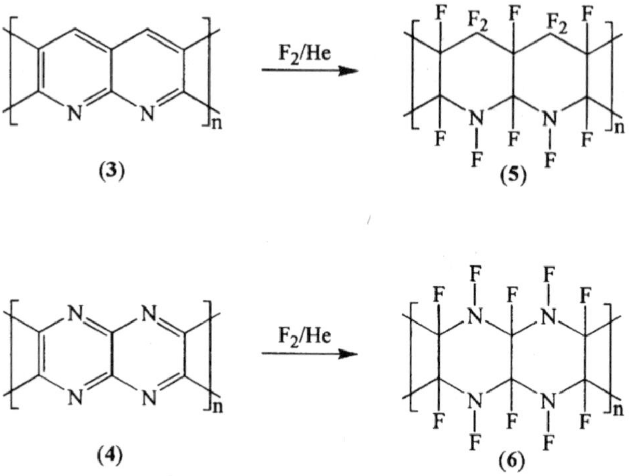

Figure 14.9. Synthesis of perfluoro nitrogen-containing ladder polymers.

Fluorination of **3** and **4** appears to have produced the saturated analogues **5** and **6**, respectively. Characterization of the perfluorinated products provides evidence that the fused polycyclic structures were retained.

The products of perfluorination are both white, very different from the original black hydrocarbon polymers. Both materials are moisture-sensitive powders and slowly degraded by atmospheric moisture, **6** more quickly than **5**. The materials oxidize iodide ion to iodine owing to the presence of the N—F moiety. A series of iodometric titrations showed that **6** required twice the number of equivalents of titrant as did **5**. This result supports the proposed structures **6** having twice as many N—F moieties as **5**.

Both IR spectra are dominated by the strong, broad absorbance between 1100 and 1300 cm^{-1} that is due to C—F stretching. A less prominent absorbance observed just below 1000 cm^{-1} is assigned to N—F stretching. The absence of absorbance in the 1500–2000 cm^{-1} region confirms that no unsaturation, either C=C or C=N, remains in the final perfluorinated products. The results of mass spectral analysis of **5** showed m/z(formula), 69(CF$_3$), 81(C$_2$F$_3$), 95(C$_2$NF$_3$), 100(C$_2$F$_4$), 114(C$_2$NF$_4$), 119(C$_2$F$_5$), 131(C$_3$F$_3$), 152(C$_2$NF$_6$), and **6** showed m/z(formula), 69(CF$_3$), 76(C$_2$NF$_2$), 83(CNF$_3$), 95(C$_2$NF$_3$), 114(C$_2$NF$_4$), 121(C$_3$N$_2$F$_3$), 135(C$_3$N$_3$F$_3$), 159(C$_3$N$_2$F$_3$), The absence of the characteristic C$_n$F$_{2n+1}$ fragments of a perfluorinated straight-chain compound precludes the fused ring structure of products. The results of the elemental analysis of **5** (Calcd: C, 24.83. N, 9.65. F, 65.52; Found: C, 24.87. N. 9.34. F, 65.33) are in excellent agreement with the calculated values, those for **6** (Calcd: C, 18.75. N, 21.87. F, 59.37; Found: C, 22.02. N, 20.30, F, 53.66) show some deviation. This may reflect the increased difficulty in maintaining the difunctional structure during fluorination or may simply be a result of greater sensitivity of **6** to decomposition after fluorination.

The thermal stability of these new materials was of special interest in regard to their possible use as high-performance lubricants. Material **5** melted sharply at 310°C producing a liquid that began to yellow at 340°C. Material **6** proved to be less thermally robust than **5**, exhibiting a melting point of 175°C with the liquid yellowing above 200°C. TGA confirmed that both these materials exhibit inferior thermal properties to poly(carbon monofluoride). Approximately 50% of the original weight of **6** is lost by 250°C compared to **5** having lost only 20% at that temperture. Material **5** has a region of maximum weight loss in the range 275–350°C, while **6** shows an area of comparable degradation at 160–300°C.

These two new materials may find use as more reactive forms of solid graphitic fluoride compounds for fluorine-containing cathodes in high-energy lithium batteries. New synthetic methods to prepare higher-molecular-weight paracyanogen and pyrolyzed poly(acrylonitrile) may be a route to increase the thermal stability of the subsequent perfluorinated polymers. However, it can be expected that these fluoropolymers will always be less stable than graphite fluoride, $(CF_{1.12})_n$, which does not contain a carbon–nitrogen bond. Another possible application for these new fluoropolymers is as a solid source of fluorine. Making use of the reactive fluorine bound to nitrogen, these fluoropolymers act as mild fluorinating agents. Recent work has shown molecular N—F compounds to be useful selective fluorinating agents.[58–61]

14.5. SURFACE FLUORINATION OF POLYMERS

The best known aspect, and the first one to find commercialization in the direct fluorination area, was the fluorination of polymer surfaces. This Lagow–Margrave invention, trademarked "Fluorokote," involved many types of polymeric materials in various forms: e.g., polyethylene bottles, polypropylene objects, and rubber gloves. Polyethylene bottles are easily given fluorocarbon surfaces (> 0.1 mm), and this has been commercialized. Air Products has at least 20 licenses for what is known as their "Aeropak" process and Union Carbide has a "Linde Fluorination" process as well. Applications in chemical, pharmaceutical, and cosmetic storage are widespread.

Direct fluorination of polymer or polymer membrane surfaces creates a thin layer of partially fluorinated material on the polymer surface. This procedure dramatically changes the permeation rate of gas molecules through polymers. Several publications in collaboration with Professor D. R. Paul[62–66] have investigated the gas permeabilities of surface fluorination of low-density polyethylene, polysulfone, poly(4-methyl-1-pentene), and poly(phenylene oxide) membranes.

14.6. CONCLUSION

It is clear that many new fluoropolymers can be synthesized by direct fluorination technology that cannot be obtained through other routes. The information in this chapter should serve as a strong indication that perhaps the best and ultimate synthetic method for fluoropolymers on both laboratory and manufacturing scales in the future will be direct fluorination reactions.

ACKNOWLEDGMENT: We are grateful for support of this work by the U.S. Department of Energy, the Air Force of Scientific Research, the Office of Naval Research, and the Army Research Office (Durham).

14.7. REFERENCES

1. R. J. Lagow and J. L. Margrave, *J. Inorg. Nucl. Chem.* 35, 2084–2085 (1973).
2. J. L. Adcock, and R. J. Lagow, *Inorg. Chim. Acta. 10*, 9–11 (1974).
3. E. S. K. Liu and R. J. Lagow, *J. Am. Chem. Soc. 98*, 8270 (1976).
4. E. S. K. Liu and R. J. Lagow, *J. C. S. Chem. Commun. 1977*, 450–451.
5. J. P. DeYoung, H. Kawa, and R. J. Lagow, *J. C. S. Chem. Commun. 1992*, 811–812.
6. W. H. Lin and R. J. Lagow, *J. Fluorine Chem. 50*, 348–358 (1990).
7. H. N. Huang, H. Roesky, and R. J. Lagow, *Inorg. Chem. 30*, 789–794 (1991).
8. H. C. Wei, S. Corbelin, and R. J. Lagow, *J. Org. Chem. 61*, 1643–1644 (1996).
9. R. J. Lagow and J. L. Margrave, *Proc. Natl. Acad. Sci. U.S.A. 67* (4), 8A (1970).
10. R. J. Lagow and J. L. Margrave, *Progress in Inorganic Chemistry, Vol. 26*, John Wiley and Sons, New York (1979), pp. 161–210.
11. T. R. Bierschenk, R. J. Lagow, T. J. Juhlke, and H. Kawa, U.S. Patent 5,093,432 (1992).
12. R. J. Lagow and J. L. Margrave, *Polym. Lett. 12*, 177–184 (1974).
13. A. J. Otsuka and R. J. Lagow, *J. Fluorine Chem. 4*, 371–380 (1974).
14. D. T. Clark, W. T. Feast, W. K. R. Musgrave, and I. Ritchie, *J. Polym. Sci. Polym. Chem. Ed. 13*, 857–890 (1975).
15. D. T. Clark, W. T. Feast, W. K. R. Musgrave, and I. Ritchie, *Advances in Friction and Wear, Vol. 5A* (L. H. Lee, ed.), Plenum Press, New York (1975), p. 373.
16. O. Ruff, D. Bretschneider, and F. Elert, *Z. Anorg. Allg. Chem. 217*, 1 (1934).
17. W. Rudorff and G. Rudorff, *Z. Anorg. Allg. Chem. 253*, 281 (1947).
18. W. Rudorff and G. Rudorff, *Chem. Ber. 80*, 417–423 (1947).
19. A. K. Kuriakase and J. L. Margrave, *Inorg. Chem. 4*, 1639–1641 (1965).
20. R. J. Lagow, R. B. Badachhape, J. L. Wood, and J. L. Margrave, *J. Am. Chem. Soc. 96*, 2628–2629 (1974).
21. R. J. Lagow, R. B. Badachhape, J. L. Wood, and J. L. Margrave, *J. C. S. Dalton Trans. 1974*, 1268–1273.
22. R. L. Fusaro and H. E. Sliney, NASA Technical Memorandum NASA TMX 5262Y (1969).
23. N. Gisser, M. Petronio, and A. Shapiro, *J. Am. Soc. Lubric. Eng. 61* (May 1970).
24. K. Braner, Technical Report ECOM-3322, U.S. Army Electronics Command, Fort Monmouth, N.J.
25. H. F. Hunger, G. J. Heymark, Technical Report ECOM-4047, U.S. Army Electronics Command, Fort Monmouth, N.J.
26. G. E. Gerhardt and R. J. Lagow, *J. C. S. Chem. Commun. 1977*, 259–260.

27. G. E. Gerhardt and R. J. Lagow, *J. Org. Chem. 43*, 4505–4509 (1978).
28. G. V. D. Tiers, *J. Am. Chem. Soc. 77*, 4837–4840 (1955).
29. G. V. D. Tiers, *J. Am. Chem. Soc. 77*, 6703–6704 (1955).
30. G. V. D. Tiers, *J. Am. Chem. Soc. 77*, 6704–6706 (1955).
31. J. T. Hill, *J. Macromol. Sci.—Chem. 8*, 499–520 (1974).
32. H. S. Eleuterio, *J. Macromol. Sci.—Chem. 6*, 1027 (1972).
33. D. Sianesi, G. Benardi, and F. Moggi, French Patent 1,531,092 (1968).
34. D. Sianesi and R. Fontanelli, British Patent 1,226,566 (1971).
35. Y. Ohsaka, T. Tohzuka, and S. Takaki, European Patent Application 841160039 (1984).
36. R. J. Lagow and S. Inoue, U.S. Patent 4,113,772 (1972).
37. G. E. Gerhardt and R. J. Lagow, U.S. Patent 4,523,039 (1985).
38. D. F. Persico and R. J. Lagow, U.S. Patent 4,675,452 (1987).
39. R. J. Lagow, U.S. Patent Application Serial No. 147,173 (1988).
40. G. E. Gerhardt and R. J. Lagow, *J. C. S. Perkin Trans. 1 1981*, 1321–1328.
41. D. F. Persico and R. J. Lagow, *Macromolecules 18*, 1383–1387 (1985).
42. D. F. Persico, G. E. Gerhardt, and R. J. Lagow, *J. Am. Chem. Soc. 107*, 1197–1201 (1985).
43. D. F. Persico and R. J. Lagow, *J. Polym. Sci.: Polym. Chem. Ed. 29*, 233–242 (1991).
44. T. R. Bierschenk, T. J. Juhlke, and R. J. Lagow, U.S. Patent 4,760,198 (1988).
45. T. R. Bierschenk, T. J. Juhlke, and R. J. Lagow, U.S. Patent 4,827,042 (1989).
46. W. R. Jones, Jr., T. R. Bierschenk, T. J. Juhlke, H. Kawa, and R. J. Lagow, *Ind. Eng. Chem. Res. 27*, 1497–1502 (1988).
47. R. J. Lagow, T. R. Bierschenk, T. J. Juhlke, H. Kawa, U.S. Patent 4,931,199 (1989).
48. W. H. Lin, W. I. Bailey, and R. J. Lagow, *J. C. S. Chem. Commun. 1985*, 1350–1352.
49. W. D. Clark, T. Y. Lin, S. D. Maleknia, and J. R. Lagow, *J. Org. Chem. 55*, 5933–5934 (1990).
50. T. Y. Lin and R. J. Lagow, *J. C. S. Chem. Commun. 1991*, 12–13.
51. H. C. Wei, V. M. Lynch, and R. J. Lagow, *J. Org. Chem. 62*, 1527–1528 (1997).
52. J. S. Brodbelt, S. D. Maleknia, C. C. Liou, and R. J. Lagow, *J. Am. Chem. Soc. 113*, 5913–5914 (1991).
53. J. S. Brodbelt, S. D. Maleknia, R. J. Lagow, and T. Y. Lin, *J. C. S. Chem. Commun. 1991*, 1705–1707.
54. J. S. Brodbelt, C. C. Liou, S. D. Maleknia, T. Y. Lin and R. J. Lagow, *J. Am. Chem. Soc. 115*, 11069–11073 (1993).
55. T. Y. Lin, W. H. Lin, W. D. Clark, R. J. Lagow, S. D. Larson, S. H. Simonsen, V. M. Lynch, J. S. Brodbelt, S. D. Maleknia, and C. C. Liou, *J. Am. Chem. Soc. 116*, 5172–5179 (1994).
56. H. C. Wei and R. J. Lagow, Unpublished results (1997).
57. J. J. Kampa and R. J. Lagow, *Chem. Mat. 5*, 427–429 (1993).
58. G. Resnati and D. D. DesMarteau, *J. Org. Chem. 56*, 4925–4929 (1991).
59. Z. Wu, D. D. MesMarteau, and M. Witz, *J. Org. Chem. 57*, 629–635 (1992).
60. G. Resnati, W. T. Pennington, and D. D. DesMarteau, *J. Org. Chem. Soc. 57*, 1536–1539 (1992).
61. G. Resnati and D. D. DesMarteau, *J. Org. Chem. Soc. 57*, 4281–4284 (1992).
62. C. L. Kiplinger, D. F. Persico, R. J. Lagow, and D. R. Paul, *J. Appl. Polym. Sci. 22*, 2617–2626 (1986).
63. J. M. Mohr, D. R. Paul, T. E. Mlsna, and R. J. Lagow, *J. Membrane Sci. 55*, 131–148 (1991).
64. J. M. Mohr, D. R. Paul, Y. Taru, T. E. Mlsna, and R. J. Lagow, *J. Membrane Sci. 55*, 149–171 (1991).
65. J. D. Le Roux, D. R. Paul, J. J. Kampa, and R. J. Lagow, *J. Membrane Sci. 90*, 21–35 (1994).
66. J. D. Le Roux, D. R. Paul, J. J. Kampa, and R. J. Lagow, *J. Membrane Sci. 94*, 121–141 (1994).

15

Surface Fluorination of Polymers Using Xenon Difluoride

GEORGY BARSAMYAN and
VLADIMIR B. SOKOLOV

15.1. INTRODUCTION

Surface engineering by means of fluorination is an effective way to change surface properties, and is used for both polymer surfaces and inorganic substrates. Polymer surface fluorination has been around a long time. The first patent we know of dates back to 1938,[1] but it was only in the 1970s that the introduction of several major industrial applications led to a rapid acceleration in development.

Surface fluorination is in popular use because most of the desirable properties of fluoropolymers are largely the result of surface phenomena. Surface fluorination enables us to modify the surface properties of a polymer while retaining others often useful bulk properties (e.g., mechanical strength, elasticity, and ease of processability).

Surface-fluorinated polymers have a number of desirable inherent properties, including:

1. Decreased gas permeability (increased barrier properties).
2. Decreased friction (decreased wear and long lifetime).
3. Enhanced wettability (better adhesion).
4. Enhanced chemical and ozone resistance.
5. Enhanced hardness.

GEORGY BARSAMYAN · Samsung Corning Co., Ltd, Dong-Suwon, Kyeongii-Do, Korea, 442-600. Present address: Solvay S.A. Moscow Office, Moscow, 123310, Russia.   VLADIMIR B. SOKOLOV · Russian Research Center "Kurchatov Institute," Moscow 123182, Russia

Fluoropolymers 1: Synthesis, edited by Hougham *et al.*, Plenum Press, New York, 1999.

However, the fact that industrial applications of polymer surface fluorination employ a fluorine/nitrogen mixture as the fluorinating agent complicates matters because fluorine gas is toxic, may explode when brought into contact with organic substances, and causes severe burns on human tissue. Moreover, the use of fluorine requires highly qualified personnel and special safety systems.

The idea of utilizing xenon difluoride (XeF_2) as a fluorinating agent for surface engineering was first suggested in the late 1980s, based primarily on three considerations. First, direct fluorination with noble gas fluorides is a well-known, effective method used extensively in inorganic preparative chemistry. Second, direct fluorination of polymeric surfaces with a F_2/N_2 mixture has proved to be a very effective tool for modifying surface properties. Third, owing to the method of low temperature thermocatalytic synthesis developed in the Russian Research Center "Kurchatov Institute" (RRC "KI"),[2] the XeF_2 production process was considerably simplified and shortened, which resulted in a significant decrease in the cost of XeF_2. Thus the general idea was to try to effect direct polymer surface fluorination using the less hazardous and less toxic solid XeF_2 instead of the conventional gaseous F_2/N_2 mixture fluorinating agent.

XeF_2 is a colorless crystalline compound stable up to 500°C, m.p. −129°C, with considerable vapor pressure for solids—4.5 mm Hg (20°C). It is a linear symmetrical molecule. The mean thermochemical energy of the Xe—F bond in XeF_2 is 132 kJ/mol,[3] which is less than the F—F bond energy—157.3 kJ/mol.[4]

It is appropriate at this point to recall that XeF_2 was originally considered to be exotic and was used only as laboratory chemical in very small quantities. While direct polymer surface fluorination with a gaseous F_2/N_2 mixture was already in use for several significant commercial applications.

15.2. FLUORINATION OF SUBSTANCES AND SURFACES

Fluorine was isolated by Henri Moissan at the end of June 1886 during an electrolysis of liquefied anhydrous hydrogen fluoride, containing potassium fluoride, at −23°C. The gas, produced at the anode, was fluorine. This achievement earned Moissan the 1906 Nobel prize in chemistry. Thousands of tons of fluorine are being produced today by essentially the same, albeit slightly improved, electrolytic method. Obviously, this scale of fluorine production means that fluorine chemistry has turned into an important branch of industry. This development can be understood if we look at fluorine from a chemist's point of view.

So, what is so unusual in this pale yellow highly toxic gas that condenses to a pale greenish-yellow liquid at −188°C, solidifies to a yellow solid at −220°C, has a strong ozone-like odor, and is easily detectable at concentrations about 0.1−0.2 ppm ($\approx$ ml/m^3).[5]

Let us consider some notorious experimental facts. Fluorine combines readily with most organic and inorganic materials at or below room temperature. Many organic and hydrogen-containing compounds can burn or explode when exposed to pure fluorine. Owing to its extremely high oxidizing ability fluorine forms compounds with all elements except helium, neon, and argon. In all compounds fluorine always shows the same oxidation degree of -1.

Generally speaking, direct fluorination is a well-known, effective method in the applied chemistry of fluorine. The continuing interest in inorganic fluorine chemistry is connected with the possible use of fluorine and fluorocompounds as powerful oxidants for rocket fuels and in the recovery and processing of metals such as molybdenum, tungsten, and rhenium; recovery of residual uranium from spent nuclear fuel elements by vaporizing it, forming UF_6, and separating it; and creation of new materials with unique properties or applications, such as, e.g., SF_6, an excellent electrical insulator used in large electrical transformers; WF_6 and ReF_6, in production of rhenium–tungsten alloys; BF_3, CoF_3, CrO_2F_2, and OsF_6, as catalysts in organic chemistry and in processing of isotope-enriched elements and compounds; AgF_2, CoF_3, MnF_4, SbF_5, HgF_2, ClO_3F, NO_2BF_4, IF_5, NF_3, and $NOBF_4$, as fluorinating agents in organic synthesis; $LiBF_4$ and $LiPF_6$, as components for lithium batteries; NF_3, as a source of fluorine for HF/OF high-energy chemical lasers; CF_4, an important gas for silicon etching in microelectronics, IF_5, an impregnating agent used in the textile industry for water- and oil-repellency; and LiF, CaF_2, BaF_2, and ZrF_4, materials for ceramics and optics.[9]

To illustrate the extreme reactivity of fluorine, one can note that it is the only element that reacts directly with the heavier noble (inert) group gases, xenon and radon, to form fluorides. It is surprising that the reaction between fluorine and xenon gas occurs even at room temperature.[6] Most noble (inert) gas compounds have been synthesized either by direct fluorination or with the help of fluoro-compounds, e.g., an electric discharge in a mixture of krypton and fluorine can produce KrF_2. In 1962 Neil Bartlett synthesized the first noble gas compound: deep red-brown PtF_6 vapor in the presence of xenon gas at room temperature produced a yellow-orange compound then formulated as $Xe^+PtF_6^-$.[7] Within a few months of this discovery, XeF_4 was prepared as was XeF_2. Since that time, apart from the simple fluorides mentioned above, more complex fluorides have been synthesized, including among many others, KrF_2, $XeFSbF_6$, $KrFAuF_6$, XeF_5AsF_6, XeO_3, XeF_5SO_3F, $XeOF_4$, $FXeOClO_3$, and Na_4XeO_6.[3] It is interesting to note that oxygen-containing noble gas compounds can be obtained only through reactions with xenon fluorides. Thus, because of its extreme reactivity, fluorine gave rise to new branch of chemical science—chemistry of noble gas compounds.

Another, more recent, fluorine contribution to the development of new branches of chemical science is connected with fullerenes, the first members of this family of close-caged molecules being obtained only about a decade ago. Fullerenes, having the composition C_{60}, C_{70}, C_{76}, $C_{84}\ldots$, represent the new (third) form of carbon. Each fullerene consists of 12 pentagonal rings and a number of hexagonal ones that have double bonds in the pentagonal rings. Nowadays only C_{60} and C_{70} are synthesized on a relatively large scale—by vaporizing graphite, followed by chromatographic extraction of the fullerene mixture. Chemically C_{60} and C_{70} behave like typical electron-deficient alkenes: the maximum number of functional groups that can be combined with C_{60} in methylation, chlorination, bromination, and other such reactions is 24, apart from the reaction with hydrogen ($C_{60}H_{36}$). By contrast, fluorination of fullerenes with F_2, KrF_2, or XeF_2* forms stable derivatives with considerably larger numbers of combined functional groups: $C_{60}F_{48}$ and $C_{60}F_{60}$. Furthermore, it was shown that fluorofullerenes can react with ammonia to form fluoro-containing amines, e.g., $C_{60}F_{30}$ $(NH_2)_{18}$.[8] Hence, in this case, as well, the higher fluorides of fullerenes, which were obtained using noble gas fluorides, become the basic substances for new theoretical and applied directions in organic chemistry.

Thus one can draw the conclusion that there is no element in the periodic table, including other halogens and oxygen, that possesses stronger oxidizing properties than fluorine, which is really at the head of the oxidants series:

$$F_2 > O_2 > Cl_2 > Br_2 > I_2$$

Another important conclusion is that fluorine atoms determine the chemical bonding and consequently the formation of the chemical compound. In other words the general energy apportionment in a reacting system, the rate of energy apportionment (chemical reaction rate), the direction of the chemical reaction, and the composition of the final products are determined by the oxidizing ability and reactivity of the fluorine atoms. As a radical, atomic fluorine, which has high reactivity ($F > [O] > Cl > Br > I$) over a broad temperature range (reactions occur practically without activation energy), interacts with all metals and metal-loids and their compounds, with the formation of ionic salts and covalent fluorides characterized by the highest bonding strength among halogen analogues and other elements, e.g., C—F (450–500), C—Cl (340), C—Br (280), C—H (410), C—O (360), H—F (560), C—S (270), C—N (300), C—C (360), but not C=C (610), C=O (750), C≡C (840) in kJ/mol. Under certain conditions a chain reaction is initiated in fluorine/halogen, fluorine/hydrogen, and fluorine/hydrocarbon systems, which is accompanied by thermal explosion. Initiation is due to the action of the fluorine atoms.

The molecules participate in the chemical reactions as "carriers" of the atoms and radicals. According to such an interpretation one can consider molecular fluorine as the simplest chemical compound of atoms—FF (like ClF,

---

* Fluorides of "simple" carbon are well-known: they have the general formula $(CF_x)_n$ and are widely used as cathodic material in high-energy primary batteries.

BrF, AgF, FOF, FKrF, FXeF) and its chemical properties as the result of oxidizing ability and reactivity of the fluorine atoms. The real conditions of the reaction system (temperature, pressure, agent content, reaction environment, the composition and properties of the initial agents and reaction products...) have a great influence on the chemical behavior of molecular fluorine, limiting or sharply increasing its chemical activity.

The high ("anomalously" high) oxidizing ability and reactivity of molecular fluorine in chemical reactions can be understood and explained by noting the principal properties and characteristics of fluorine, which significantly determine and characterize its chemistry, including its reactions with organic polymers, namely low energy of molecular fluorine dissociation, 157.3 kJ/mol, the highest electronegativity F of all elements, 4.10 (Ollred and Rochow scale), high electron affinity F, 350 kJ/mol, high stability of C—F, H—F, and other bonds, weak polarizability of fluorine (0.82×10^{-24} cm^3), small atomic, covalent, and ionic radii, 0.64, 0.72, and 1.33, respectively (in Å). Standard values of enthalpy of formation and free energy of formation for atomic fluorine are: $\Delta H^{\circ}_{298}F = 75$ kJ/mol, $\Delta G^{\circ}_{298}F = 60$ kJ/mol.

A low dissociation energy and low activation dissociation energy (145.5 kJ/mol) indicates the presence of considerable amounts of atomic fluorine in molecular fluorine, even at moderate temperatures and pressures. A high molecular fluorine dissociation rate in combination with a low recombination velocity ensures that at every moment there is a sufficiently high concentration of atomic fluorine in the reaction system and, consequently, a high chemical reaction rate. The interaction of atomic fluorine with saturated hydrocarbons is accompanied by the formation of HF and its corresponding radical, followed by the formation of saturated polymers containing $>$ CHF and $>$ CF$_2$ groups:

$$C_{n-1}(CH_2)H_{2n} \xrightarrow{F} C_{n-1}(C^*H)H_{2n} + HF \xrightarrow{F}$$

$$C_{n-1}(CFH)H_{2n} \xrightarrow{F} C_{n-1}(C^*F)H_{2n} + HF \xrightarrow{F} C_{n-1}(CF_2)H_{2n} \ldots$$

Hydrogen abstraction by fluorine is thermodynamically advantageous, since C—H bond strengths are about 410 kJ/mol compared to 560 kJ/mol for H—F and 450–500 kJ/mol for C—F bonding.

In contrast to saturated hydrocarbons, the unsaturated hydrocarbons react with atomic fluorine by two pathways, i.e. (atomic fluorine addition at $>$ C$=$C $<$ double bond and hydrogen substitution by fluorine atoms. The reaction of fluorine with aromatic hydrocarbons proceeds with the formation of F-derivatives and hydrogen atoms break off:

It is assumed that all similar fluorination reactions proceed via an intricate radical chain-reaction mechanism. The overall reactions for the substitution of hydrogen by fluorine (RH + F$_2$ → RF + HF, $\Delta H_{298} \approx -430$ kJ/mol per carbon atom) are more exothermic than the reactions for adding fluorine to the double bonds (R$_2$C=CR$_2$ + F$_2$ → R$_2$FC − CFR$_2$, $\Delta H_{298} \approx -350$ kJ/mol per carbon atom). This is the reason that the carbon skeleton in the first case is often fractured ($E_{C-C} = 360$ kJ/mol, $E_{C=C} = 610$ kJ/mol). Thus it is necessary to minimize and control the energy of the process for successful fluorination of organic compounds,[10] including organic polymers.

The principal laws for the fluorination of polymeric hydrocarbons are the same as those described above for the simple case. Direct fluorination has been used extensively in organic chemistry (but only since the early 1970s) in low-temperature methods, where the fluorine is strongly diluted with some inert gas (helium, argon, nitrogen, krypton). One can note the La Mar, aerosol-based, and liquid-phase fluorination methods.

While on the subject of fluoroorganic compounds, one cannot overstate the importance of fluoropolymers in modern industry and science. There is hardly anyone around today who has never heard of polytetrafluoroethylene (PTFE) [CF$_2$—CF$_2$]$_n$. Housewives who know nothing about fluorine use PTFE-coated frying pans or pots, and know that PTFE makes washing up easier, because virtually nothing sticks to a PTFE-coated utensil.

Nonstickiness, low friction, low wettability, and high thermal- and chemical-resistance are the major properties of PTFE, which was accidentally discovered in the DuPont laboratories in 1938, and these properties are more or less typical of other fluoropolymers that have been developed since.

In discussing fluoropolymers one has to mention also the fluororubbers, which are mostly copolymers of fluoromonomers and common rubbers. They are also much in demand by industry, owing to the fact that while they retain the traditional features of rubber—elasticity and tensile strength—they have high chemical- and thermal-resistance and low friction.

The useful properties of fluoropolymers noted above are mostly attributed to a high C—F bond strength (456 kJ/mol) and the atomic arrangement in the macromolecules. The large fluorine atoms in PTFE are so neatly packed and close to the C—C chain that it makes the carbon chain inaccessible to other atoms and molecules. Besides providing protection from chemical attack, the tightly packed nonpolarizable fluorine atoms are responsible for the PTFE's low surface energy (18.6 mN/m).[10]

Thus, the strong C—F bond, the special arrangement of atoms in macromolecule, and low surface energy impart some unique physical properties to PTFE and other fluoropolymers: high chemical and thermal resistance, nonstick character, low friction coefficient, and low wettability. This combination of properties naturally makes fluoropolymers very attractive to industry and they are produced by the thousands of tons all over the world.

However, one should not forget that apart from the complexity of the synthesis fluoropolymers are very expensive. For example, the price of fluoro-rubber is more than 30-fold that of an ordinary rubber such as butadiene-styrene (SBR) or ethylene-propylene (EPDM). Cost was one of the factors that gave impetus to research polymer surface fluorination, with the object of imparting the properties of fluoropolymers to the surfaces of less expensive polymers without changing their bulk properties.

The violent nature of reactions between fluorine and hydrocarbon compounds has already been noted here, and the direct fluorination of organic polymers is not a exception: it is so exothermic that if the reaction is not controlled, it generally leads to fragmentation and charring of the substrate. Moderation of the reaction rate can be effected by:

1. Diffusion control of the gas phase through the presence of diluent gases.
2. Thermal control of the gas or substrate.
3. Diffusion control in the solid phase.

It is believed that polymer surface fluorination proceeds via a free radical mechanism, where fluorine abstracts hydrogen atoms from the hydrocarbon, and fluorine atoms are substituted.[11] Of course, the precise conditions depend on the nature of the polymer in question and the surface properties required.

Today, controlled polymer surface direct fluorination is used in a number of specific applications. According to Annand[22] there are primarily two different methods by which elemental fluorine can be applied to the surfaces of polymeric materials:

1. Postforming exposure (posttreatment).
2. Simultaneous treatment and exposure (*in situ* treatment).

These techniques differ not only in terms of their principles of operation but also in regard to economic flexibility, applicability, and the ultimate properties of the product.

In the post-treatment of films or fibers, fluorine-containing gas is con-tinuously injected into the reactor and gaseous by-products are removed. The other type of reactor for posttreatment is the batch process reactor, which consists of a suitably sized vacuum chamber provided with means of evacuation and injection.

In-situ treatment, on the other hand, uses existing polymer processing equipment to apply the desired fluorine-containing gas to the polymer in question. Of course, there has to be some modifications of the processing equipment. The

Aeropak® process—simultaneous blow-molding and inner-surface fluorination of polymeric containers—is the most well-known example of an *in-situ* treatment process.

Surface fluorination changes the polymer surface drastically, the most commercially significant use of polymer surface direct fluorination is the creation of barriers against hydrocarbon permeation. The effectiveness of such barriers is enormous, with reductions in permeation rates of two orders of magnitude. Applications that exploit the enhanced barrier properties of surface-fluorinated polymers include: (1) Polymer containers, e.g., gas tanks in cars and trucks, which are produced mostly from high-density polyethylene, where surface fluorination is used to decrease the permeation of fuel to the atmosphere; and perfume bottles. (2) Polymeric membranes, to improve selectivity; commercial production of surface-fluorinated membranes has already started.[13]

While enhancing barrier properties is the most commercially significant use of direct surface fluorination it is not by any means the only example of surface modification by direct fluorination. Another example worthy of note is the change of surface energy and adhesion. It was discovered that direct surface fluorination of polyolefins generally raises the surface energy.[14] Several ways have been suggested to exploit this phenomenon but it is the effect on adhesive and coating properties that has generated the greatest interest.

The resulting improvement in polymer adhesion can be significant and is being explored on a commercial scale for: (1) enhancement of the paint receptivity of molded plastic items for automotive and other applications; (2) enhanced adhesion between polyester yarn, cord, or fabric and rubber in automobile tire production; (3) increased resistance to delamination in coated flexible film; (4) improved strength in fiber-reinforced composite parts through better matrix-to-fiber energy transfer; (5) increased coating integrity on molded thermoplastic containers.[15]

The third major application of polymer surface fluorination is improvement in tribological characteristics of rubber and rubber articles. It is commonly known that fluorination of natural or synthetic rubber creates a fluorocarbon coating, which is very smooth. Rubber articles, such as surgical gloves, O-rings, gaskets, and windshield wiper blades can be surface-fluorinated, without the bulk material losing its elastic and flexible properties.[16] Fluorinated O-rings have longer lifetimes and in some cases can even be used without lubricants. Surgical gloves that have been fluorinated from the inside can be put on without talc, but the outer surface remains rough so that instruments will not slip out of the surgeon's hands.[17]

In closing of this general introduction it may be said that the development of the applied chemistry of fluorine, which started with fluorination of substances has led to surface fluorination of materials, which is now an effective tool in changing of surface properties.

15.3. POLYMER SURFACE FLUORINATION WITH XENON DIFLUORIDE

As fluorine has such an extremely high reactivity alternative agents are being used for the fluorination of organic compounds, in particular milder XeF_2 ($\Delta H^\circ{}_{298}XeF_{2(gas)} = -108.4\,kJ/mol$; $\Delta G^\circ{}_{298}XeF_{2(gas)} = -74.7\,kJ/mol$). Furthermore XeF_2 is "pure" fluorinating agent, with its only reduction product being the noble gas xenon.

The comparative accessibility and desirable properties of XeF_2 enable its use for the synthesis of fluoroorganic compounds under "mild" conditions. As a fluorinating agent XeF_2 can be used to effect electrophilic fluorination of various organic compounds with good yield and it is a useful reagent for generating radical cations. According to Bartlett,[3] in many fluorination and oxidation reactions, XeF_2 is first ionized to XeF^+ and then electron transfer gives the radical $XeF^.$. In many cases XeF_2 is an attractive alternative to the conventional Balz–Schiemann reaction for the preparation of fluorinated aromatic compounds, e.g., fluorobenzenes (in CCl_4, CH_2Cl_2, CH_3CN, ...):

$$RC_6H_6 \xrightarrow{XeF_2} RC_6H_5F + Xe + HF \xrightarrow{XeF_2} RC_6H_4F_2$$

$$C_5H_5N \xrightarrow{XeF_2} C_5H_4FN + Xe + HF \xrightarrow{XeF_2} C_5H_3F_2N$$

Xenon difluoride reacts with carbon–carbon single, double, and triple bonds giving addition fluoroorganic compounds, e.g.,

$$R{-}CH = CH_2 \xrightarrow{XeF_2} RCH_2 - CHF_2 + Xe$$

and

$$RCH_2{-}CH_3 \xrightarrow{XeF_2} RCHF{-}CH_3 + Xe + HF \xrightarrow{XeF_2} RCF_2{-}CH_3 + Xe + HF$$

In this respect XeF_2 is similar to molecular fluorine. The relatively low oxidizing ability and reactivity of XeF_2 in the same type of chemical reactions is explained by its greater dissociation energy. The activation energy of the dissociation ($XeF_2 \rightarrow XeF + F$) is 190–210 kJ/mol. The XeF radicals are very weakly bound ($E_{XeF} = 58$–$80\,kJ/mol$) and can be an effective source of F atoms.[3]

15.4. EXPERIMENTAL

The procedure was a very simple one. The sample and a certain amount of XeF_2 were placed together in a stainless steel reactor ($V = 6$ liters). The reactor was then closed and evacuated if needed. The system was left undisturbed for some time. Owing to the high vapor pressure, the fluorination proceeded via a chemical reaction between the solid polymer surface and the geseous XeF_2. After a certain amount of time had elapsed the reactor was evacuated and opened, and the samples were put under a hood for 24 h for degassing.

15.5. RESULTS AND DISCUSSION

In the first stage in order to test the process of various types of polymer films were surface-fluorinated. From 1990 to 1994 it was shown that XeF_2 could be used effectively for surface fluorination of a variety of plastics. Polyethylene film and plates,[18] aromatic polysulfone,[19] polyvinyltrimethylsilane,[20] and polycarbonate,[21] among other polymeric materials, were fluorinated successfully.

The fact that fluorination had taken place was established by various analytic methods: ESCA, IR and FTIR spectroscopy, bulk anlaysis, NMR, and DSC. The presence of chemically bonded fluorine in the surface layer of treated samples were uniquely determined by analysis data. Further details can be found elsewhere.[22]

Here we would like to emphasize that treatment with XeF_2 fluorinates the polymer surfaces, with the depth and degree of fluorination depending upon the reaction conditions. We never observed the formation of a perfluorinated surface layer after treatment with XeF_2. Results of XPS analysis of XeF_2-treated polyethylene film $-(-CH_2-CH_2-)_n-$ showed that the maximum degree of fluorination corresponded to the composition $-(-CF_2-CH_2-)_n-$. After that, visible destruction of the polymer surface, i.e., cleavage of C—C bonds in the polymer macromolecule, would start. These results differ from those of the fluorination of polymer powders; it was reported by Lagow and Margrave in the early 1970s that polymer powders were fully converted into perfluorinated polymers under the action of a F_2/N_2 mixture.[23]

Apart from the whole range of physicochemical analyses of XeF_2-treated polymers, some other potentially interesting and useful surface characteristics were tested and measured.

First, it is interesting to note that the increased surface energy of the XeF_2-treated polymer surfaces enhances their wettability. So, e.g., the water contact angle for polyethylene and polyurethane drops after treatment up to three- to fourfold and two- to threefold, respectively. Increased wettability can be very useful in biomedical applications; and, of course, it improves adhesion. Treated polypropylene films that were glued together showed a substantially increased adhesion strength, which was found to be greater than its cohesion strength.

Another interesting and possibly useful result of fluorination of XeF_2-treated polymer surfaces is decreased gas permeability. For example, fluorinated polyvinyltrimethylsilane film showed a fourfold decrease in the CO_2 permeability coefficient, as determined by gas chromatography.

PVC tubes and pipes used in microchip manufacturing for supplying deionized water, where purity is a very important factor, were XeF_2-treated. Tests showed that the treated tubes did not contaminate the deionized water with organic compounds below 90°C. Certain types of polymers showed significant improvement in swelling in organic solvents after XeF_2 treatment. Aromatic polysulfone showed no swelling in kerosene (80°C, 7 h) while the control (untreated) sample showed a 4% increase in weight. A similar effect was found with fluororubbers. However, the most significant results were obtained with XeF_2 treatment of rubbers.

15.6. RUBBER SURFACE FLUORINATION WITH XENON DIFLUORIDE

In 1993 it was established that surface fluorination of rubbers with XeF_2 led to a significant decrease in the friction coefficient. Table 15.1 compares friction coefficients before and after treatment with XeF_2. These results generated considerable interest from the former Skega AB—one of the major producers of rubber products in Sweden, now Svedala Skega AB and Skega Seals AB. Their interest centered on the decrease in the initial friction force in small (10.3 × 2.4 mm) O-rings made from a peroxide-cured ethylene-propylene rubber (EPDM). Our joint venture lasted for two years, during which time the optimum conditions for XeF_2 surface fluorination were established and the results shown in Table 15.2 were obtained.

Friction force measurements at Skega AB were carried out by the following procedure. A piston/cylinder type of test ring was used. The O-ring being tested was greatly stretched over the piston and fitted into the seal groove. The cylinder was carefully wiped and dried with a clean cloth before the piston was put into the

Table 15.1. Friction Coefficient Decrease
in Russian-Made Rubbers after XeF_2
Treatment

Rubber type	Friction coefficient	
	Initial	After treatment
SBR	2.40	0.23
IR	2.00	1.48
Revertex	1.88	0.60
NBR	0.60	0.16

Table 15.2. Mean Values of Friction Force Decrease in EPDM
O-Rings after XeF$_2$ Treatment

Type	Friction force (F), N	$F_{fluorinated}/F_{initial}$
Initial (nonfluorinated)	127.00	—
Fluorinated, mode 1	17.40	0.14
Fluorinated, mode 2	99.06	0.78
Fluorinated, mode 3	28.00	0.22
Fluorinated, mode 4	10.31	0.08
Fluorinated, mode 5	5.56	0.04
Fluorinated, mode 6	6.14	0.05

cylinder bore. The test ring was then placed in the compression test machine (L&W TCT 50) and the piston was pushed into the cylinder to a distance of approximately 17 mm at a constant speed of 400 mm/min. three tests were made on each O-ring. After each test the piston was removed from the cylinder and the bore was carefully cleaned as described above.

It is very important to know how surface fluorination affects the stability of tribological characteristics. In the case of Russian-made rubbers we used a standard friction machine (SMT), where abrading was done with a stainless steel instead of a conventional brass grid so that the test would be more rigorous. The results obtained are shown in Table 15.3.

The friction force endurance of EPDM O-rings after XeF$_2$ treatment in optimum mode can be illustrated by a graphic presentation of friction force measurements during 50 cycles in the compression test machine (L&W TCT 50) (Figure 15.1).

To validate our idea that treating rubber products with gaseous XeF$_2$ causes them to become surface-fluorinated, i.e., that chemically bonded fluorine atoms are incorporated in the surface layer, we did ESCA analyses of the samples treated in various modes. The results are shown in Table 15.4.

Table 15.3. Times to the Initial Values of the
Friction Coefficients of XeF$_2$-Treated Samples[a]

Rubber type	Time, h
SBR	4.5
NBR	5.2
IR	2.4
Revertex	1.8

[a] The nontreated rubber samples began losing the initial friction
coefficient value during this test after 0.5 h

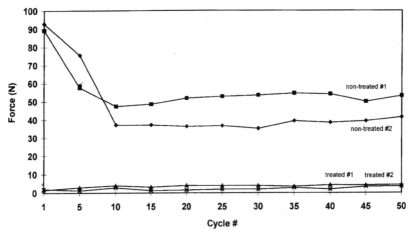

Figure 15.1. Friction force endurance of EPDM O-rings.

Although we made no attempt to elucidate the mechanism of friction decreases in rubbers after surface fluorination, it seems to us that apart from the substitution of H atoms to F in the polymer macromolecule, which forms a fluoropolymer on the surface, there is another phenomenon that makes a significant contribution to the friction decreases, i.e., fluorination of carbon black, which is used in rubber recipes for reinforcement. It appears that when the carbon black in the surface of the rubber is fluorinated it produces a lubricating effect, followed by "blooming" on the surface of the treated rubber while it is under a friction load. So, in our opinion, two effects contribute to friction decrease of carbon-filled rubbers: fluorination of the rubber macromolecules and fluorination of the carbon black; rubbers that do not contain carbon black show a much smaller decrease in friction after XeF_2 treatment.

Rubber (NBR) samples treated with XeF_2 in an optimum mode, i.e., showing the lowest friction force while retaining all the other bulk properties of NBR

Table 15.4. ESCA Analysis Results of the Rubber (SBR) Samples Treated in Different Modes

Sample No.	Elemental composition, at. %		
	Carbon	Oxygen	Fluorine
1	82.9	15.5	1.6
2	82.5	15.0	2.5
3	82.7	15.2	2.1

rubber, were analyzed in Sweden to detect surface-incorporated fluorine and the depth of fluorination with scanning electron microscope (SEM) equipped with EDS. The samples were vacuum-treated before they were sputtered with gold.

In Figure 15.2 Pictures 1.1 and 2.1 are SEM images of nonfluorinated (reference) material and fluorinated samples (Batch 15A20PR7), and Pictures 1.2 and 2.2 are the corresponding X-ray spectra. The EDS spectrum shows that Batch 15A20PR is fluorinated, since a fluorine peak can be seen in the left part of the spectrum (Picture 1.2).

Figure 15.3 shows the results of the second SEM analysis designed to give a qualitative determination of the depth of fluorination. A carefully prepared cross section of the O-ring was analyzed using the so-called "mapping technique." The light streak across Picture 3.2 shows where the fluorine is concentrated. With the aid of the microscale shown it can be estimated that the fluorination is approximately 5 μm deep.

15.7. INDUSTRIAL APPLICATION OF RUBBER SURFACE FLUORINATION WITH XENON DIFLUORIDE

After two years of extensive research and testing, in 1955 Skega AB placed an order with the RRC Kurhatov Institute for a specially designed plant for XeF_2 treatment with a capacity of 5000 O-rings per shift. The design of the plant is very simple and it was not expensive to build. The plant has been installed at Skega Seals AB and is operating successfully.

The simple calculations presented below will show that surface fluorination of rubbers with XeF_2 is economically advantageous as compared with fluororubbers. The amount of XeF_2 needed for batch treatment of 5000 O-rings made from EPDM rubber (diam. 10.2 × 1.5 mm) is approximately 20 g. That means that the cost of the XeF_2 needed for one batch treatment, is 20 × $1.5 = $30 for 5000 O-rings or about $0.0006 each. It should be kept in mind that, as mentioned above, the price for fluororubber is thirty times that of EPDM.

Although we have no specific data on the economics of rubber surface fluorination with the traditional F_2/N_2 mixture, in our opinion the following facts make the XeF_2 technique very attractive for rubber manufacturers:

1. Low cost, based on the quantity of XeF_2.
2. Equipment and technique are simple and inexpensive.
3. Unlike using traditional methods there are no health, explosion, or fire hazards.

Another argument in favor of the XeF_2 method is the evidence that since 1995 it has been used by a manufacturing company that had no previous experience with direct fluorination or fluorine at all.

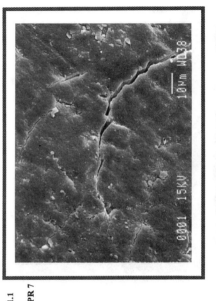

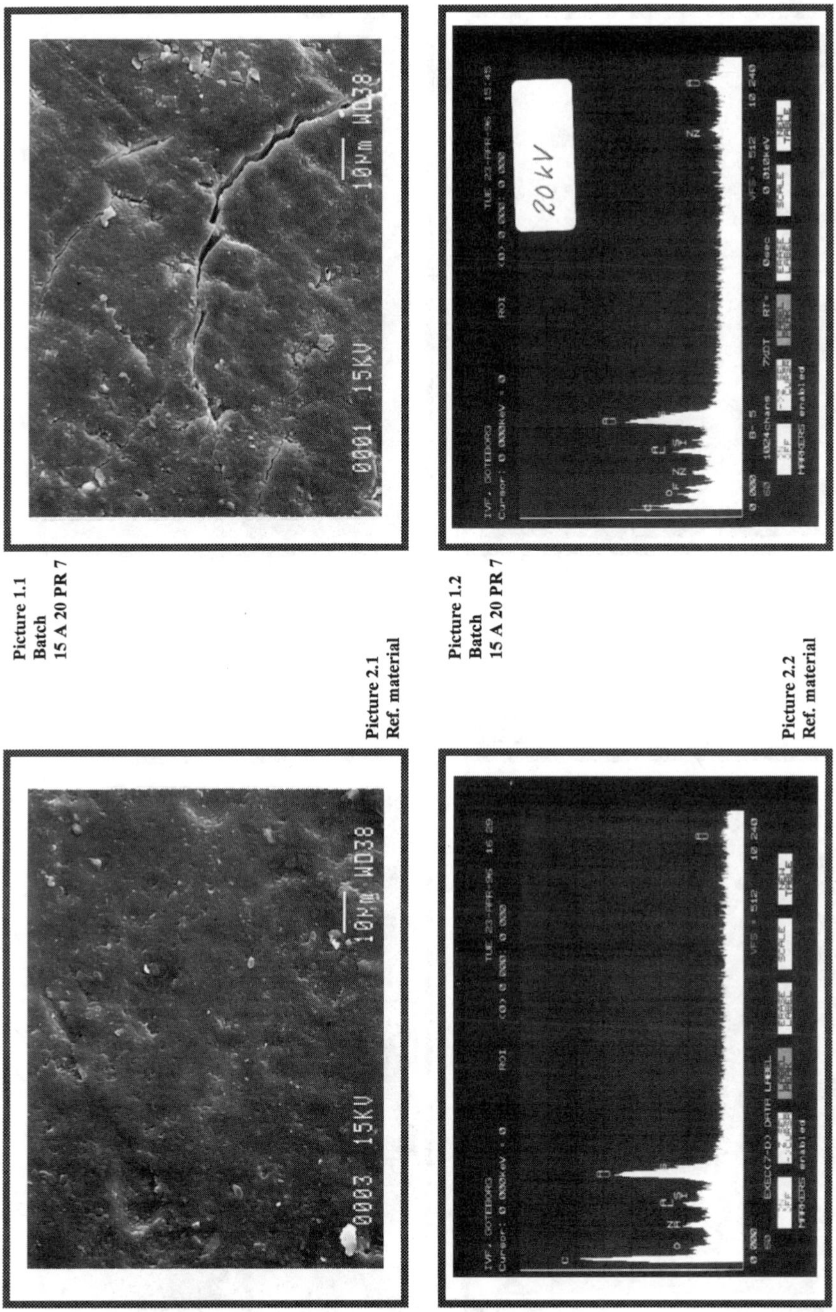

Picture 1.1
Batch
15 A 20 PR 7

Picture 2.1
Ref. material

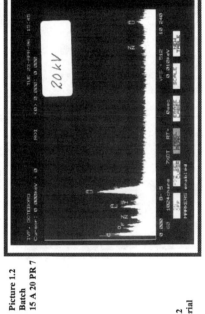

Picture 1.2
Batch
15 A 20 PR 7

Picture 2.2
Ref. material

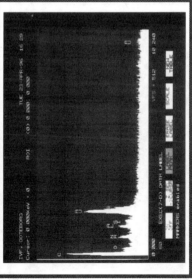

Figure 15.2. SEM Images (1.1 and 2.1) and X-ray spectra (1.2 and 2.2) of nonfluorinated reference materials and fluorinated samples.

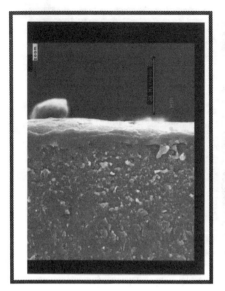

**Picture 3.1
Cross-
section**

**Picture 3.2
"Mapping"**

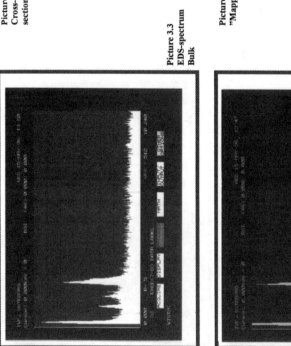

**Picture 3.3
EDS-spectrum
Bulk**

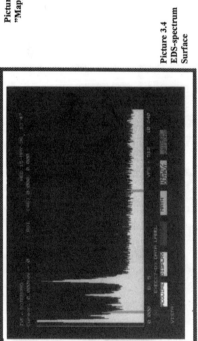

**Picture 3.4
EDS-spectrum
Surface**

Figure 15.3. SEM images reflecting the depth of fluorination.

15.8. RECENT DEVELOPMENT OF THE XENON DIFLUORIDE METHOD

An interesting advance in XeF_2 polymer surface fluorination was made in 1995. Polyethylene (PE) film was treated with XeF_2. The reaction chamber was evacuated and blown with argon, after which a fluoromonomer—tetrafluoroethylene (TFE) (C_2F_4) was added to the reaction chamber. This second stage formed a new surface layer on the PE film. The IR spectrum of the treated PE film shows the presence of a C—F bond, but differs a lot from PE samples treated just with XeF_2, which also show the C—F bonding. One of the most reasonable assumptions here is postpolymerization of TFE on the XeF_2-activated PE film surface. The potential of this method is evident and needs no comment. Another very attractive idea is to substitute TFE for another perfluoromonomer compound that is easier to handle.

Recent research indicates that the XeF_2/TFE method can also be used for inorganic surfaces,[24] which would be a very important development.

15.9. CONCLUSIONS

1. Xenon difluoride has proved to be an effective fluorinating agent for surface fluorination of polymers.
2. The technique of surface fluorination with XeF_2 is much less sophisticated, expensive, and hazardous than the conventional F_2/N_2 treatment.
3. The XeF_2 polymer surface fluorination technique can be easily applied at manufacturing companies that have no experience in dealing with fluorine.
4. The laboratory results that have been obtained point to several interesting possible applications of XeF_2/TFE method.

15.10. REFERENCES

1. J. Soll, U.S. Patent 2,129,289 (1938).
2. B. B. Chaivanov, V. B. Sokolov, and S. N. Spirin, Preprint IAE–4936/13, Moscow, Atominform, 1989.
3. N. Bartlett and F. O. Sladky, in *Comprehensive Inorganic Chemistry*, Pergamon Press, Oxford and New York (1978).
4. R. E. Banks and J. C. Tatlow, in *Organofluorine Chemistry: Principles and Commercial Applications* (R. E. Banks, B. E. Smart, and J. C. Tatlow, eds.), Plenum Press, New York (1994), p. 16.
5. M. Howe-Grant (ed.), *Fluorine Chemistry: A Comprehensive Treatment*, Encyclopedia Reprints Series, John Wiley and Sons, New York (1995), p. 2.
6. A. B. Neiding and V. B. Sokolov, *Usp. Khim.* 12, 2146–2194 (1974).

7. *Kirk-Othmer Encyclopedia of Chemical Technology, 4th Ed., Vol. 13*, John Wiley and Sons, New York, (1995), p. 38.

8. V. B. Sokolov and G. Barsamyan, *Proceedings of XI European International Symposium on Fluorine Chemistry*, Bled, Slovenia, 1995, pp. 132–135, 181–182.

9. R. E. Banks, D. W. A. Sharp, and J. C. Tatlow (eds.), *Fluorine: The First Hundred Years* (1886–1986), Elsevier Sequoia, Lausanne and New York (1986).

10. N. Isikava (ed.), *Soedinenya Ftora*, Mir, Moscow (1990), p. 16.

11. A. Kharitonov and Yu. L. Moskvin, in Proceedings of the Second International Conference on Fluorine in Coatings, Salford, England, 28–30 Sept. 1994.

12. M. Annand, J. P. Hobbs, and I. J. Brass, in *Organofluorine Chemistry: Principles and Commercial Applications* (R. E. Banks, B. E. Smart, and J. C. Tatlow, eds.) Plenum Press, New York (1994), pp. 469–480.

13. V. Frolov, V. V. Teplyakor, and A. P. Kharitonov, in Proceedings of the Second International Conference Fluorine in Coatings, Salford, England, 28–30 Sept. 1994.

14. W. J. Koris, W. T. Stannet, and H. B. Hopfenberg, *Polym. Eng. Sci. 22*, 738 (1982).

15. M. Annand, J. P. Hobbs, and I. J. Brass, in *Organofluorine Chemistry: Principles and Applications* (R. E. Banks, B. E. Smart, and J. C. Tatlow, eds.), Plenum Press, New York (1994), p. 474.

16. R. J. Lagow and J. L. Margrave (to DAMW Associates), Canadian Patent 1,002,689 (Dec. 28, 1976). R. J. Lagow and J. L. Margrave (to DAMW Associates), U.S. British Patent, 440,605 (Oct. 20, 1976).

17. R. B. Badachhape, C. Homsy, and J. L. Margrave (to Vitek Inc. and MarChem, Inc.), U.S. Patent 3,992,221 (Nov. 16, 1976).

18. G. B. Barsamyan, V. B. Sokolov, B. B. Chaivanov, and S. N. Spirin, *J. Fluorine Chem. 54*, 85 (1991).

19. G. B. Barsamyan, V. B. Sokolov, and B. B. Chaivanov, *J. Fluorine Chem. 58* (N2–3), 220 (1992).

20. G. B. Barsamyan, N. A. Vargasova, S. D. Stavrova, and V. P. Zubov, *Zhurnal Prikl. Khimi 67*, 610–612 (1994).

21. G. B. Barsamyan, V. B. Sokolov, and B. B. Chaivanov, in *Proceedings of XI European International Symposium on Fluorine Chemistry*, Bled, Slovenia, September 1995.

22. G. B. Barsamyan, K. C. Belokonov, N. A. Vargasova, V. B. Sokolov, B. B. Chaivanov, and V. P. Zubov, Preprint IAE-5748/13 (1994).

23. J. L. Margrave and R. J. Lagow, *J. Polym. Sci. Polym. Lett. Ed. 12*, 177 (1974).

24. Korean Patent Application 96-47649, filed October 23, 1996.

16

New Surface-Fluorinated Products

P. A. B. CARSTENS, J. A. DE BEER, and J. P. LE ROUX

16.1. INTRODUCTION

In 1938 Söll filed a patent[1] covering the treatment of rubbers with diluted F_2 to improve their resistance with regard to F_2 reactivity. Interestingly, he also suggested the possibility of liquid-phase fluorination, a principle that was later further explored and patented for surface fluorination.[2]

The two most important properties induced by surface fluorination of plastics, i.e., permeation and adhesion, were patented by Joffre[3] in 1957. His results proved that surface fluorination created excellent liquid as well as gas barrier properties.

Today it is claimed that the surface fluorination of polymers using F_2 gas mixtures enhances a wide range of properties, e.g., low permeability to nonpolar liquids,[4] improved permselectivity,[5,6] excellent wettability and adhesion,[7] low friction coefficient (especially for elastomers),[8] and chemical inertness.[9] Obviously, these properties depend on the chemical composition of the fluorinated layer, which in turn is determined by the chemical structure of the base polymer, the composition of the F_2 gas mixture, and the fluorination parameters.

Commercially the most exploited uses of surface fluorination are fluorinated HDPE fuel tanks for automotive vehicles and to a lesser extent fluorinated HDPE containers for the packaging of solvent-based crop protection chemicals. These products can be fluorinated using either in-line or postfluorination processes.[10]

P. A. B. CARSTENS, J. A. DE BEER, and J. P. LE ROUX · Atomic Energy Corporation of South Africa, Ltd., Pretoria, 0001, South Africa.

Fluoropolymers 1: Synthesis, edited by Hougham *et al.*, Plenum Press, New York, 1999.

Owing to the product diversity of the South African market, the AEC has developed its surface fluorination technology based on the more versatile post-fluorination process. This decision has led to numerous novel applications and improvements to existing products, some of which will be discussed in this chapter.

16.2. PERMEATION-BASED PRODUCTS

16.2.1. Containerization of Crop Protection Chemicals

The packaging of liquid insecticides and herbicides in specially treated metal containers is problematic owing to corrosion and other problems. The natural choice would be HDPE, but many formulations contain solvents such as xylene and other hydrocarbons that permeate through HDPE, leading to product loss[11] and, in many cases, to severe paneling of the container.[12,12a]

A generally accepted method for studying solvent permeation involves high-temperature exposure of the contained fluid over a set time period to determine the amount of fluid loss from permeation through the container walls.

To determine the suitability of fluorinated HDPE containers as packaging material for the solvent-based formulations, the containers were placed in well-ventilated ovens at 50°C for 28 days. During this period the mass loss of the contents was monitored. To date more than 130 liquid crop protection chemical formulations have been evaluated at the AEC by this method with at least 91 of them having serious permeation problems through the untreated HDPE containers. In only two cases fluorinated containers did not have acceptable barrier properties.

The performance of a number of representative formulations in fluorinated and untreated natural HDPE containers is summarized in Table 16.1. It must be kept in mind that owing to the larger ratio of volume to exposed surface area, the mass loss for larger containers would be significantly lower, as indicated by the results in brackets in the table.

However, when these products were packed into pigmented HDPE fluorinated containers, there was severe paneling of the containers. Intensive research at the AEC solved this problem with pigmented containers, and today more than 13 colors are available that are suitable for fluorination. Indications are that the AEC is the only organization that has solved the paneling problem experienced with pigmented HDPE fluorinated containers used for packing of a broad range of solvent-based crop protection chemicals. To assist smaller companies, the AEC, together with a local blow-molding company, has developed a special range of fluorinated containers, called Fluorotainers, which conform to SABS 0229 and are thus UN-certified.

Table 16.1. Performance of Crop Protection Chemicals in Natural 100 ml HDPE Fluorinated and Unfluorinated Containers at 50°C after 28 Days

Product	Mass loss (%)[a]		Improvement	Container appearance[b]	
	Fluorinated	Control		Fluorinated	Control
Alachlor	0.39 (0.05)	21.50	55.13	NP	NP
Bromoxynil	0.25	12.62	50.48	NP	SP
Chlorpyrifos	0.52	19.79	38.06	NP	SP
Lorsban	0.73	18.02	24.68	NP	NP
Synthetic pyrethroids	1.57 (0.09)	55.90	35.61	NP	SP
Trifluralin	1.13 (0.02)	20.89	18.49	NP	SP
Wenner	0.06	0.90	15.00	NP	NP

[a] Results of larger, pigmented fluorinated containers are included in brackets.
[b] NP = no paneling; SP = severe paneling.

16.2.2. Fluorinated HDPE Containers for Other Applications

It is known that fluorinated HDPE containers are highly suitable for the packaging of numerous other products such as adhesive systems, oil-based paints, polishes, hand cleaners, and hydrocarbon solvents.[12]

It has also been reported that some products with distinct flavors can be successfully contained in fluorinated containers.[13-15] Work done at the AEC on the containment of concentrated flavorants in HDPE containers showed excellent containment for butter LR12390,* aniseed oil, and dimethyl sulfide. The unacceptable barrier properties found for OR onion, OR garlic, and banana LR4186* led to a massive slump in popularity of one of the authors from colleagues who accused him of poisoning them!

A further interesting property that is induced by surface fluorination is a roughening[7,16] of the surface, leading to a matt or frosted texture. Currently this is a highly desirable surface finish for certain cosmetic containers.

Another benefit of the fluorinated layer is its excellent adhesion properties. Printing (including water-based inks) and labels will normally adhere tenaciously to a fluorinated layer. In some cases it is even possible to fluorinate printed containers without discoloration of the printing. In these cases the printing will be far more insoluble to a large spectrum of solvents and will then only be removed with difficulty.

* Formulation numbers used by Haarmann and Reimer.

16.2.3. Rotational-Molded LLDPE Automotive Fuel Tanks

In-line and nowadays postfluorinated HDPE fuel tanks have been used for many years without any problems in Europe. These tanks comply with European 1.ECE Regulation 34 Appendix 5 as well as with the SHED (Sealed House Emission Determination) tests.[17]

In South Africa, owing to the diversity and small production volume of specific motor car models, it is not economically feasible to blow-mold HDPE fuel tanks. The only cost-effective plastic tank manufacturing process used in South Africa is rotational molding, using LLDPE with a density of $0.939 \, g/cm^3$. Some selected properties of LLDPE before and after petrol exposure are compared to HDPE in Table 16.2. It is clear that owing to weaker mechanical properties, LLDPE is far less suitable than HDPE for petrol containment, mainly because of its lower crystallinity.

By a combination of tank redesign and surface fluorination the problems were overcome and today a number of local motor car manufacturers are using surface-fluorinated LLDPE petrol tanks. These tanks performed exceptionally well in petrol permeation tests in which control tanks lost about 16 g/day, while fluorinated tanks lost less than 0.5 g/day at 40°C. With advanced fluorination techniques mass loss values of less than 0.1 g/day have already been achieved for some of these fluorinated tanks.

16.2.4. HDPE Pipes[18]

Owing to corrosion problems experienced in using underground mild steel pipes for petrol conveyance between reservoirs and pumps at filling stations, the possibility of using HDPE pipes was considered. The obvious disadvantage in using HDPE is its high permeation for nonpolar liquids such as petrol.

This problem was most effectively solved by fluorination of the HDPE pipes. A fluorinated layer thickness, one-fifth of that used for fluorinated LLDPE fuel tanks, was used initially. As can be seen from Figure 16.1, a significant

Table 16.2. Comparison of Some Properties of LLDPE and HDPE before and after Exposure to Leaded, Alcohol-Free Petrol

Polymer grade	Yield strength (MPa)		% Mass swell	% Dimensional swell
	Before	After		
LP 3180[a] (LLDPE)	20.0	14.3	12.0	4.1
GM 7650[b] (HDPE)	29.2	21.7	8	2.1

[a] Product from Polifin, South Africa.
[b] Product from Plastomark, South Africa.

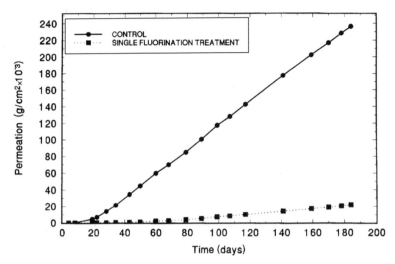

Figure 16.1. Petrol permeation through HDPE pipe at 50°C: Pipe = Type 4 Class 10; petrol grade = 90 octane high altitude.

improvement was experienced. The effect of increasing the fluorinated layer thickness is depicted in Figure 16.2. A quadruple fluorination treatment resulted in an 80-fold improvement when comparing the steady state permeation rate of the control pipes with the fluorinated pipe.

As far as can be ascertained, no performance standards exist for this product. In the absence of such standards, the existing standards for automotive vehicles were used as guidelines. By using the most stringent standard, the SHED test, a petrol permeation rate of approximately $3.3\ g/m^2$ for 24 h at 40°C can be estimated. With a single-fluorination treatment a pipe already exceeds this standard with a steady state permeation rate of $1.7\ g/m^2$ per 24 h at 50°C. Since it is a known fact that permeability increases drastically with a rise in temperature, a permeability of less than $0.17\ g/m^2$ per 24 h at 30°C is expected for a single fluorination treatment.

16.2.5. Benefits of Surface Fluorination for Flexible Plastics

Apart from the benefits of excellent barrier properties with regard to the permeation of nonpolar liquids and good adhesion, little is known about the influence of surface fluorination on gas barrier properties. The permeability of only a few inorganic gases through surface-fluorinated plastics such as PE,[19] PP, and PET[20] have been evaluated to some extent.

The influence of fluorine mixtures with or without reactive gases on the permeability of O_2 and N_2 through LLDPE was investigated.[20a] It was found, as

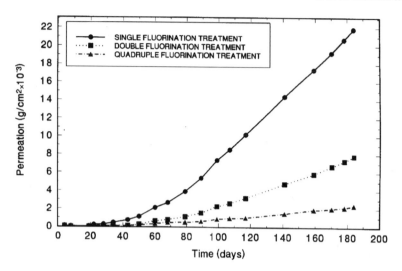

Figure 16.2. Petrol permeation through HDPE pipe at 50°C: Pipe = Type 4 Class 10; petrol grade = 93 octane high altitude.

seen in Table 16.3, that the presence of other reactive gases led to the formation of a significantly better barrier layer than with fluorination alone. From Table 16.3 it is also clear that this new gas mixture incorporates far less fluorine into the surface of the film. This newly discovered F_2/reactive gas mixture[21] was then used to treat a number of different plastic films used in the packaging industry. The results of this study are summarized in Table 16.4.

Table 16.3. Permeabilities (in Barrer) for Oxygen and Nitrogen Gases Through a 70 μm Low-Density PE Film Fluorinated in the Absence (F_2/N_2) and Presence of a Reactive Gas (F_2/X)[a]

Polymer film	Treatment time (min)	Neutron activation		O_2 permeability		N_2 permeability	
		F_2/N_2	F_2/X	F_2/N_2	F_2/X	F_2/N_2	F_2/X
Control LDPE	0	—	—	6.1383	6.1383	1.9715	1.9715
LDPE	1	10.2 ± 1.1	2.65 ± 0.44	5.2538	2.7006	1.4415	0.4155
LDPE	3	12.3 ± 0.8	4.31 ± 0.51	5.6305	2.2399	1.4660	0.3447
LDPE	5	18.8 ± 1.6	5.58 ± 0.57	5.3805	1.9533	1.4564	0.2822
LDPE	10	26.3 ± 2.1	7.91 ± 0.75	5.0132	1.5360	1.3413	0.2034
LDPE	20	37.7 ± 2.8	7.91 ± 0.76	4.9130	1.1720	1.2925	0.1752
LDPE	30	45.9 ± 3.3	13.4 ± 1.10	4.7346	1.0688	1.2568	0.1668
LDPE	60	68.4 ± 4.7	14.4 ± 1.20	4.5530	0.7850	1.1747	0.1248

[a] Total surface fluorine concentration values (μg F/cm^2) were determined using neutron activation analysis.

Table 16.4. Oxygen Permeability of F_2/Reactive-Gas-Treated Polyolefin Films

Plastic film	Film thickness (μm)	Permeation (cm^3/m^2 per 24 h)		Permeability decrease (%)
		Control	Treated	
LDPE	70	8273	174	98
HDPE	100	928	117	87
PP	40	3352	185	94
PS	60	3000	300	90
PET	20	188	10	94.7

Similar reductions in permeability were found for CO_2 and N_2. Apparently this treatment method creates a universal gas barrier and is seemingly universal for polymers with a high hydrocarbon content. It is anticipated that by optimizing this approach, O_2 permeability values for PE of less than 50 cm^3/m^2 per 24 h will be achieved.

The above results suggest a potential for utilizing surface fluorinated plastics for modified atmosphere and vacuum packaging applications. If needed, the barrier properties can be improved further by vacuum metalization. Excellent adhesion was found for vacuum-deposited aluminum on this new fluorinated layer, achieving an adhesion index of 10 using the classical cross-hatch method. With this treatment method it was also possible to heat-seal PP or PE onto PET.

16.3. PRODUCTS BASED ON IMPROVED ADHESION

16.3.1. Introduction

Polyolefins are preferred materials in many applications owing to their low cost, easy processability, chemical inertness, and mechanical properties. The problem with their use in many industrial applications is that, owing to their low surface energy and nonpolar surface, most adhesives do not bond well to their surfaces. One possible solution is roughening of the surface to induce mechanical interlocking, but this is undesirable because it involves a potentially labor-intensive additional processing step and does not usually work very well, as it is known that mechanical interlocking plays a minor role in most adhesive applications.[22] The other solution is to change the surface of polyolefins chemically in order to increase the surface energy and/or to introduce polar functional groups on the surface. This can be done in many ways, including wet chemical reactions such as oxidation with chromic acid.[23] However, wet chemical treatments lead to the problem of producing liquid waste (usually toxic), which can be difficult to dispose of. The ideal way is to treat the surface with a gaseous reagent, the waste of which (if toxic) can be treated with scrubber systems before being released into the atmosphere. Treatment with mixtures of fluorine and nitrogen and/or oxygen gas meets this criterion.

Treatment with fluorine/nitrogen mixtures (fluorination) increases the polar component of the surface energy initially, but longer treatment times result in a drop in both the polar component and the total surface energy[24,25] (see Figure 16.3). At very long treatment times the surface energy of the treated surface can be lower than that of the polyolefin itself.

Treatment with fluorine/oxygen mixtures (oxyfluorination), however, results in a large increase in both the polar component of the surface energy and the total surface energy. The latter does not decrease again with longer treatment times (see Figure 16.4).[24] The surface thus produced exhibits good adhesive bonding with most adhesives, the bond often being stronger than the material itself. It has been shown that this treatment has a long-lasting effect,[7] as opposed to other treatment methods such as corona discharge or flaming. Products can therefore be treated and then stored for a considerable length of time before adhesive bonding is done. This also implies that the treatment does not have to be undertaken on the same fabrication site as the adhesive bonding.

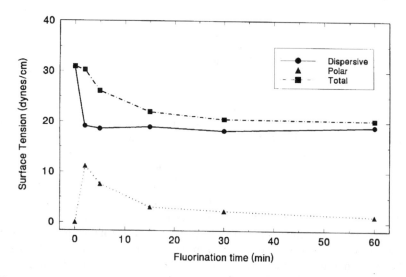

Figure 16.3. Effect of fluorination time on surface tension of PE (10% F_2/N_2, 30°C, constant flow of 20 cm^3/min).

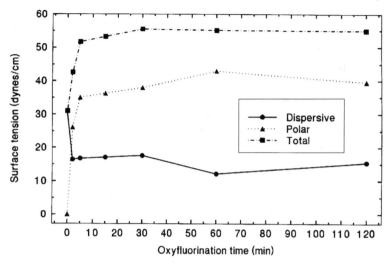

Figure 16.4. Effect of oxyfluorination time on the surface tension of PE (10% F_2/O_2, 30°C, constant of 20 cm^3/min).

16.3.2. Sheet Cladding

Generally, cladding is needed to protect materials from corrosion in hostile environments, e.g., concrete in chemical plants. Two of the most widely used construction materials today are concrete and mild steel. There are, however, two things known for certain, namely concrete will crack, and steel will rust.

Polymers are favored as cladding materials since there is a wide range of substrates to choose from, with different chemical resistances, operating temperatures, and thicknesses. There is the added advantage that a complete liquid-proof lining can be constructed in new or old structures. Applications include bunded* areas in chemical plants and linings for sewage pipes.

Owing to the poor adhesive properties of polymeric substrates, especially polyolefins, these claddings are attached by physical methods, which include melting polyester fleecing onto one side of the polymer to promote adhesion, attaching rubber to the polymer for the same reason, or anchored linings (lugs on one side of the polymer). The fleece-backed and rubber-lined materials are used in adhesion applications, whereas the anchored linings are cast into the concrete structure. Cladding that is applied by these physical methods is very costly. A more affordable method applicable to a number of cladding materials, including

---

* A bunded area is a containment area where potentially harmful and dangerous chemicals (mostly liquid) can be temporarily contained safely, if they are spilled or released from process containers, and can then be treated and be done away with safely.

polyolefins, is to activate the surface with surface fluorination. The activated cladding can then be glued to basically any substrate, by using suitable adhesives, e.g., epoxies.[26]

It has been found that when surface-fluorinated PE or PP was adhered to concrete using epoxy adhesives, efforts to remove it caused breakages in the concrete. Typical activation conditions used are 50 kPa of a 20% F_2 gas mixture for 30 min at 50°C. The butt test was used to measure adhesion to concrete, and an adhesion strength of 2 MPa was measured before a cohesive break started in the concrete. T-peel tests (ASTM 1876-92) returned a value greater than 5 N/mm when surface-fluorinated PE was glued to surface-fluorinated PE using epoxy adhesives (see Table 16.5). In cases where the joint was fully immersed in water at 50°C for 60 h, no decrease in bond strength was measured using certain epoxy adhesives. Similar results were obtained for PP and UHMWPE with epoxy adhesives.

Two factors are very important for adhesion to polyolefins with epoxy adhesives, i.e., (a) the surface activation method, and postactivation cleaning of the surface.

The surface activation method, especially the gas mixture used, determines the chemistry of the surface and can thus promote the adhesion properties with epoxy adhesives. The following example illustrates the importance of the correct activation gas mixture. As is clearly illustrated by Table 16.6, oxyfluorination induces better adhesive properties than fluorination on PE with Pro-Struct 30/71.

Postactivation cleaning of the surface can be vital in cases where the polyolefin sheeting is transported and subjected to contamination. The contaminants, e.g., oils and dust, have to be successfully removed prior to adhesion to ensure a strong bond. The effect of certain organic solvents on the adhesion to PE with Pro-Struct 30/71 can be seen in Table 16.7.

No adhesion was obtained when alkaline washes were used; thus neutral aqueous degreasers/cleaners are recommended.

Table 16.5. Peel Strengths Obtained for Surface Fluorinated PE with Epoxy Adhesives[a]

Adhesive	Peel strength (N/mm)
Pro-Struct 30/71	> 7.0
Pro-Struct 632	> 7.0

[a] All results are based on the average of quintuplet samples. Cross-head speed = 200 mm/min. Pro-Struct adhesives are manufactured by Pro-Struct, Pinetown, South Africa.

Table 16.6. Lap Shear Strengths of Oxyfluorinated and Fluorinated PE with Pro-Struct 30/71[a]

Sample No.	Lap shear strength (MPa)	
	Oxyfluorination	Fluorination
1	11.00	1.786
2	15.12	1.392
3	9.79	1.464
Average	11.97	1.547

[a] All results are based on the average of quintuplet samples. Cross-head speed = 5 mm/min.

Table 16.7. Lap Shear Strengths of Surface Fluorinated PE after Cleaning with Pro-Struct 30/71[a]

Solvent	Lap shear strength (MPa)
Control	10.750
Trichloroethylene (TCE)	8.416
Acetone	10.520
Ethanol	6.071
Methyl ethyl ketone (MEK)	6.083
Xylene	6.947

[a] All results are based on the average of quintuplet samples. Cross heat speed = 5 mm/min.

16.3.3. Pipe Products Manufactured with Fluorination Technology

In many industrial fluid conveyance applications, e.g., chemical plants, polyolefin pipes are preferred, owing to their light weight, chemical resistance, and cost advantage. The disadvantage of these pipes is their poor flexural rigidity and relatively low pressure ratings. These disadvantages are to a large extent addressed by steel encapsulation or by reinforcing the pipes on the outside with resin-impregnated glass fiber. An important design criterion for these pipes is that axial and radial strain in the plastic liner be minimized so as to obviate premature failure from fatigue. This in turn implies that excellent adhesion should exist between the plastic liner and the cladding reinforcement. Owing to the poor adhesive susceptibility of polyolefin surfaces, three competitive technologies have evolved to facilitate the manufacturing of glass-fiber-reinforced polyolefin pipes:

1. Grooving of the external surface of the polyolefin pipe: These grooves serve as a mechanical interlock with the reinforcement. Strain is thus primarily limited to the axial direction only. Grooving is an expensive mechanical operation, and also requires the use of thick-walled liners.

2. Heat-bonding process: Glass fiber mat is embedded into the exterior surface of the pipe liner. This serves as an adhesion key for the reinforcement resin. This is an expensive process owing to the necessity for strict control over the surface fusion process.

3. Activation of the surface by fluorination for adhesion with the reinforcement resin[27]: This is a cost-effective mass production technology, which does not interfere with the overwrapping production processes, e.g., filament winding (see Figure 16.5).

16.3.4. Adhesive Performance of Surface-Activated Polyolefin Surfaces with Respect to Reinforcement Resins

16.3.4.1. Lap Shear Test Based on an Adaptation of ASTM D1002-72

The lap shear test involves measuring the adhesive shear strength between two surface fluorinated polyolefin sheet tokens that are adhesively secured with a reinforcement resin. The tokens are individually reinforced with steel backing plates to eliminate flexural distortion in the shear joint. Lap shear tests carried out with various reinforcing polyester-type resins, contrasting fluorination and oxyfluorination as surface treatment, are shown in Table 16.8.

From the above it can be deduced that, in general, oxyfluorinated PP surfaces give slightly improved adhesion over oxyfluorinated HDPE surfaces. This can be attributed to the greater ease with which PP forms the acyl fluoride functionality during treatment.[28]

Our experience indicates that fluorination generally results in decreased adhesion strength compared to oxyfluorination (see Tables 16.6 and 16.8). This is in contrast to reported findings in the literature where epoxy adhesives were used.[7,29] This discrepancy can be attributed to the surface energy implications of the increased acyl fluoride to the C—H—F functionalization ratio (compare Figures 16.3 and 16.4). However, as indicated by our data, certain resin formula-

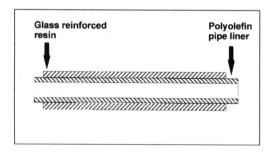

Figure 16.5. Cross section of a plastic-lined GRP pipe.

Table 16.8. Typical Lap Shear Strengths for HDPE and PP Activated by Fluorination and Oxyfluorination[a]

Polyester resin	Oxyfluorinated HDPE (MPa)	Oxyfluorinated PP (MPa)	Fluorinated HDPE (MPa)
NCS Crystic 392	9.9	12.5	6.1
NCS Crystic 600	10.4	12.5	3.5
Dion 9100	11.6	12.0	10.9
NCS N7384	10.4	12.3	4.6

[a] All results are based on the average of quintuplet samples. Cross-head speed = 5 mm/min.

tions, e.g., DION 9100 (epoxy vinyl ester) do not demonstrate a strong dependency on the treatment gas mixture. The generalization (oxyfluorination gives better adhesion than fluorination) should thus be treated with caution.

16.3.4.2. BS6464 Tests

Flat laminated panels are manufactured using fluorinated sheet substrates. The manufacturing of these test panels simulates the production process used to manufacture the fiberglass-reinforced pipes. The BS6464 test prescribes a minimum adhesive shear and peel strength of 7 MPa and 7 N/mm, respectively. Representative values for both tests are given in Tables 16.9 and 16.10 below.

16.3.4.3. Coaxial Push-Out Tests

Coaxial push-out tests (DIN 53769-A) are typically performed on production run pipe samples for quality-control purposes. Typical results using oxyfluorinated HDPE as liner material are shown in Table 16.11 below.

Table 16.9. BS6464 Lap Shear Strength Tests[a]

Substrate	Average shear strength (MPa)
HDPE	10.1
PP	9.9

[a] Crystic 392 (NPG resin) was used to manufacture the fiberglass reinforcement. All results are based on the average of quintuplet samples. Cross-head speed = 5 mm/min.

Table 16.10. BS6464 Peel Strength Tests[a]

Substrate	Average peel strength (N/mm)
HDPE	10.2
PP	13.3

[a] Crystic 392 (NPG resin) used to manufacture fiberglass reinforcement. All results are based on the average of quintuplet samples. Cross-head speed = 5 mm/min.

Table 16.11. Coaxial Push-Out Shear Tests Performed on Production Run Samples

Polyester	Shear strength of 20-mm-wide composite pipe section (MPa)
NCS Crystic 600	10.7
SB Crystic 600	12.3
DION 6694	10.1
NCS Crystic 392	11.8

16.3.4.4. Technology-Related Products for the Liquid Containment Industry

Polyester cloth-backed flat PP sheets (fleece-backed sheets) are normally used to manufacture large reservoirs for liquid containment purposes. Typically, such reservoirs are reinforced on the outside with fiberglass. The polyester mat backing serves as an adhesion key between the PP and the fiberglass resin. Fluorine surface activation can be used as a substitute for the fleece-backed material at a considerable cost-saving. Table 16.12 shows a comparison between the adhesive performance of the materials with two polyester resins.

Table 16.12. Adapted ASTM D1002-72 Lap Shear Tests on Sheet Materials[a]

Sample description	Shear strength using NCS Crystic 600 (MPa)	Shear strengh using NCS Crystic 392 (MPa)
Commerically available fleece-backed PP material	10.6	8.9
Surface fluorinated PP	11.1	12.5

[a] All results are based on the average of quintuplet samples. Cross-head speed = 5 mm/min.

A typical oxyfluorination treatment that is sufficient to activate polyolefin surfaces with respect to polyester adhesion includes the following:

- Gas mixture: 20% F_2/O_2.
- Gas pressure: 20 kPa absolute.
- Reaction time: 30 min.
- Reaction temperature: 50°C.

16.3.5. Surface-Fluorinated Fibers in Cementitious Mixtures

Concrete is an inherently brittle material with low tensile strength and strain capacities. Its brittle characteristics lead to easy nucleation and propagation of cracks, thus restricting its range of applications. To address this deficiency, fibers of different materials such as asbestos, glass, metal, and synthetics[30,31] are used as additives, with the following results[30,31]:

1. Improvement in tensile or flexural strength.
2. Improvement in impact strength.
3. Control of cracking and failure mode by means of postcracking ductility.
4. Change in the rheology or flow characteristics of the cementitious mixture in the fresh state.
5. Improved durability on exposure to freeze–thaw cycling, especially synthetic fibers.

Of the synthetic fibers, polypropylene fibers are most commonly used, owing to their low cost, alkali resistance, and high melting point. One of the most important negative attributes of PP is its hydrophobic surface, which results in poor bonding with the concrete matrix. Owing to weak interfacial bonding, the fiber-matrix adhesion must rely on mechanical anchoring. To improve the mechanical bonding or interlocking of PP fibers, the shape and surface structure of the fibers are changed, e.g., twisting, micropegging through fibrillations in the fiber, or by wedge action through the open or button end.[32,33]

To improve the interfacial bonding between the PP fiber and a cementitious matrix, commercially available PP fibers with different geometries were fluorinated and the effect evaluated with regard to crack control, impact resistance (ACI Committee 544 Report), and water absorption (ASTM C-948).[34,34a] In all cases, as seen in Table 16.13, fluorination improved the induced properties.

It is well known that addition of PP fibers to concrete increases water absorption. It is, however, interesting to note that with fluorinated fibers, the water absorption is very similar to unfilled concrete, suggesting an improvement in wettability and chemical adhesion between fiber and matrix.

Table 16.13. Effect of Fluorinated and Unfluorinated Fibrillated PP Fibers on Certain Properties of Concrete[a]

Sample No.	Crack control (%)	Impact resistance (blows to cause failure)	Water absorption (%)
No fiber	—	16	5
A (unfluorinated)	86	61	6.5
A (fluorinated)	92	64	5.1
B (unfluorinated)	87		6.7
B (fluorinated)	92		5.4
C (unfluorinated)	75	59	6.6
C (fluorinated)	79	80	4.8

[a] Fiber content = 0.1 vol.%

Pull-out tests[34a] done with PP monofilaments having a rectangular cross section that ranges from 0.5×1.0 to 0.65×1.4 mm, indicated that oxyfluorination readily doubles the shear bond strength (see Table 16.14).

It is not yet known if the improvement in adhesion strength is a result of an improvement in chemical adhesion between the fiber and matrix or whether it is due to a change in the concrete microstructure surrounding the fiber or both.

A totally unexpected observation was made when the compression strength of (0.1% by volume) fluorinated fiber concrete increased from about 31 MPa (unfilled) to about 38.6 MPa (filled with fluorinated fibrillated tape). This observation is in contrast to published results[31] that PP fibers do *not* enhance compressive strength.

These observations were later confirmed in field tests where significant increases in initial and final concrete strengths were measured (see Table 16.15). As yet, no satisfactory explanation can be offered for these results.

Table 16.14. Shear Bond Strengths between Concrete Matrix and Oxyfluorinated Monofilament PP Fibers with Different Oxyfluorination Treatment Levels[a]

Oxyfluorination treatment level	Bond strength (MPa)
0	0.19
1	0.45
2	0.41
8	0.30
20	0.32
30	0.36
60	0.36

[a] Cement : fly ash : water : sand : 6 mm crushed stone = 0.8 : 0.2 : 0.42 : 1.5 : 1.5.

Table 16.15. Cube Test Results of
Casted Slabs[a]

% Fiber dosage	Compressive strength (MPa)	
	7 days	28 days
0	23.2	35.2
0	23.9	36.7
0	22.1	33.6
1.2	26.9	37.5
1.2	25.2	37.7
1.2	30.4	40.7
1.8	30.5	40.4
1.8	32.2	39.8
1.8	31.1	42.2

[a] The mix design was as follows: strength: 30 MPa, cement OPC: 280 kg, fly ash: 120 kg, water: 140 kg, stone, 19 mm: 1110 kg, stand: 245 kg, crushed dust: 585 kg, W/C ratio: 0.5.

16.4. CURRENT STATUS AND FUTURE DEVELOPMENTS

In the development of the discussed fluorinated products a lot of fundamental insight has been gained into permeation and adhesion properties of surface-fluorinated polymers. Although this insight is not discussed in this paper, it may be of interest to mention a few significant fundamental discoveries and also to suggest directions for future development.

1. To decrease the permeation of nonpolar liquids using fluorination, a number of factors must be addressed: (a) Fluorination should be performed at temperatures lower than the melting point of the polymer[16] and preferably in a very controlled manner, control can be exercised, for instance, by the stepwise introduction of fluorine[35]; this control is required in order to decrease surface roughness[16] and to suppress carbon–carbon polymer backbone scission; in the most extreme, charring of the surface owing to the highly exothermic nature of the fluorination reaction can take place. (b) The incorporation of oxygen in the fluorinated layer must be minimized.[36,37] (c) The influence of the chemical composition of the fluorinated layer on the barrier properties should be investigated further. (d) The formation of a laminated barrier structure where the fluorinated layer is one of the layers will most probably be needed for mixtures of polar and nonpolar solvents. The optimization of some of these has already resulted in a marked reduction of the surface fluorine concentration required. This

has been accompanied by a drastic improvement of the barrier properties of the fluorinated layer.

2. It may be possible to predict the permeation characteristics of the fluorinated layer by using the acid–base theory of surface energy together with solubility theory.[38]

3. The role of additives in plastics in the creation of low-permeability fluorinated layers is totally underestimated/ignored in the open literature.

4. The conclusion that might be drawn from published results,[7,29] that adhesive/coating systems adhere just as well to fluorinated and oxyfluorinated surfaces, is not supported by our experience based on activation conditions normally encountered in commercial postfluorination plants.

5. We have found enough exceptions to refrain from predicting that a specific adhesive system, e.g., epoxy or polyester, will always adhere better to a fluorinated or oxyfluorinated surface.

6. Some postactivation liquid degreasers will totally destroy the adhesion of epoxy adhesives to a fluorinated plastic surface, while having no effect on the adhesion of polyester adhesives.

7. Adhesive strength may possibly be predicted by using the acid–base theory.[39] We have already found linear relationships between adhesive strength and either the acid or base component of the fluorinated surface.[40]

8. To decrease the risk of kilogram quantities of high-pressure fluorine on site of companies engaging in surface fluorination, on-site fluorine generation would be a very attractive alternative. This will decrease the amount of fluorine present on site at any given moment to a few kilograms with a maximum pressure of about 100 kPa.

ACKNOWLEDGMENT: The authors would like to thank the R&D fluorination team, and especially S. A. Marais, and Dr. J. B. Wagener in helping to prepare this paper.

16.5. REFERENCES

1. J. Söll, U.S. Patent 2,129,289 (1938).
2. C. Bliefert, H.-M. Boldhaus, and M. Hoffman, U.S. Patent 4,536,266 (1985).
3. S. P. Joffre, U.S. Patent 2,811,468 (1957).
4. K. A. Goebel, V. F. Janas, and A. J. Woytek, *Polym. News 8*, 37–40 (1982).
5. J. M. Mohr, D. R. Paul, T. E. Mlsna, and R. J. Lagow, *J. Membrane Sci. 55*, 131–148 (1991).
6. J. D. Le Roux, D. R. Paul, J. Kampa, and R. J. Lagow, *J. Membrane Sci. 94*, 121–141 (1994).
7. G. Kranz, R. Lüschen, T. Gesang, V. Schlett, O. D. Hennemann, and W. D. Stohrer, *Int. J. Adhes. Adhes. 14* (4), 243–253 (1994).
8. A. D. Roberts and C. A. Brackley, *J. Nat. Rubb. Res. 4* (1), 1–21 (1989).
9. T. Nagura, Japanese Patent 01 81,802 [CA 111: 195664 (1989)].
10. M. Anand, J. P. Hobbs, and I. J. Brass, in *Organofluorine Chemistry: Principles and Commercial Applications* (R. E. Banks, B. E. Smart, and J. C. Tatlow, eds.), Plenum Press, New York (1994), Ch. 22.

11. R. L. Sandler and R. A. Verbelen, *ASTM Spec. Tech. Publ., STP 1183* (13), 121–133 (1993).
12. D. M. Buck, P. D. Marsh, and K. J. Kallish, *Polym.-Plast. Technol. Eng. 26* (2), 71–80 (1987).
12a. J. P. Hobbs, M. Anand, and B. A. Campion, in *Barrier Polymers and Structure* (W. J. Koros, ed.), ACS Symposium Series 423 (1990), pp. 280–294.
13. R. K. Mehta and D. M. Buck, *Packaging August 1988,* 100–101.
14. M. Anand, *Proceedings of the 4th Annual High Performance Blow-Moulding Technical Conference,* October 16–17, 1989.
15. C. J. Farrell, *Ind. Eng. Chem. Res. 27,* 1946–1951 (1988).
16. M. Eschwey, R. van Bonn, and H. Neumann, U.S. Patent 4,869,859 (1989).
17. M. Eschwey, *Gas Aktuell 40,* 11–16.
18. P. A. B. Carstens, South African Patent 95/8968 (1995).
19. C. L. Kiplinger, D. F. Persico, R. J. Lagow, and D. R. Paul, *J. Appl. Polym. Sci. 31,* 2617–2626 (1986).
20. A. P. Kharitonov, Yu. L. Moskvin, and G. A Kolpakov, *Sov. J. Chem. Phys. 4* (4), 877–885 (1987).
20a. P. A. B. Carstens, South African Patent 96/D555 (1977).
21. P. A. B. Carstens, South African Patent application 97/0498 (1997).
22. F. Garbassi, M. Morra, and E. Occhiello, *Polymer Surfaces,* John Wiley and Sons, Chichester (1994), Ch. 10.
23. D. M. Brewis, *J. Adhesion 37,* 97–107 (1992).
24. F. J. du Toit, R. D. Sanderson, W. J. Engelbrecht, and J. B. Wagener, *J. Fluorine Chem. 74,* 43–48 (1995).
25. J. D. Le Roux, D. R. Paul, M. F. Arendt, Y. Yuan, and I. Cabasso, *J. Membrane Sci. 90,* 37–53 (1994).
26. P. A. B. Carstens, G. A. B. M. G. Boyazis, and J. A. de Beer, South African Patent 95/4900 (1995).
27. P. A. B. Carstens and G. A. B. M. G. Boyazis, South African Patent 95/4903 (1995).
28. J. L. Adcock, S. Inoue, and R. J. Lagow, *J. Am. Chem. Soc. 100* (6), 1948–1950 (1978).
29. I. Brass, D. M. Brewis, I. Sutherland, and R. Wiktorowicz, *Int. J. Adhes. Adhes. 11* (3), 150–153 (1991).
30. D. J. Hannant, *Mater. Sci. Technol. 11* (9), 853–861 (1995).
31. Z. Zheng and D. Feldman, *Prog. Polym. Sci. 20,* 185–210 (1995).
32. B. Curie and T. Gardiner, *Int. J. Cem. Compos. Lightweight Concr. 11* (1), 3–9 (1989).
33. A. Bentur, S. Mindess, and G. Vondran, *Int. J. Chem. Compos. Lightweight Concr. 11* (3), 153–158 (1989).
34. P. A. B. Carstens, South African Patent 95/1800 (1995); U.S. Patent 5,744,257 (1998).
34a. L. Tu, D. Ing. Thesis, Rand Afrikaans University, Johannesburg (1988).
35. J. P. Hobbs, U.S. Patent 5244 615 (1993).
36. P. Wouters, J.-J. Van Schaftingen, P. Dugois, and M. Obsomer, European Patent 657493 (1994).
37. L. Böhm, M. Eschwey, R. Kaps, E. Raddatz, and R. Van Bonn, European Patent 738751 (1966).
38. C. J. van Oss, *Coll. Surf. A: Physicochem. Eng. Aspects 78,* 1–49 (1993).
39. R. J. Good, in *Contact Angle, Wettability and Adhesion* (K. L. Mittal, ed), VSP, Utrecht (1993), pp. 3–36.
40. L. Tu, D. Kruger, J. B. Wagener, and P. A. B. Carstens, *J. Adh. 62,* 187–211 (1977).

17

Modified Surface Properties of Technical Yarns

MARCUS O. WEBER and DIEDERICH SCHILO

17.1. INTRODUCTION

Modifications of the surface properties of man-made fibers play a dominant role in several end-uses. Via a variety of application processes it is possible to treat man-made fibers with fluorine gas. The electronegativity of flouorine is 4.0 eV, the highest of all elements. In particular, the surface activity of yarns can be increased considerably by incorporating fluorine atoms. As part of this investigation some polymers and fiber types for diverse end-uses were tested (Table 17.1). All end-use applications for these yarns have important surface function requirements, e.g., for tire yarns the adhesion to rubber is very important while for tarpaulins it is the adhesion to PVC. Textile and medical applications with different end-use characteristics were also considered. The modifications attempted to this investigations did not influence other properties such as tenacity, elongation, or shrinkage.

Industrial use of fluorine gas treatment started at the end of the 1980s. Surface modifications with fluorine offer improved reactivity for subsequent demands. One of the major advantages of fluorine gas treatment is the fact that modifications can be carried out under standard pressure and temperature (Figure 17.1), so that it can be used for on-line processes at low cost. Areas of application include, plastic fuel containers, gluing, dyeing or printing preparations on plastics.[1–3]

MARCUS O. WEBER and DIEDERICH SCHILO · Acordis Research GmbH Obernburg, 63784 Obernburg, Germany.

Fluoropolymers 1: Synthesis, edited by Hougham *et al.*, Plenum Press, New York, 1999.

Table 17.1. Effects of Fluorination on Fiber Properties in Diverse Applications

Product	Target	Result[a]
Polyester yarn (Diolen)	Fiber/rubber adhesion	+
	Fiber/PVC adhesion	−
	Wettability	+
Aramid yarn (Twaron)	Fiber/rubber adhesion	+
	Fiber/PVC adhesion	+
Dialysis membranes	Complement activation	+/+
(Cuprophan/SPAN)	Thrombin-antithrombin III	−/−
	Platelets	+/+

[a] (+) Better properties; (−) worse properties.

Many industrial yarns have specific surface function requirements. For technical yarns the market share for composites or coated fabrics is almost 70%. Furthermore, textile applications also can benefit from a special surface treatment in order to improve the water repellency. Capillary membranes for dialysis, however, have totally different requirements; enhanced biocompatibility of the membranes is needed.[4–6]

Two methods of applying fluorine gas to fibers have been used during these investigations. For fibers and woven fabrics winding and unwinding units transport the yarn through a fluorine gas atmosphere. In this atmosphere the concentration of gas can be varied up to 6% F_2 gas. The treatment chamber was ventilated at both ends, and separate entry ports for the mixture F_2/N_2 and for N_2 gas alone made it possible to adjust the specific setting of the F_2 content (Figure 17.2). To clean the apparatus N_2 alone was blown through the treatment chamber. The used gas mixture was cleaned in a calcium carbonate tower, where the fluorine was absorbed.

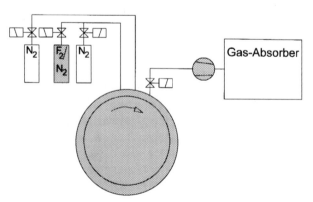

Figure 17.1. Batch process.

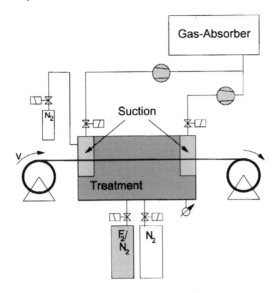

Figure 17.2. Continuous process.

To treat hollow dialysis fibers the fluorine gas was passed through the inside of the capillaries (Figure 17.3). By flow stream measurements the exact amount of gas that entered the capillaries could be determined. The treatment time and the fluorine concentration was measured to determine possible effects on the main biocompatibility properties.[7,8]

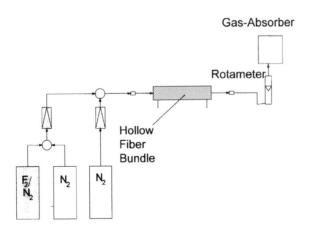

Figure 17.3. Hollow fiber fluorination.

17.2. ADHESION TO RUBBER AND PVC

Tire yarns are mostly made of polyester and viscose, but aramid yarns offer advantages for high-performance applications. Besides strength and shrinkage behavior the fibers need a good adhesion to rubber. As viscose yarns already have good adhesion properties to rubber, this test concentrated only on polyester (Diolen) and aramid (Twaron). The adhesion is measured against peel testing standards ASTM D-4393-85. While both fibers (Figures 17.4 and 17.5) showed significantly improved adhesion, compared to the conventional two-bath dip process the values achieved were not sufficiently good. However, as the use of epoxy predips will become increasingly more problematic as environmental legislation becomes more stringent, alternative low-cost treatments will have to be developed to meet the increasing demands.[9,10]

Strong adhesion of Twaron and Diolen to PVC is also required for different applications. In particular, truck and flat tarpaulins and roofing/textile constructions represent the biggest markets for PVC-coated fabrics. The adhesion of Twaron to PVC can be improved significantly (Figure 17.6), but compared to the specially activated yarn-type Twaron 1001, the improvement is nevertheless minimal. The PVC adhesion of fluorinated Diolen was found to be rather poor (Figure 17.7); the test results are much below the values with the cohesion agent Vulcabond. In particular, the Diolen fluorination together with Vulcabond gave worse results than when Vulcabond was used alone.

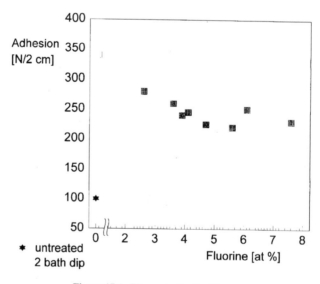

Figure 17.4. Tire-cord-adhesion Twaron.

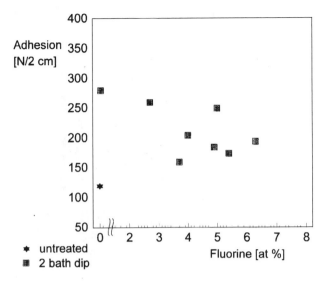

Figure 17.5. Tire-cord-adhesion Diolen.

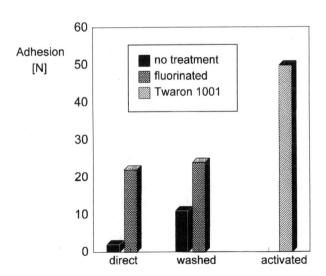

Figure 17.6. PVC-adhesion Twaron.

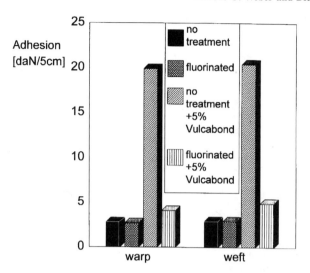

Figure 17.7. PVC-adhesion Diolen.

17.3. WETTING BEHAVIOR OF FLUORINATED POLYESTER

Polyester fibers are common in many textile end-uses. For the applications in apparel the properties concerning wetting behavior are important in two ways. On the one hand, for the dyeing process it is necessary to have a good penetration of the dyestuff; on the other hand, for outer wear it is advantageous to have a good water repellency. In this investigation we tested water-drop penetration and dissipation (Figures 17.8 and 17.9). The drop penetration was determined by measuring the time a drop needs to penetrate into the woven fabric. The drop dissipation was measured as the spreading length for both the wrap and weft directions. In Figures 17.9 the mean values from both directions are indicated.

The penetration time for a woven fabric in warp 76 dtex f 24 gl pr x 1 and weft 100 dtex f 144 x 2 texturized could be reduced by fluorination. The effect diminished with the extent of the washing, which was a regular machine washing with detergent at 40°C. The drop dissipation improved only slightly after fluorination. The fluorination is an approach to improve water repellency in a narrow range. For an optimum in dyeing performance, fluorination should not be applied before dyeing. Printing behavior of fluorinated fabrics was not investigated.

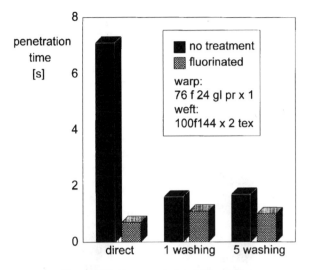

Figure 17.8. Drop-penetration Diolen fabric.

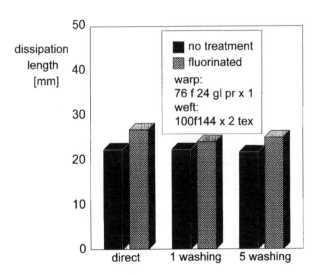

Figure 17.9. Drop-dissipation Diolen fabric.

17.4. INFLUENCES ON BIOCOMPATIBILITY

Various membranes are in common use for the filtration of blood during dialysis. In this investigation cellulosic (Cuprophan) and synthetic (acryl nitrile, SPAN) capillary membranes were tested. The fluorine gas treatment was performed as described before. Three parameters are chosen for the assessment

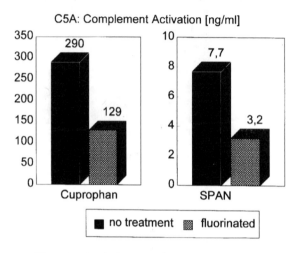

Figure 17.10. Biochemistry of dialysis hollow fibers I.

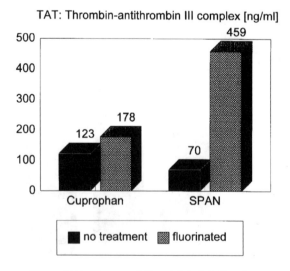

Figure 17.11. Biocompatibility of dialysis hollow fibers II.

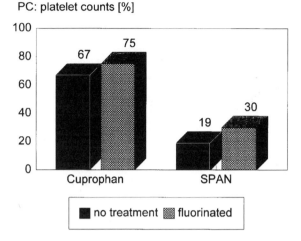

Figure 17.12. Biocompatibility of dialysis hollow fibers III.

of the biocompatibility: complement activation (C5a), thrombin-antithrombin III complex (TAT), and platelet counts were measured. The complement activation gives an indication of the immunological stimulation of blood components, while the TAT value describes the activation of the blood coagulation process. Both values have to be as low as possible. The platelets that can be counted after testing with blood have to be the highest possible; otherwise they remain on the membrane with possible negative effects on filtration and on the patient.

Fluorination improved the complement activation and reduced the loss of platelets (see Figures 17.10 and 17.12), but there was a deterioration in the coagulation properties (Figure 17.11). It is interesting to note that these trends were independent of the fiber raw material (cellulosic or synthetic) used to make the membrane.

ACKNOWLEDGMENTS: We thank Mr. B. Möller, Fluortec GmbH for the technical support in these investigations and Mr. Bowry and Dr. Budgell for their contributions to this article.

17.5. REFERENCES

1. R. Milker and A. Koch, *Maschinenmarkt 9*, 42–46 (1992).
2. B. Möller, *Paper+Kunstst.-Verarb. 7*, 7–8 (1992).
3. W. Götz, *Dicht- und Oberflächentechnik 1* S. 4, (1991).
4. M. Weber and D. Schilo, Paper presented at International Man-Made Fibers Congress in Dornbirn, Austria, 25–27 September 1996.

5. M. Weber and D. Schilo, Paper presented at Fluorine in Coatings II in Munich, Germany 24–26 February 1997.
6. M. Weber and D. Schilo, *J. Coated Fabrics* **10** (1996).
7. D. A. Lane and S. K. Bowry, *Nephr. Dial. Transplant.* **9 (Suppl. 2)**, 18–28 (1993).
8. R. J. Johnson, *Neph. Dial. Transplant.* **9 (Suppl. 2)**, 36–45 (1994).
9. B. Rijpkema and W. E. Weening, Paper presented at Adhesion '93, York (UK) 6–8 September 1993.
10. M. Doherty, B. Rijpkema, and W. E. Weening, Paper presented at the Rubber Division of the American Chemical Society, Philadelphia, May 1995.

III

Vapor Deposition

18

Vapor Deposition Polymerization as a Route to Fluorinated Polymers

J. A. MOORE and CHI-I. LANG

18.1. INTRODUCTION

Microelectronics has become a dominant influence in our lives. This latest industrial revolution was originally driven by the need for very small and lightweight electronic circuits for military and aerospace applications. With the development of, first, the transistor and, later, the integrated circuit (IC), microelectronics has now grown into a multibillion dollar industry, and its applications are ubiquitous. Device miniaturization has brought us from small-scale (SSI), medium-scale (MSI), and large-scale (LSI) integration to very large-scale integration (VLSI) with 10^5 or more components per chip.

There are three main elements in the fabrication of computer chips: the integration of devices, the wiring (or the interconnects), and the packaging. Research has been going on intensively in all three areas to provide ever higher performance (speed) and density. Active research foci include the scaling of devices and the search for novel materials and processing technologies for interconnects and packaging. Because speed is now limited largely by packaging technology rather than state-of-the-art chip construction, the bottleneck to computer speed improvement resides in further developments of packaging schemes and materials. The development of new insulators for interconnects and packaging is one approach to increase the speed of pulse propagation.[1]

Advances in the speed and complexity of integrated circuits and multichip modules (MCMs) have created a demand for the development of high-density

J. A. MOORE and CHI-I. LANG · Department of Chemistry, Rensselaer Polytechnic Institute, Troy, New York 12180-3590.

Fluoropolymers 1: Synthesis, edited by Hougham *et al.* Plenum Press, New York, 1999.

interconnects with controlled impedance.[2] Compatibility with thin-film fabrication techniques; low-dielectric losses; good adhesion to a variety of substrate materials; and thermal, mechanical; and chemical stability have made polymer materials attractive choices as the interlayer dielectric (ILD). In its simplest form, the multilevel interconnect structure comprises a metal pattern on a substrate, a polymer layer on top with metalized vias, and a second metal pattern on top of the polymer.

As the operating frequencies of electronic devices enter the gigahertz range and as the dimensions of electronic devices approach the submicron level, dielectric media with low dielectric constants (<3) become increasingly more important for the reduction of signal coupling among transmission lines. Therefore, there is a clear need to understand the chemistry of materials used in these applications to evaluate better their effects on the electrical and physical performance of devices. There is also a need to develop a new materials that address the problems and requirements of IC manufacturing.

Organic materials exhibit high bulk and surface resistivities. Table 18.1 presents a comparison of the bulk resistivities of some relevant materials. The response of a dielectric material to a suddenly imposed voltage can be quite complex. Polar groups become oriented and local environments and chain stiffness affect the rate of this conformational reorientation. A distribution of response times characterizes this process. Mobile ions also begin to migrate. The response generally outlasts the polar group orientation that accounts for the initial surge of current. This current dissipates quickly as the orientation process nears completion. The mobility of ions and other impurities gives rise to a pseudo-steady-state current dictated by the rate at which trapped ions become mobilized.

These postulated mechanisms[3] are consistent with the observed temperature dependence of the insulator dielectric properties. Arrhenius relations characterizing activated processes often govern the temperature dependence of resistivity. This behavior is clearly distinct from that of conductors, whose resistivity increases with temperature. In short, polymer response to an external field comprises both dipolar and ionic contributions. Table 18.2 gives values of dielectric strength for selected materials. Polymers are considered to possess

Table 18.1. Bulk Resistivity of Selected Materials

Resistivity range (ohm-cm)	Material
$10^{-6} - 10^{-3}$ (conductor)	Cu, Ag, Al, Ni
$10^{0} - 10^{8}$ (semiconductor)	Si, Ge, Se, CuO
$10^{10} - 10^{11}$ (insulator)	Urea-formaldehyde resin
$10^{13} - 10^{14}$	Polyethylene
$10^{14} - 10^{15}$	Ceramics
$10^{16} - 10^{17}$	Teflon

Table 18.2. Intrinsic Dielectric Strength
(Breakdown Voltage)

Material	V (V/mil)
Sodium chloride	3,800
Minerall oil	5,000
Polyimide	8,000
Polyethylene	16,500
Poly(methyl methacrylate)	25,000

adequate strengths against breakdown in packaging applications.[4] However, moisture absorption lowers the breakdown threshold.

For a material to be suitable as an interconnect dielectric, it should have a low dielectric constant and be able to withstand temperatures higher than the 500°C necessary for the subsequent heating steps without the evolution of volatile by-products. Other important considerations are compatibility with other materials; long-term thermal, chemical, and electrochemical stability; ease of fabrication; and low cost. SiO_2, which has a dielectric constant of 3.5–4.0, has been used as the interlayer dielectric material in the industry. Further breakthroughs in high-performance chips hinge critically on the development of new insulators with dielectric constants much lower than that of SiO_2. It is generally believed that to achieve such a low dielectric constant organic polymeric materials must be considered instead of the traditional inorganic materials. An examination of Table 18.3 points to some of the advantages and disadvantages of organic polymers over inorganic dielectrics.

Apart from new low-dielectric materials, a clean method to deposit the dielectric as a uniform film is also required. Owing to the way fabrication technology in the microelectronics industry has developed and because larger silicon wafers (≥ 8 in.) are being used (currently, the technical difficulty of

Table 18.3. Comparison of Properties between Polyimide and Selected Inorganic Dielectrics

Physical properties	Polyimide	SiO_2	Si_3N_4
Process temperature, °C	300–350	350–450	700–900
Decomposition temperature, °C	500	1710	1900
Dielectric strength, MV/cm	3–4	5–8	5–10
Volume resistivity, ohm-cm	$10^{15}-10^{16}$	$> 10^{16}$	$10^{14}-10^{16}$
Dielectric constant	3.2–3.6	3.5–4.0	7–10
Expansion coefficient 10^{-6}/°C	20–70	0.3–0.5	4
Thermal conductivity, W/cm $-$ °C	0.0017	0.021	0.12

controlling thin-film uniformity for larger wafers has been a challenging area for spin-coating), a methodology with the ability to deposit polymers with appropriate properties as conformal films from the gas phase would be a useful advance, yielding high purity, uniform films on larger wafers.

Organic polymers such as polyimides and poly(benzocyclobutenes) are applied by spin-coating, and these materials can planarize the underlying topography to provide a planar surface for the next metal deposition. Senturia et al.[5] reported that planarization between 48 and 69% can be achieved with a single coating of polyimide. The degree of planarization is determined by the ratio of the step height with the polyimide coating to the initial step height of the metal pattern.

While these organic dielectrics have excellent planarizing properties, there are problems associated with their use: occasionally, polymer films are contaminated with solvent; polymer film curing processes are not complete; polymer films may be hygroscopic; and metal bubbling caused by water vapor trapped at the metal/polymer interface during high-temperature processing can, at times, be severe. Additionally, the nonuniformity when larger wafers are used is an unavoidable difficulty for the spin-coating technique. Among these problems, an important consideration is that of environmental safety, which strongly suggests that future technology should adopt a solvent-free process.

Although commercial materials such as DuPont's Vespel polyimide and Teflon AF, Union Carbide's Parylenes, or Dow's poly(benzocyclobutenes) have lower dielectric constants than inorganic insulators, some of these materials still fall short of certain requirements, such as thermal stability. For instance, Parylenes (vide infra), depending on the type, have dielectric constants ranging from 2.38 to 3.15, and Teflon AF has an even lower dielectric constant of 1.89 but poor thermal stability (360°C in air). However, VLSI interconnection and packaging applications also require high thermal stability of the films being used. A further consideration is that diffusion in and adhesion of metal to polymer films rely strongly on the thermal stability of the polymer film. For instance, Cu diffusion in Parylene-N starts at a temperature of 300–350°C, which corresponds roughly to the onset of thermal degradation. Adhesion failure between Cu and Parylene-N also starts at 300°C. The thickness of the film begins to shrink at 350°C while it is being annealed in nitrogen.[6]

The development of low-dielectric-constant materials as ILDs is crucial to achieve low power consumption, reduce signal delay, and minimize interconnect cross-talk for high-performance VLSI devices. In one of the multilevel interconnect process routes, metal lines (e.g., Al–Cu or Cu) are patterned through reactive ion etching, and then dielectric films are filled in the trenches formed between these lines. These trenches can have widths in the sub-0.5 μm range and aspect ratios greater than 3. Therefore, small gap-filling capability is also required for such dielectrics.

The main objective of the work to be described here is to design and synthesize new organic dielectric materials, and to develop new techniques to deposit these materials as thin, thermally stable films with very low dielectric constants, for use as ILDs. New dielectric materials must not only possess proper electrical, thermal, and mechanical properties but should also minimize or eliminate the use of solvents in this era of concern for environmental pollution.

A detailed consideration of the physical and chemical parameters that affect the magnitude of the dielectric constant of organic polymers as well as modulate thermal stabilities and mechanical properties[7] reveals that an ideal material for a low-dielectric-constant medium would be a rigid, (partially) fluorinated hydrocarbon polymer. Adding the requirements of good to excellent thermal stability leads to the narrower choice of rigid, perhaps lightly cross-linked, nonconjugated aromatic fluoropolymers. If we make the standards even more stringent, we would include a requirement of freedom from the need for solvents and a film-generation process that leads to conformal, dense films. One might say that the described task is an impossible one. However, we believe that we may have found a promising process in vapor deposition polymerization, as described in this brief overview of our work to date.

18.2. VAPOR DEPOSITION POLYMERIZATION

Chemical vapor deposition (CVD) is a process whereby a thin solid film is synthesized from the gaseous phase by a chemical reaction. It is this reactive process that distinguishes CVD from physical deposition processes, such as evaporation, sputtering, and sublimation.[8] This process is well known and is used to generate inorganic thin films of high purity and quality as well as form polyimides by a step-polymerization process.[9–11] Vapor deposition polymerization (VDP) is the method in which the chemical reaction in question is the polymerization of a reactive species generated in the gas phase by thermal (or radiative) activation.

18.2.1. Parylene N

The best developed example of a material produced by VDP is poly(p-xylylene) designated as Parylene-N by the Union Carbide Corporation. Poly(p-xylylene) was discovered by Szwarc[12] in 1957 and then commercialized by Gorham at Union Carbide.[13,14] (Scheme 1). Gorham has reported that di-p-xylylene is quantitatively cleaved by vacuum vapor-phase pyrolysis at 600°C to form two molecules of the reactive intermediate p-xylylene, which subsequently polymerizes on the cold substrate. In a system maintained at less than 1 Torr, p-xylylene spontaneously polymerizes on surfaces below 30°C to form high-molecular-weight, linear poly-*p*-xylylene. The facile cleavage of di-*p*-xyly-lenes to *p*-xylylene is, in all probability, due to the high degree of steric strain in the dimeric species and to the comparatively stable nature of *p*-xylylene.

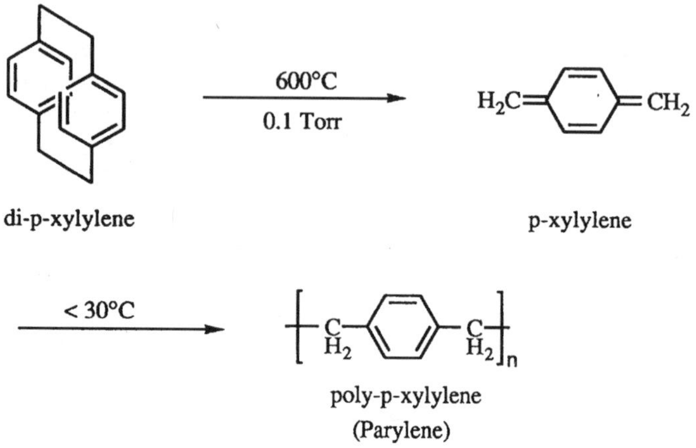

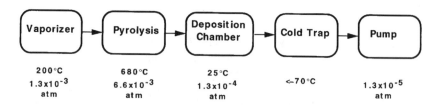

di-p-xylylene   p-xylylene

poly-p-xylylene
(Parylene)

Scheme 1. Synthesis of Parylene.

The VDP process takes place in two stages that must be physically separated but temporally adjacent. Figure 18.1 is a schematic of typical Parylene deposition equipment and indicates the approximate process operating conditions.

The Parylene-N process has certain similarities with vacuum metalizing. The principal distinction is that truly conformal Parylene coatings are deposited even on complex, three-dimensional substrates, including on sharp points and into hidden or recessed areas. Vacuum metalizing, on the other hand, is a line-of-sight coating technology. Whatever areas of the substrate cannot be "seen" by the evaporation source are "shadowed" and remained uncoated. This evidence provides insight into the differences between the two processes.

The major drawback of poly(*p*-xylylene) is that it reverts to a monomer when thin films are heated above ca. 400°C and it cracks when the films are annealed at 300–350°C in nitrogen. During module assembly the chip-joining (soldering)

Vaporizer	Pyrolysis	Deposition Chamber	Cold Trap	Pump
200°C	680°C	25°C	<-70°C	1.3x10^-5
1.3x10^-3	6.6x10^-3	1.3x10^-4		atm
atm	atm	atm		

Figure 18.1. Parylene-N deposition apparatus.

process causes short exposure to temperatures between 300–450°C and therefore precludes use of this material under these conditions.

18.2.2. Parylene F

Although Parylene-N possesses an outstanding combination of physical, electrical, and chemical properties, the benzylic C—H bonds present are potential sites for thermal and oxidative degradation. It is well known that replacing a C—H bond with a C—F bond not only enhances the thermal stability of the resulting polymer, but also reduces the dielectric constant. Because incorporation of fluorine is known to impart thermal and oxidative stability, it became of interest to prepare poly($\alpha,\alpha,\alpha',\alpha'$-tetrafluoro-$p$-xylylene), Parylene-F. Joesten[15] reported that the decomposition temperature of poly(tetrafluoro-p-xylylene) is ca. 530°C. Thus, it seemed that the fluorinated analog would satisfy many of the exacting requirements for utility as an on-chip dielectric medium.

The comparison of physical and chemical properties of Parylene-N and Parylene-F is shown in Table 18.4. Parylene-N is considerably less stable in air than in nitrogen as a result of oxidative degradation. However, the similarity between its behavior in air and in nitrogen suggests that Parylene-F has very good thermal oxidative stability, which is most likely the result of the high stability of the C—F bond, and provides evidence that oxidative attack starts at the benzylic C—H bonds in Parylene-N.[15]

Hertler[16] was the first to report the preparation of poly(tetrafluoro-p-xylylene) by a multistep synthesis as shown in Scheme 2. Pyrolysis (330°C, 0.025 Torr) of dibromotetrafluoro-p-xylene ($Br_2F_4C_8H_4$) over zinc led to deposition of the polymer film in a cold trap.

Table 18.4. Several Properties of Parylene-N and Parylene-F

Polymer	Parylene N	Parylene F
Decomposition Temperature	430°C in N_2 300°C in air	530°C in N_2 450°C in air
Dielectric constant	2.65 ± 0.5	2.36 ± 0.5
UV stability	Stable in N_2 Degraded in air	Stable in N_2 Stable in air

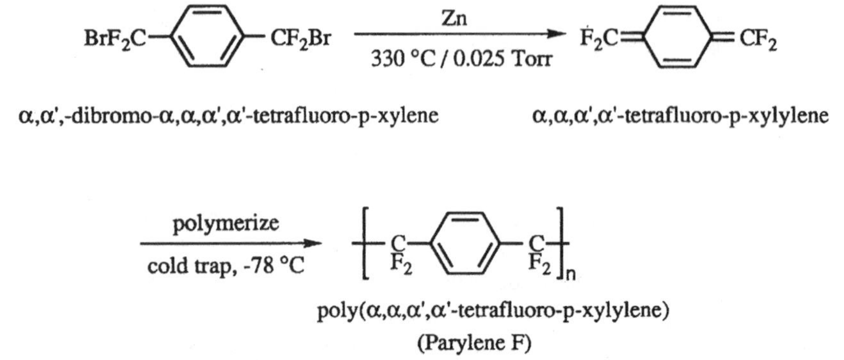

α,α',-dibromo-α,α,α',α'-tetrafluoro-p-xylene   α,α,α',α'-tetrafluoro-p-xylylene

poly(α,α,α',α'-tetrafluoro-p-xylylene)
(Parylene F)

Scheme 2. Synthesis of poly(tetrafluoro-p-xylylene.

Fuqua and co-workers[17] tried to develop a much shorter route from α,α,α',α'-tetrafluoro-p-xylene to poly(tetrafluoro-p-xylylene) but were unsuccessful in generating a polymer because they conducted their pyrolyses at 820–925°C/3–5 Torr, and under those conditions instead of losing H_2 to form α,α,α',α'-tetrafluoro-p-xylylene, α,α,α',α'-tetrafluoro-p-xylene lost HF and underwent rearrangement to form β,β,p-trifluorostyrene [Eq. (1)].

(1)

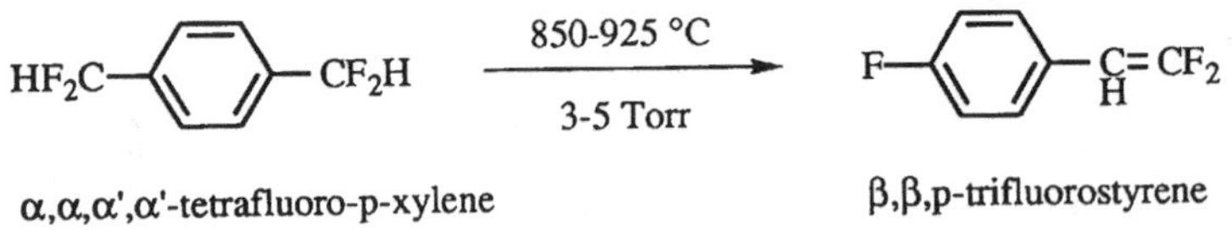

α,α,α',α'-tetrafluoro-p-xylene   β,β,p-trifluorostyrene

Chow and co-workers[18] developed a multistep synthesis for the commercial production of α,α,α',α'-tetrafluoro-p-xylene that uses octafluoro[2.2]paracyclophane (PA-F dimer) as the precursor to polymer. PA-F dimer was cracked at 720–730°C and polymer was deposited on a substrate at −25 to −35°C [(Gorham method) Eq. (2)]. Chow[19] also attempted to pyrolyze $Br_2F_4C_8H_4$ at very high temperatures. The film that was deposited was of poor quality compared to that prepared from dimer.

(2)

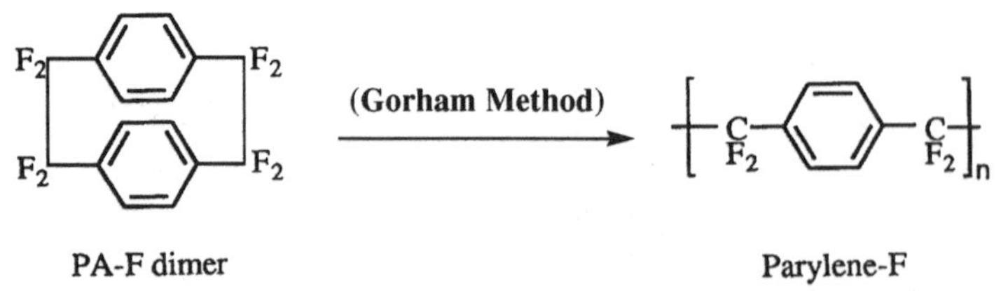

PA-F dimer   Parylene-F

In spite of its potential commercial utility, PA-F dimer has not been extensively used as a Parylene-F precursor because the only reported preparative methods for PA-F dimer involve pyrolysis of different precursors at very high temperatures 600–950°C and yields are very low. The conventional way of synthesizing PA-F dimer involves a pyrolysis process as shown in Eq. (3).

(3)

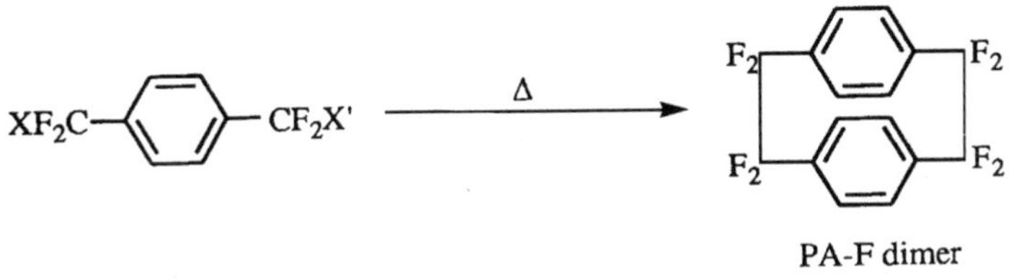

PA-F dimer

The special apparatus required for the synthesis makes laboratory preparation unattractive, while the required high temperatures make commercialization impracticable. In principle, any method that could generate $\alpha,\alpha,\alpha',\alpha'$-tetrafluoro-$p$-xylylene would be a potential source of dimer.

Recently, Dolbier[20,21] reported a new, nonpyrolytic method for the synthesis of PA-F dimer [Eq. (4)]. This method allows easy laboratory preparation of gram quantities of PA-F dimer but still presents difficulties with regard to commercialization because of the high-dilution methodology that is required for a good yield of high-purity product.

(4)

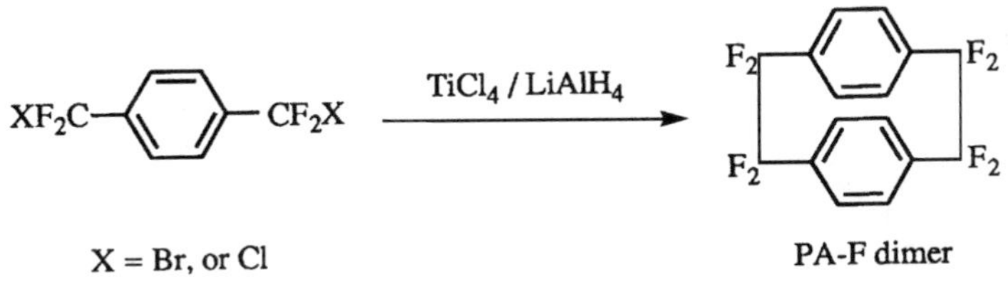

X = Br, or Cl

PA-F dimer

One advantage of VDP is that it is compatible with the deposition of metal on silicon wafers. Cracking dimer does provide a clean and efficient process for depositing film but the main hindrance in producing PA-F thin films in large quantities by this method are the extremely high cost and the short supply of PA-F dimer. Therefore this situation provided the impetus to develop an alternative process to make PA-F film. In principle, a process that could generate $\alpha,\alpha,\alpha',\alpha'$-tetrafluoro-$p$-xylylene directly from the vapor state in a conventional vacuum system would be a potential method for depositing poly(tetrafluoro-p-xylylene) film.

As noted above, the synthesis of dimer involved a complicated synthetic procedure and produced very low yields. Alternative routes for deposting poly-(tetrafluoro-p-xylylene) thin film were studied: the precursor method and a new method (Scheme 3). "Parylene-F" is a trade name and is only used for material produced by the Gorham method from dimer. The polymer produced by any other process is called poly(tetrafluoro-p-xylylene).

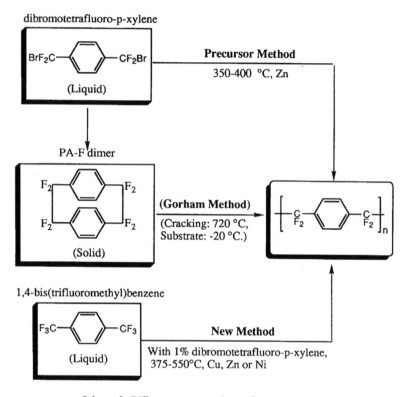

dibromotetrafluoro-p-xylene

Precursor Method

350-400 °C, Zn

PA-F dimer

(Gorham Method)

(Cracking: 720 °C, Substrate: -20 °C.)

1,4-bis(trifluoromethyl)benzene

New Method

With 1% dibromotetrafluoro-p-xylene, 375-550°C, Cu, Zn or Ni

Scheme 3. Different routes to poly(tetrafluoro-p-xylylene).

Three different techniques, the Gorham method, the precursor method, and a new method, were used to deposit polymer. Films ranging in thickness from 2000–8000 Å were deposited. Thicker films could be deposited by using more source material or a longer reaction path. FTIR spectroscopy indicates that the films, as deposited, are essentially identical to those prepared by the conventional route. The average dielectric constant of the films deposited by this technique was 2.6 ± 0.1 (presursor method), while the value measured on films generated by cracking the dimer was 2.38 ± 0.5 (Gorham method). X-ray photoelectron spectroscopy (XPS) measurements of the as-deposited film from the precursor method revealed the presence of O (5.86–9.00 at.%), Zn (4.03 at.%), and Br (5.4 at.%) contaminants.[22] Annealing the films in nitrogen at 530°C for 30 min removed Br and and reduced the O and Zn levels to 3.63 and 3.23%, respectively.

Further improvements in the design of the apparatus and optimization of the deposition conditions are expected to remove the remaining discrepancies.

To simplify the synthetic effort required to deposit such films, attempts were made to deposit films by pyrolyzing tetrafluoro-p-xylene ($F_4C_8H_6$). Under similar reaction conditions, a polymer film was deposited that was different from poly(tetrafluoro-p-xylylene) as the FTIR spectrum indicates that it contains more hydrogen and less fluorine. Presumably HF is preferentially eliminated rather than H_2.

Attempts were made not only to find an alternative way to replace dimer and to deposit high-quality poly(tetrafluoro-p-xylylene) film, but also to eliminate the dibromide as the precursor because of the difficulty of synthesis. Therefore, the deposition of poly(tetrafluoro-p-xylylene) film by using hexafluoro-p-xylene as the precursor instead of dibromotetrafluoro-p-xylene was tried. However, no polymer film was deposited on the wafer. Effort was expanded and other metal reagents such as nickel or copper were used to react with 1,4-bis(trifluoromethyl)-benzene to generate $\alpha,\alpha,\alpha',\alpha'$-tetrafluoro-$p$-xylylene to deposit poly(tetrafluoro-p-xylylene) film. However, the result showed that no film was deposited, which was not unexpected, because a C—X bond that is weaker than C—F bonding might be necessary to initiate the formation of the desired intermediate.

Accidently, using hexafluoro-p-xylene with the contaminated copper wire obtained from the precursor method experiments, a polymer film was deposited on the silicon substrates. Obviously, some dibromotetrafluoro-p-xylene from the precursor method that adhered to, or reacted with, the metal could somehow initiate this VDP process. However, a complete explanation of these results is not yet available. As an extension of this discovery, commercially available 1,4-bis(trifluoromethyl)benzene in conjunction with a catalyst/initiator has proved to be a potential alternative by which to deposit poly(tetrafluoro-p-xylylene) film successfully.[23]

The copyrolysis of 1 wt% dibromotetrafluoro-p-xylene with commercially available hexafluoro-p-xylene (Aldrich) with metals was examined and it was found that it was indeed possible to prepare films that were spectroscopically indistinguishable from those deposited from dimer. The PA-F films obtained are of excellent quality, having dielectric constants of 2.2–2.3 at 1 MHz and dissociation temperatures up to 530°C in N_2. A uniformity of better than 10% can be routinely achieved with a 0.5-µm-thick film on a 5-in. silicon wafer with no measurable impurities as determined by XPS. During a typical deposition, the precursor was maintained at 50°C, the reaction zone (a ceramic tube packed with Cu or Ni) was kept at 375–550°C, and the substrate was cooled to -10 to $-20°C$. The deposited film had an atomic composition, $C:F:O = 66:33:1 \pm 3$ as determined by XPS. Except for O, no impurities were detected. Within instrumental error, the film is stoichiometric. Poly(tetrafluoro-p-xylylene) has a theoretical composition of $C:F = 2:1$. Figure 18.2 illustrates the XPS of the binding energy

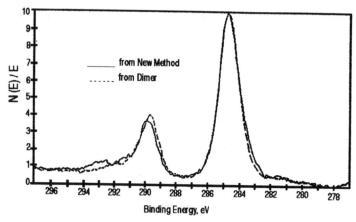

Figure 18.2. XPS of poly(tetrafluoro-*p*-xylylene) from the new method.

of C1*s* electrons of poly(tetrafluoro-*p*-xylylene) deposited by the new method. The two peaks in the spectrum represent two groups of C atoms in different chemical environments, which correspond to the six C atoms of the benzene ring and the two CF_2 groups of the poly(tetrafluoro-*p*-xylylene) repeating unit. The stronger peak in Figure 18.2 is assigned to the C1*s* binding energy of $C{=}C$ and $C{-}H$, which have almost the same C1*s* binding energies of approximately 285 eV. The other peak in the spectrum is from the CF_2 group with a C1*s* binding energy of 290 eV.

A source of difficulty here remains the metal reagent used to generate the reactive intermediate. It became deactivated after the surface of the metal had been in contact with the dibromide precursor and hexafluoro-*p*-xylene, and the film growing process was interrupted as a result. An interesting result was observed during annealing of the polymer films from cracking dimer (Gorham method) and from the new method. Some 30% of the films deposited by the Gorham method developed cracks when the films were annealed at 510°C in nitrogen but no cracks were observed in any of the films deposited by the new method.

18.2.3. Poly(Benzocyclobutenes)

The use of cross-linked polymers as engineering materials has been extensive owing to their rigidity, strength, mechanical and thermal stability, and chemical resistance. Typically, these materials are synthesized using step-growth methods where polymer growth and cross-linking arise from the same chemical reaction.[24] These materials are usually processed as low-molecular-weight prepolymers because of the intractability of the final cross-linked material.

Another approach adopts a two-step process: processable prepolymers are synthesized, and then the prepolymers are cured to drive the polymerization reaction to higher conversions, initiating network formation. Chemically, this scheme has the problem that complete conversion is rarely achieved and defects from unreacted chain ends and intramolecular cyclization are common.

18.2.3.1. Benzocyclobutene Chemistry

The Parylene family has very attractive properties for use as dielectric materials as was noted above, but their thermal stability at the temperatures used in the fabrication of electronic devices is less than optimum. When considering alternatives as possible precursors for VDP, the isomeric *ortho*-xylylene (*o*-quinodimethane) is a likely candidate (Scheme 4). This approach involves the thermolysis of benzocyclobutene derivatives to generate a reactive dieneoid intermediate (*o*-quinodimethane),

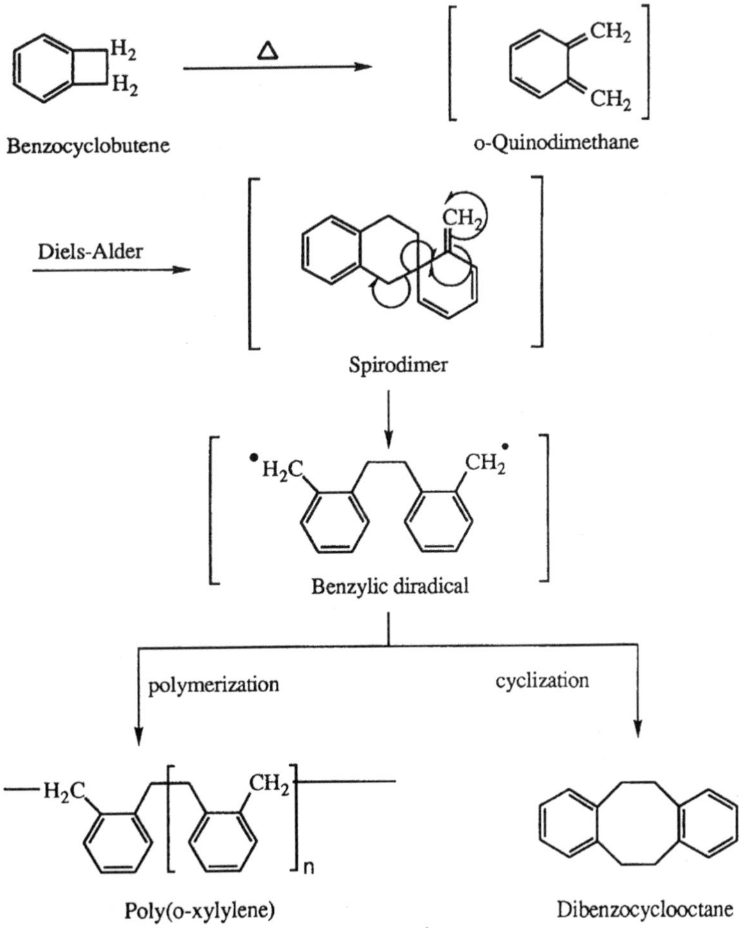

Scheme 4. Ring-opening reaction of benzocyclobutene.

which can then undergo a Diels–Alder reaction between two *o*-quinodimethane molecules, one as a diene and one as a dienophile, to yield an intermediate spirodimer. The spirodimer then fragments to give a benzylic diradical species, which may undergo intramolecular coupling to give dibenzocycloocta-1,5-diene (cyclo-di-*o*-xylylene) or oligomerize to provide poly(*o*-xylylene).[25,26] If two benzocyclobutene units are joined together, they should undergo polymerization upon heating to yield an insoluble, cross-linked system without evolving by-products.

18.2.3.2. Polymers Containing Benzocyclobutenes

In the late 1970s, Kirchhoff at Dow Chemical Company developed the use of benzocyclobutenes in polymer synthesis and modification. These efforts culminated in 1985 with the issuance of the first patent describing the use of benzocyclobutene in the synthesis of high-molecular-weight polymers.[27] Similar work that involved a thermosetting system based on Diels–Alder cycloaddition between terminal benzocyclobutene and alkyne groups,[28,29] was reported separately and independently by Tan and Arnold.[28] Since these initial discoveries, the field of benzocyclobutene polymers has expanded rapidly and benzocyclobutene chemistry constitutes the basis of a new and versatile approach to the synthesis of high-performance polymers for applications in the electronics and aerospace industries.[30]

The basic benzocyclobutene technology involves a family of thermally polymerizable monomers that contain one or more benzocyclobutene groups per molecule.[31–33] Depending on the degree and type of additional functionality, these monomers can be polymerized to yield either thermosetting or thermoplastic polymers. For monomers of the class that contains only benzocyclobutene moieties as reactive groups, the *o*-quinodimethane groups react rapidly with one another to give a highly cross-linked polymer.

Although the moisture insensitivity of poly(benzocyclobutenes) is superior to polyimides, unlike polyimide films, poly(benzocyclobutenes) are very sensitive to oxygen and must be cured in a nitrogen atmosphere. In general, the thermal stabilities of poly(benzocyclobutenes) are also lower than those of the polyimide family. If too much oxygen is incorporated during the cure cycle, several important properties are adversely affected, including dielectric constant, chemical resistance, and flexibility.[34] Introduction of fluorine would be expected to enhance thermal stability significantly.

(*a*) *Poly(Octafluorobisbenzocyclobutene)*. To date, the systems reported have used R groups which are oligomeric and are, therefore, not volatile [Eq. (5)]. If, however, *R* were to be made small enough that the mass of the

(5)

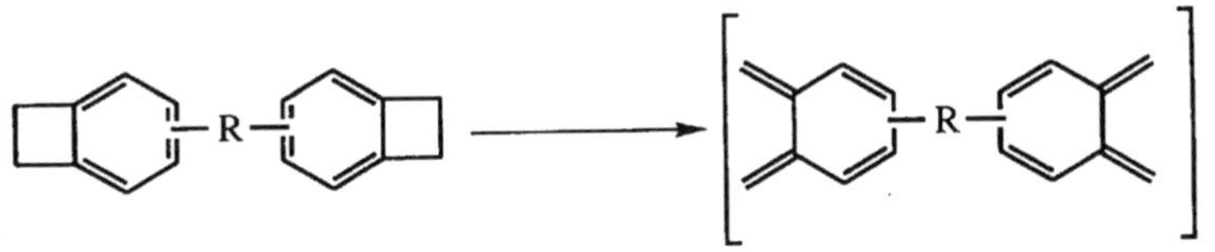

monomer was close to that of paracyclophane, it should be possible to use VDP to prepare polymers from such monomers (Figure 18.3).

Although the mass of the fluorinated benzocyclobutene analogue is significantly higher than that of the hydrocarbon it should be noted that the incorporation of fluorine often enhances the volatility of derivatives beyond the expected levels, e.g., tetrafluorobenzocyclobutene with a mass of 176 boils only 10° higher than o-xylene with a mass of 106 (153°C vs. 143°C).

Using a fluorinated benzocyclobutene-based monomer (Figure 18.4) should provide at least one advantage over the already promising properties of fluorinated poly(p-xylylene). All the desirable properties such as low dielectric constant and low affinity for water should remain but the thermal stability should be enhanced because of the cross-linking that would accompany the generation of these films. The synthesis and polymerization paths for poly(octafluorobisbenzocyclobutene) are depicted in Scheme 5.

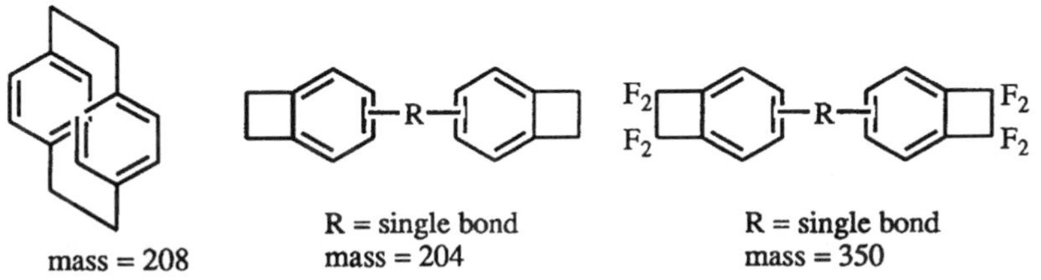

mass = 208

R = single bond
mass = 204

R = single bond
mass = 350

Figure 18.3. Di-p-xylelene variants based on benzocylobutene.

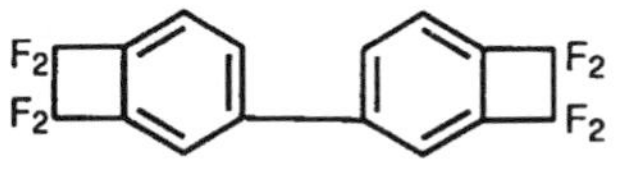

F$_8$-bis-Benzocyclobutene

Figure 18.4. Octafluorobisbenzocyclobutene.

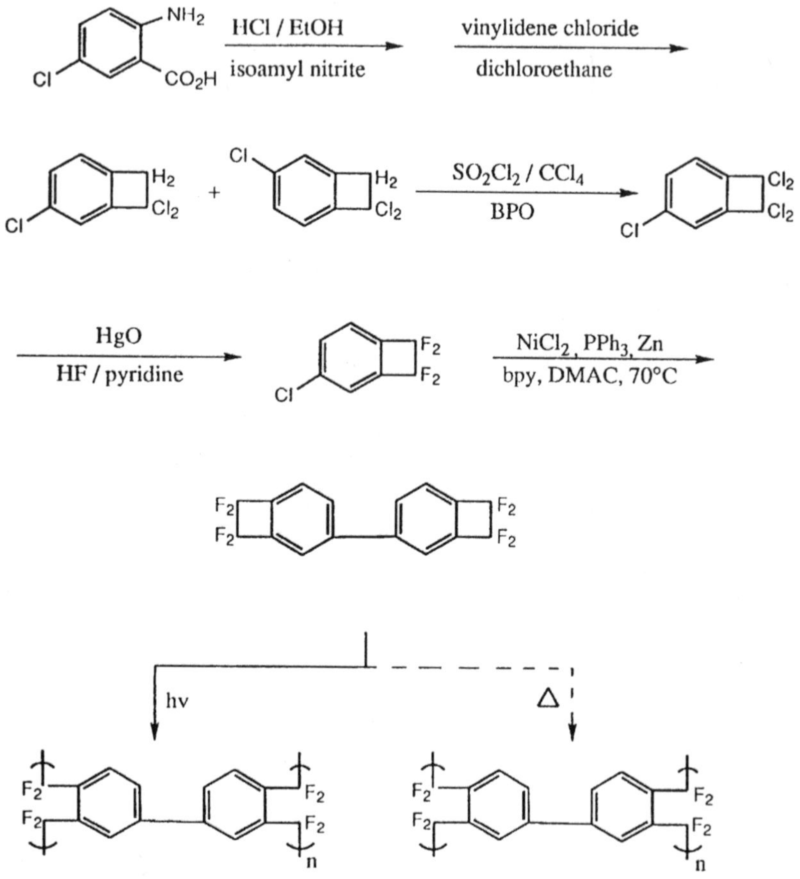

Scheme 5. Synthesis of poly(octafluorobisbenzocyclobutene).

Initially the deposition conditions used for Parylene-F were used in attempts to deposit polymer film on glass or silicon substrates. Reaction temperature, pressure, and retention time of the fluorinated precursor in the reaction zone were varied from 600 to 850°C and from 0.25 to 760 Torr in packed and unpacked tubes, to no avail. The cure chemistry of these systems is primarily based upon the fact that under appropriate thermal conditions, the strained four-membered ring of benzocyclobutene undergoes an electrocyclic ring opening. The temperature at which such a concerted process occurs depends principally on the substitutents at the alicyclic, rather than on the aromatic positions. It has been reported by Kirchhoff[35] that an electron-donating substitutent at C_7 and/or C_8 will favor ring-opening, but electron-withdrawing groups at those positions will make the ring-opening energetically more demanding. 7,7',8,8'-Octafluoro-4,4'-bis(1,2-dihydrobenzocyclobutene) has four fluorine atoms on each ring and it might be expected that the reaction temperature would be higher than for an unfluorinated system. However, the electronic effect is so overwhelming that this compound did not polymerize or undergo Diels–Alder reactions with added dienes or dienophiles at the temperatures or conditions that were useful for benzocyclobutene.

(*b*) *UV-Assisted Vapor Deposition Polymerization.*[36] In this approach, a monomer is activated in the gas phase by UV radiation. Subsequently, the activated monomer polymerizes on the substrate. Despite the fact that octafluorobisbenzocyclobutene did not polymerize efficiently within a convenient temperature range, it was found that it was possible to activate it by UV irradiation in the vapor state or in solution [Eq. (6)]. A polymer was obtained as a film that was not soluble in common laboratory solvents.

(6)

7,7',8,8'-Octafluoro-4,4'-bis-   Poly(octafluorobenzocyclobutene)
(1,2-dihydrobenzocyclobutene)

The film that was obtained was very thin and it was not possible to grow thicker films. This result was most probably caused by absorption of the incident radiation by the film formed on the interior of the quartz reactor, thereby blocking the incoming UV light and preventing the activation of the monomer and continous polymerization. The UV absorption of the monomer and of polymer film reside in the same region. Figure 18.5 and 18.6 show the UV absorption spectra of the precursor and the polymer film as deposited on the quartz surface, respectively.

18.2.3.4. Properties of Poly(Octafluorobisbenzocyclobutene)

(*a*) *Infrared Spectroscopy.* The IR spectrum of poly(octafluorobenzocyclobutene) is the most informative in confirming the structure of the proposed polymer[37,38] at the present time. Figure 18.7 presents the IR spectrum of poly(octafluorobenzocyclobutene). The repeating unit of the polymer contains an absorption band attributable to a 1,2,4-trisubstituted phenyl pattern. The bands at 981–831 cm^{-1} are assigned to the bending of a lone hydrogen atom at the 3-position, and the bands at 905–757 cm^{-1} arise from the bending of two adjacent hydrogen atoms. Additionally, the ring deformation absorption of 1,2,4-trisub-

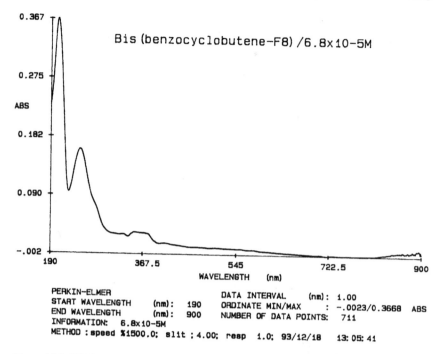

Figure 18.5. UV-absorption spectrum of 7,7′,8,8′-octafluoro-4,4′-bis(1,2-dihydrobenzocyclobutene).

stituted benzene appears in the region of 756–697 cm^{-1}. An investigation of the polymeric structure by solid state NMR has not yet been successful because of the strong fluorine coupling, which complicates the ^{13}C spectrum.

(*b*) *X-Ray Photoelectron Spectroscopy.* The composition of the polymer obtained by photopolymerization, was studied by XPS. The chemical formula of the repeating unit of poly(octafluorobenzocyclobutene) is $C_{16}H_6F_8$; therefore, the theoretical ratio of different types of carbon atoms can be represented as the ratio of the areas of different binding energies is C1s electrons. Two types of C1s binding energies from XPS of the polymer (Figure 18.8) were expected to be observed, and the area ratio was expected to be C (C—C or C—H) : C (F—C—F) = 3 : 1. However, the results showed that the ratio of different C1s peaks of deposited polymer was not consistent with the theoretical value (Table 18.5).

By comparision of the calculated carbon ratios and those determined from XPS data, it can be seen that the number of fluorine atoms is lower than the theoretical value, and is approximately C (C—C or C—H) : C (F—C—F + C – F) = 4 : 1.

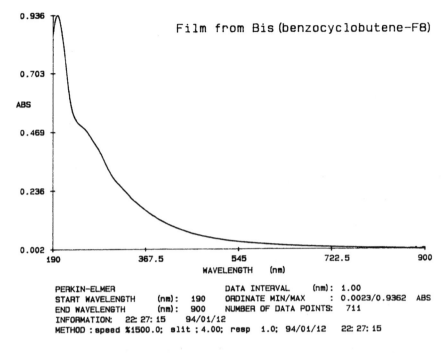

Figure 18.6. UV-absorption spectrum of a thin film of poly(fluorinatedbisbenzocyclobutene) as-deposited.

Ideally, there should be twelve carbon atoms with a $C1s$ binding energy of 286 eV and four carbon atoms with an energy of ca. 291 eV for each repeating unit of the polymer. However, the XPS data showed that the defect structure of as-deposited film was best described as roughly eight carbon atoms with $C1s$ at 286.26 eV (C—C and C—H), and tail to high binding energy because of the presence of carbon atoms bearing only one fluorine atom [$C1s$ at 288.27 (C—F)] and a small tail at the end of the spectrum [$C2s$ at 291.57 eV (F—C—F)]. Obviously some of the fluorine atoms on the aliphatic linkages were missing from the polymer structure. This deficit might be caused by fluorine atoms of C—F

Table 18.5. The XPS Results of Poly(Octafluorobisbenzocyclobutene) from UV-Assisted Polymerization

Parameter	C—C and C—H (C1s)	C—F (C1s)	F—C—F (C1s)
Peak position (eV)	286.26	288.27	291.57
Area	3820	524	365
% of total area	81.12	11.13	7.75

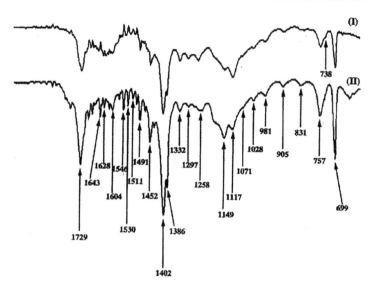

Figure 18.7. IR spectrum of poly(octafluorobisbenzocyclobutene).

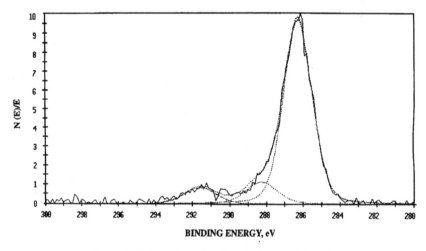

Figure 18.8. XPS of poly(octafluorobisbenzocyclobutene).

bonds being replaced by hydrogen atoms during photopolymerization by an, as yet, unclarified process or by reaction of the fluoropolymer with the quartz substrate to form Si-F bonds. These and other possibilities are under investigation.

(c) *Thermal Stability.* The thermal stability of films of the polymer as reported here is given as the temperatures of 5 and 10% sample weight loss. The thermal stability of the polymer synthesized by photopolymerization was only 268°C (5% TGA weight loss) in nitrogen and 314°C (10% TGA weight loss) as shown in Figure 18.9.

(d) *Dielectric Constant.* While this work was in progress, there was an independent effort by Kudo et al.[39] at the Fujitsu Research Center (Japan) to fluorinate poly(bisbenzocyclobutene) by using an NF_3 plasma. The hydrocarbon polymer poly(bisbenzocyclobutene) was dissolved in an organic solvent, spin-coated onto a silicon wafer, and cured to cross-link the polymer. The cross-linked film was then exposed to an NF_3 plasma in a microwave downstream plasma system to fluorinate the aliphatic C—H bonds as an approach to poly(octafluoro-benzocyclobutene) [Eq. (7)].

Poly(bis-benzocyclobutene)   Poly(octafluorobenzocyclobutene)

(7)

Before fluorination, the dielectric constant of poly(bisbenzocyclobutene) was 2.8, and this value was reduced to 2.1 after plasma treatment. No data were reported in the paper on characterization of structure or properties, except for the dielectric constant of the modified poly(bisbenzocyclobutene). The authors did report that the thermal stability of fluorinated poly(vinylidenefluoride) was inferior to the original poly(vinylidenefluoride) when treated in a similar way. One of the probable reasons for the low thermal stability is that the NF_3 plasma degraded the polymer. According to their results, the thickness of fluorinated poly(bisbenzo-cyclobutene) was reduced by 30%. The same phenomenon was observed for other hydrocarbon polymers subjected to the NF_3 plasma process. A remaining question is whether plasma treatment can modify more than a thin surface layer of the cured polymer? Additionally, one of the side products generated was hydrogen fluoride, which is a serious drawback to this approach.

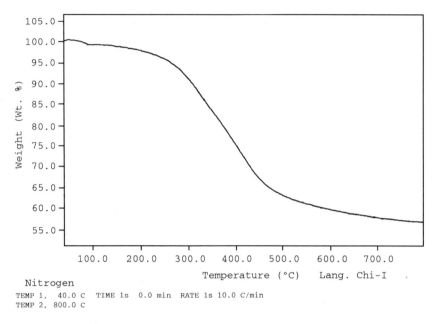

Figure 18.9. Thermogravimetric anlaysis of poly(octafluorobisbenzocyclobutene) in nitrogen.

Although we were unable to grow thicker films of poly(octafluorobenzocy-clobutene), 7,7′,8,8′-octafluoro-4,4′-bis(1,2-dihydrobenzocyclobutene) appears to be a promising compound, which may eventually lead to poly(octafluorobenzo-cyclobutene) under the proper experimental conditions. A different reactor geometry and the use of intense laser sources could prove to be a promising way to grow thicker films.

18.2.4. Polynaphthalenes

The formation of C—C bonds between aromatic rings is an important step in many organic syntheses and can be accomplished by chemical, photochemical, or electrochemical means. As was noted earlier, fundamental considerations of the parameters for a dielectric which must be dealt with in designing a thermally stable, low-dielectric-constant polymer naturally lead one to consider rigid-rod, nonconjugated aromatic polymers containing no "lossy" functional groups. A structure such as poly(naphthalene) is a likely candidate.

Poly(naphthalene) is chemically similar to poly(p-phenylene), which is an insoluble, infusible, low-molecular-weight polymer, all attributes that preclude application in thin-film form in microelectronics. Although these materials possess several very desirable properties, such as high glass transition temperature, insolubility, infusibility, resistance to oxidation, radiation, and thermal degradation,[46] the synthesis of rigid-rod macromolecules via conventional synthetic methods has been an ongoing problem because the poor solubility of these aromatic polymers has resulted in low molecular weight and poor processability.

Poly(naphthalene) is an insoluble material that precipitates as an oligomer as synthesized by classical methods. The first route to poly(naphthalene) was the direct polymerization, in 1965, of naphthalene using a technique developed by Kovacic and others.[41] The process is known as oxidative cationic polymerization and was conducted in o-dichlorobenzene with a catalyst such as ferric-chloride–water or aluminum-chloride–cupric-chloride. Another way to synthesize polynaphthalene is to use nickel-catalyzed polycondensation of Grignard reagents derived from dibromonophthalenes by Sato and co-workers.[42] Additionally, Banning and Jones[43] used a Grignard coupling reaction[44] to synthesize poly(1,4-naphthalene), poly(1,5-naphthalene), poly(2,6-naphthalene), and poly(2,7-naphthalene) (Figure 18.10). These polymers are generally obtained as powders of low molecular weight.

There are a few reports of poly(naphthalene) thin films. Yoshino and co-workers[45] used electrochemical polymerization to obtain poly(2,6-naphthalene) film from a solution of naphthalene and nitrobenzene with a composite electrolyte of copper(II) chloride and lithium hexafluoroarsenate. Zotti and co-workers[46] prepared poly(1,4-naphthalene) film by anionic coupling of naphthalene on platinum or glassy carbon electrodes with tetrabutylammonium tetrafluoroborate as an electrolyte in anhydrous acetonitrile and 1,2-dichloroethane. Recently, Hara and Toshima[47] prepared a purple-colored poly(1,4-naphthalene) film by electrochemical polymerization of naphthalene using a mixed electrolyte of aluminum chloride and cuprous chloride. Although the film was contaminated with the electrolyte, the polymer had very high thermal stability (decomposition temperature of 546°C). The only catalyst-free poly(naphthalene) which utilized a unique chemistry, Bergman's cycloaromatization, was obtained by Tour and co-workers recently (vide infra).

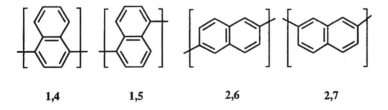

| 1,4 | 1,5 | 2,6 | 2,7 |

Figure 18.10. Structural variants of poly(naphthalene).

18.2.4.1. Bergman's Cycloaromatization

In the nearly 1970s, Bergman and co-workers[48,49] postulated that cis-hex-2-ene,1,5-diyne (A) upon thermolysis would undergo a thermal rearrangement to the benzene 1,4-diradical intermediate (or, 1,4-dehydrobenzene, B), which could revert to starting material or collapse to the rearrangement product (C) [Eq. (8)].

(8)

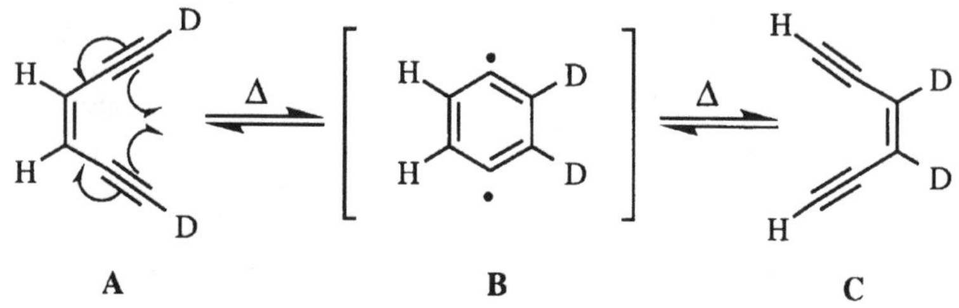

    A   B   C

Numerous synthetic and mechanistic studies were done to investigate this reaction further, and a variety of enediynes have been thermalized in the presence of radical traps such as 1,4-cyclohexadiene. Even though large excesses of radical traps were employed, the yields of the substituted benzenes were often moderate at best. Most important of all, Tour et al.[50] demonstrated that 1,4-naphthalene diradicals generated in solution couple to eventually form a polymer [Eq. (9)].

(9)

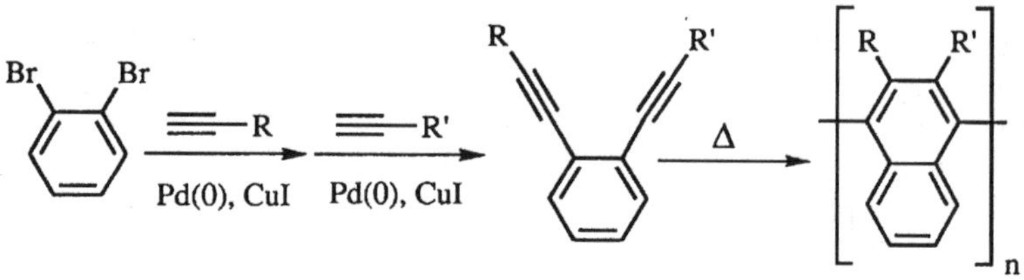

Tour's work showed that polymerization is still a preferred process even though large excesses of radical traps (1,4-cyclohexadiene) were employed. They obtained poly(1,4-naphthalene) only as an insoluble brown powder by polymerization of 1,2-diethynylbenzene in benzene solution. This route appears to be an attractive approach to a new dielectric medium:

1. It requires no catalysts or reagents other than heat for the polymerization.
2. Heteroatomic coupling sites such as halogen are not necessary in the polymerization process, and all atoms present in the monomer are also present in the polymer.
3. The monomers can be readily synthesized by Pd/Cu coupling with a variety of substitution patterns.

By carefully adapting the unique chemistry of Bergman's cycloaromatization, it seemed possible to synthesize the rigid-rod polymer as a thin film of high quality and high purity.

18.2.4.2. Poly(Naphthalene) and Poly(Tetrafluoronaphthalene)

Scheme 6 outlines the stepwise synthesis of poly(naphthalene) and poly-(tetrafluoronaphthalene).

A=H   1,2-Dibromobenzene
A=F   1,2-Dibromotetrafluorobenzene

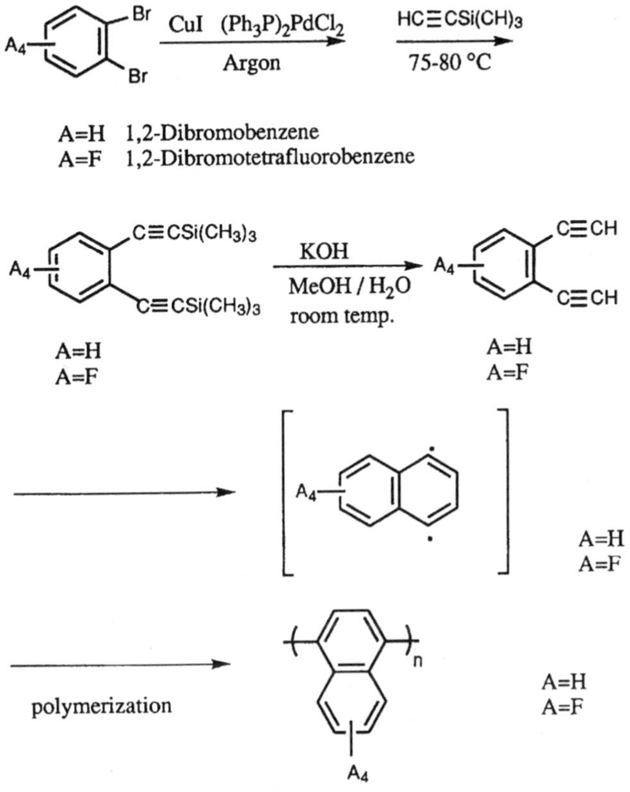

Scheme 6. Synthesis of poly(naphthalene) and poly(tetrafluoronaphthalene).

(*a*) *Solution Polymerization.* The thermal reactivities of these two diacetylenes were dramatically different when the polymerization was carried out in diphenyl ether under a nitrogen atmosphere. Aromatic solvents are expected to be unreactive toward free radical hydrogen atom abstraction and, indeed, proved to be almost completely inert toward the intermediates produced during the reaction.

Powdery PNT-N was obtained when the unfluorinated compound was refluxed in diphenyl ether solution. When the temperature was raised to 196°C for 24 h, the amount of solid increased with time and there was no observable change in the appearance of the solid. However, in the case of PNT-F, a black solid formed suddenly between 120–150°C, and no additional solid formed even when the temperature was increased to 190°C. Apparently the fluorinated derivative reacts much more rapidly and exothermically than the hydrocarbon analogue. Care should be exercised when heating such compounds in sealed glass tubes.

(b) *Vapor Deposition Polymerization.* A schematic drawing of the VDP apparatus used for the preparation of PNT-N and PNT-F thin films in this work is shown in Figure 18.11. The unit consists of four sections: a source vessel with a needle valve, a vapor introduction channel, a Pyrex or quartz tube containing silicon substrates, recycling traps, and a pumping system. 1,2-Diethynylbenzene or tetrafluoro-1,2-diethynylbenzene can be easily and safely polymerized in the deposition apparatus described. It should be carefully noted that polyacetylenic compounds are reactive materials and can be exothermically converted to graphite upon heating. In closed systems, the pressure increase generated because of the amount of heat liberated when graphite is formed can be significant. Proper shielding should be utilized at all times.

Poly(naphthalene) and poly(tetrafluoronaphthalene) were synthesized by vaporizing the appropriate precursor *in vacuo* and transporting the vapor into a hot chamber maintained at 150–400°C. In the high-temperature zone, the

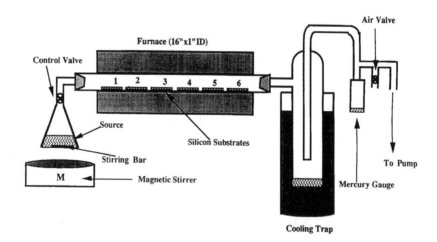

Figure 18.11. The VDP system for the deposition of PNT-N and PNT-F films.

monomer forms the intermediate and subsequently condenses and polymerizes on the substrate. The VDP was monitored by following the pressure changes accompanying the beginning of the reaction. The vapors of the monomers caused an increase in the pressure, which returned to the initial base vacuum of the system when the reaction reached completion.

The temperature distribution of the furnace of the VDP reactor is uneven, and the temperatures at positions 1, 2 and 5, 6 are lower than the temperatures at position 3, 4. For PNT-N, the film was concentrated on positions 3, 4, 5 (see Figure 18.11) and for PNT-F film, the film was concentrated at positions 2, 3, 4. This phenomenon indicates that the fluorinated compound required a lower temperature than the unfluorinated one to be polymerized. Approximately, the temperatures of positions 1 and 6, 3 and 4, and 2 and 5 are the same, and the temperature gradient is 3, 4 > 2, 5 > 1, 6. The deposition temperature is reported as the temperature at positions 3 and 4.

The diethynylbenzene analogues, depending on their individual properties, can be evaporated or sublimed into the reaction zone. The lower limit of the deposition temperature for poly(naphthalene) (PNT-N) was explored, and it was observed that deposition began at a substrate temperature somewhere in range of 150–200°C. A summary of the poly(naphthalene) and poly(tetrafluoronaphthalene) deposition conditions described here is given in Table 18.6.

From the analysis of different depositions of PNT-N films, it can be concluded that the beginning of the VDP process is of particular importance because it affects the adherence of the film to the substrate. The best results were obtained by maintaining the substrate at the desired deposition temperature for an extended time (approximately 2 h) before starting the actual deposition. The heater is also kept on at the end of the deposition while evacuating the chamber to the base vacuum during cooling to avoid contaminating the film.

Table 18.6. Deposition Conditions for Poly(Naphthalene) and Poly(Tetrafluoronaphthalene)

Polymer	Poly(naphthalene)	Poly(tetrafluoronaphthalene)
Source material	(1,2-diethynylbenzene)	(1,2-diethynyltetrafluorobenzene)
Base vacuum	0.2–0.25 Torr	0.2–0.25 Torr
Deposition pressure	0.5–1 Torr	0.5–1 Torr
Deposition temperature	350°C	350°C
Substrate temperature	350°C	350°C
Deposition rate	0.1–0.5 Å/s	0.3–1.0 Å/s
Thickness	0.5–4 μm	0.5–2 μm
Dissociation temperature	570°C in N_2	590°C in N_2
Dielectric constant	2.2–2.3	2.1–2.2

18.2.4.3. Properties of PNT-N and PNT-F Films

The properties of PNT-N and PNT-F were measured on films deposited under the conditions reported in Table 18.6. The films, generated by vapor deposition on hot surfaces such as glass or silicon substrates as previously described, are not soluble in common laboratory solvents. The best film was obtained in the absence of oxygen. Film that was deposited in an oxygen-rich ambient exhibited lower thermal stability and poor adhesion.

(a) *X-Ray Powder Diffraction.* The X-ray powder diffractogram showed that the PNT-N and PNT-F films deposited by the VDP process at 0.45 Torr are slightly crystalline (Figure 18.12). The vertical axis is the intensity and the horizontal axis is the diffraction angle.

When argon was used as a carrier gas to deposit PNT-N film at 10 Torr, amorphous PNT-N film was obtained but the mechanical properties of the film as deposited were inferior to the film deposited at 0.45 Torr without a carrier gas.

(b) *Thermal Stability.* One of the most important properties of a good dielectric material is thermal stability. The thermal stability of PNT can be very different depending on the polymerization process. According to Tour's[50] work, a brown powder of PNT-N was obtained in benzene solution in a pressure tube. The thermal stability of the PNT-N obtained was reported to be 500°C. Using the procedure described by Tour, we obtained a brown powder with a thermal stability of 420°C. We also polymerized *o*-diethynyl benzene under an argon atmosphere in diphenyl ether solution, to give a material with a thermal stability of 460°C. All

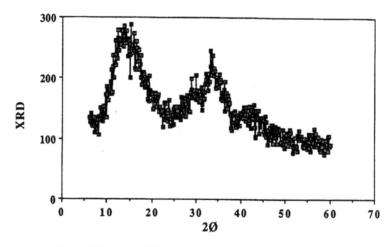

Figure 18.12. X-ray diffractogram of PNT-N film deposited by VDP.

the samples of PNT-N obtained from solution polymerizations were obtained as insoluble brown powders. The thermal stability of PNT-N film obtained by Hara[47] using an electrochemical reaction was reported to be 546°C. The thermal stability measurements of PNT-N and PNT-F films from the VDP process at 0.45 Torr are presented in Figure 18.13.

The thermal stability of PNT from different polymerization methods is presented in Table 18.7. It appears that the colored (dark brown) but transparent PNT-N film synthesized by VDP is the cleanest film among the polynaphthalenes from other polymerization processes that have been reported. These PNT-N films from VDP also have very low dielectric constants in comparison to poly(tetra-fluoro-*p*-xylylene) films. PNT-N and PNT-F films have higher dissociation temperatures ($> 570°C$) and better thermal stability ($> 530°C$), and no film cracking was observed until PNT-F was annealed at 600°C in nitrogen. Table 18.8 presents a summary of the different properties of PNT-N and PNT-F prepared by the VDP process.

It appears that the main reasons for the high thermal stability of PNT-N and PNT-F arise from their inherent rigid-rod polymer structure and the high aromatic content of the repeating unit. However, the fact that the thermal stability of PNT-F is higher than PNT-N might be the result of strong intermolecular interaction between chains because of dipole–dipole interactions.

Although there is no consistent explanation of the relationship between organic polymer morphology and electrical properties,[51] amorphous structures are generally preferred over a crystalline structure. An experiment was conducted to study the structure of the film deposited using an inert carrier gas. The PNT-N

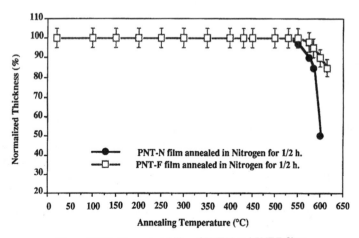

Figure 18.13. Thermal stability of PNT-N and PNT-F films.

Table 18.7. Thermal Stabilities of PNT from Different Polymerization Processes

Method	Thermal stability	$T_{50\%}$	Color
Benzene solution/170°C[50]	500°C (10% weight loss) (N_2)	> 900°C	Brown powder
Electrochemical reaction[47]	546°C	NA	Purple film
Benzene solution/170°C (this work)	420°C (10% weight loss) (N_2)	> 900°C	Brown powder
Diphenyl ether/200°C (this work)	460°C (10% weight loss) (N_2)	> 900°C	Black powder
CVD 350°C/0.4 Torr (this work)	570°C (N_2)	NA	Transparent film

Table 18.8. Properties of PNT and PNT-F Prepared by the CVD Process

Polymer	PNT-N	PNT-F
Dielectric constant	2.4 ± 0.1	2.2 ± 0.1
Electric breakdown	3×10^7 V/m	5×10^7 V/m
Structure	Semicrystalline	Semicrystalline
Dissociation temperature (°C)	570°C in (N_2)	590°C in N_2
Cracks	No cracks at 570°C in N_2	No cracks at 600°C in N_2

films, as-deposited, were amorphous in the presence of an inert vaporous diluent in the VDP process. However, this deposition condition might not be preferred because the films deposited with argon as a carrier gas did not adhere well to silicon substrates.

(c) IR Spectra of PNT-N. At the present time the IR spectra are most informative in confirming the structure of the proposed polymers. These polymers show characteristic absorption bands in the $760-940$ cm^{-1} and the $1300-1600$ cm^{-1} regions. The IR spectra of PNT-N from three different polymerization processes are collected in Figure 18.14. The top spectrum is PNT-N (powder) from solution polymerization using diphenyl ether as the solvent. The middle spectrum is PNT-N (powder) from solution polymerization in a pressure tube with benzene as the solvent, and the bottom spectrum is PNT-N (thin film) from the VDP process. Basically, these three spectra are super-imposable with slight differences among them. In the literature, there are different versions of IR characterizations of the poly(1,4-naphthalene) structure. Unfortunately, they are not exactly consistent with one another. The shoulder above 3000 cm^{-1} comes from the C—H stretching vibration. According to Tour's[50] analysis, bands at 1595 and 878 cm^{-1} (two adjacent H atoms) and 754 cm^{-1} (four adjacent H atoms) are characteristic for a 1,4-disubstituted naphthalene unit. These bands were also observed in the IR spectra of PNT-N. However, there is a very interesting peak at 699 cm^{-1} which was reported only by

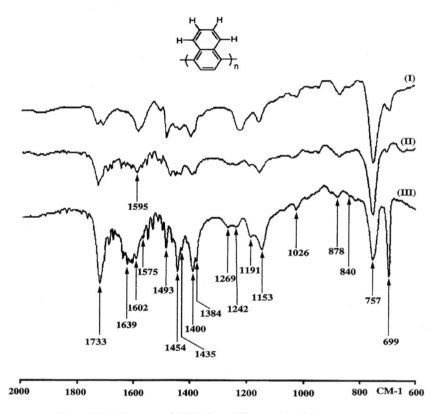

Figure 18.14. IR spectra of PNT-N from different polymerization processes.

Banning,[43] who synthesized 1,4-naphthalene oligomer using a Grignard coupling reaction with 1,4-dibromonaphthalene as the starting material. The IR spectra of PNT-N prepared in this work all have this absorption at 699 cm^{-1}. Films collected from different VDP reactions always exhibit this specific absorption, between 696–699 cm^{-1}.

(*d*) *IR Spectra of PNT-F.* For PNT-F, the IR spectrum also showed a strong absorption at 757–760 cm^{-1} and 699–700 cm^{-1}. Therefore it is reasonable to believe that the 699 and 757 cm^{-1} bands might be important indications of 1,4 linkages for polynaphthalene analogues. Figure 18.15 shows the IR spectrum of PNT-F prepared by two different polymerization processes. The top spectrum was prepared in diphenyl ether solution, and the bottom from a polymer film prepared by the VDP process.

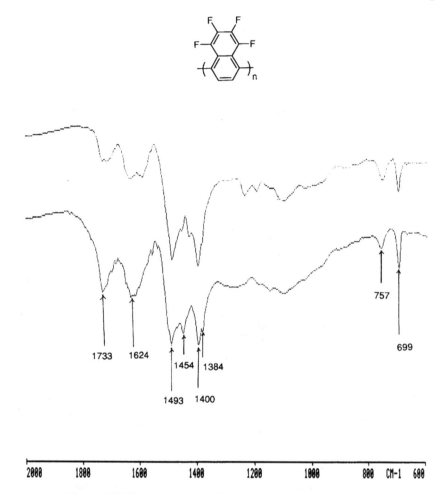

Figure 18.15. IR spectra of PNT-F from different polymerization processes.

(e) *Solid State NMR Spectrum of PNT-N.* The brown powder of PNT-N that was prepared by solution polymerization in benzene was investigated using magic-angle cross-polarization (CP/MAS) ^{13}C-NMR spectroscopy (Figure 18.16). The area ratio of the peaks for hydrogen-substituted carbon atoms and quaternary carbon atoms was ca. 55 : 45 (curve fit), which approximately corresponds to a hydrogen/carbon ratio of C : H = 10 : 5.5.

(f) *Solid State NMR Spectrum of PNT-F.* The attempts to study PNT-F were not successful because the complicated coupling pattern between carbon and fluorine atoms makes the analysis too difficult to accomplish at this time.

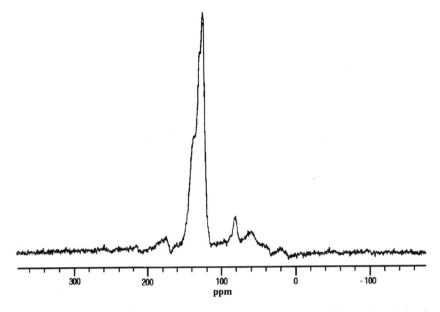

Figure 18.16. Solid State ^{13}C-NMR spectrum of PNT-N prepared by solution polymerization in benzene.

(g) *X-Ray Photoelectron Spectroscopy.* The shifts in C1*s* levels for oxygen-containing structural features in terms of the number of carbon–oxygen bonds are given below[52]:

$$O = C \underset{\diagdown O}{\overset{\diagup O}{}} > O{=}C{-}O > C{=}O \sim O{-}C{-}O > C{-}O$$

$$\sim 290.6 \text{ eV} \qquad \sim 289.0 \text{ eV} \qquad \sim 287.9 \text{ eV} \qquad \sim 286.6 \text{ eV}$$

The reference binding energy for carbon not attached to oxygen (e.g., in polyethylene) is 285.0 eV and the primary substituent effect of oxygen can be described in terms of a simple additive model. The shift in binding energy of C1*s* core level electron, subsequent to replacing carbon or hydrogen by oxygen is ca. ~ 1.5 eV per carbon–oxygen bond. This value is only half the shift induced by fluorine.

(h) *XPS of PNT-N.* We were able to link the XPS data to the thermal stability and dielectric constant of the deposited films. Because of the structural simplicity of PNT-N and PNT-F films the assignment of the signals is straightforward in the light of the foregoing discussion, in which it was noted that the primary shift induced by oxygen is ca. 1.5 eV per carbon–oxygen bond and roughly additive. PNT-N does not provide an easily interpretable spectrum because only C—C and C—H atoms exist in the polymer and their C1s binding energies are indistinguishable or extremely difficult to assign. The XPS spectra of the PNT-N films, however, clearly showed that these films contain defect structures, which apparently result from the incorporation of oxygen atoms into the polymer. Ideally, there should be only one C1s peak at ca. 285 eV (C—C and C—H) for a defect-free PNT-N film. However, two to three different types of carbon atoms were observed, depending on the concentration of the incorporated oxygen atoms. Oxygen-incorporation was observed from the additional C1s peaks at ca. 286.5 eV (C—O) and at ca. 287.9 eV (O—C—O or C=O) in the XPS spectra. The low thermal stabilities and high dielectric constants of these films also confirmed these defects, indirectly.

It is apparent that VDP of PNT-N is sensitive to the presence of oxygen and, perhaps, moisture, as would be expected if the growing chain ends bear unpaired electrons. The higher the incorporated oxygen concentration as detected by XPS, the higher the dielectric constant. When both the $-C$=O and $-C$—O peaks were detected in XPS spectra, the incorporated oxygen contents in the films were usually higher than 12%. If only a $-C$—O peak was observed, approximately 5–8% oxygen was incorporated.

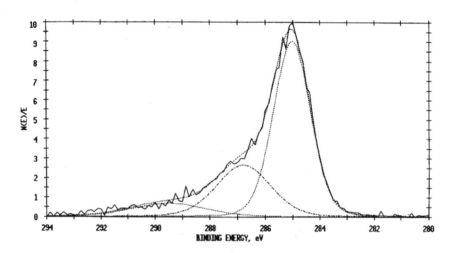

Figure 18.17. XPS spectrum of PNT-F film.

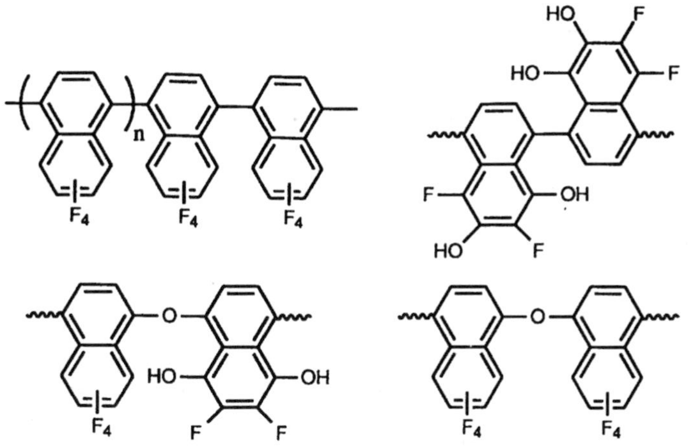

Figure 18.18. Several possible defect structures in PNT-F film.

(*i*) *XPS of PNT-F.* Figure 18.17 presents the XPS spectrum of C1s binding energy of the PNT-F sample made by VDP. The three broad peaks in the spectrum represent three groups of C atoms from different chemical environments, which correspond to the six carbon atoms of the backbone ring (*C*—C and *C*—H have almost the same C1s binding energy, approximately 285 eV), oxygenated carbon atoms of the naphthalene ring (*C*—O) and fluorinated carbon atoms (*C*—F). In Figure 18.16, the strongest group of peaks in the spectrum is assigned to the C atoms of *C*—C and *C*—H of the backbone ring, and the medium group of peaks is assigned to the C atoms of *C*—O. The smallest group of peaks in the spectrum is assigned to the *C*—F groups. Table 18.9 gives the XPS data for this PNT-F film.

The expected chemical formula of each repeating unit is ($C_{10}H_2F_4$), with a calculated ratio of C (*C*—C and *C*—H) : C (*C*—F) = 6 : 4. Thus, there should be six carbon atoms (*C*—C and *C*—H) with C1s binding energy of 285.03 eV and four carbon atoms (*C*—F) with C1s binding energy of ca. 288 eV. However, the defect structure of the as-deposited film was clearly demonstrated by the extra peaks and the stoichiometry of C atoms (the corresponding area) from the XPS analysis. Relatively, there are six carbon atoms with C1s binding energy of 285.03 eV, and the C1s spectrum possesses a tail to high binding energy owing to the presence of predominantly carbon singly bonded to oxygen, C1s 286.79 eV, and a small tail at the end of spectrum with C1s, 289.5 eV from *C*—F bonding. It seems that some fluorine atoms on the aromatic ring may have been replaced by the *C*—*OH* bond and, perhaps, further converted to diphenyl ether linkages during the deposition process. Several speculative defect structures are illustrated in Figure 18.18. The oxygen atoms within the polymer structure or formed as end

Table 18.9. XPS Results for PNT-F

Parameter	C—C and C—H (C1s)	C—O (C1s)	C—F (C1s)
Peak position (eV)	285.03	286.79	289.50
Area	1971	867	331
% of total area	62.20	27.36	10.44

Table 18.10. XPS Results for "Improved" PNT-F Film

Parameter	C—C and C—H (C1s)	C—F (C1s)
Peak position (eV)	285.57	287.66
Area	1584	1123
% of total area	58.81	41.49

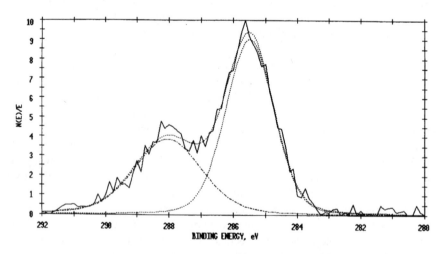

Figure 18.19. XPS spectrum of "improved" PNT-F film.

Table 18.11. Properties of Thin Films Prepared by VDP

Film	PA-N	PA-F	Teflon	Teflon AF	PNT-N	PNT-F	Bis(BCB-F$_8$)
Film deposition	Vapor	Vapor	Vapor	Vapor	Vapor	Vapor	Vapor
Source material	Dimer (solid)	Precursor (liquid)	Polymer (solid)	Polymer (solid)	Precursor (liquid)	Precursor (solid)	Precursor (solid)
Toxicity of source material	No	Hazardous	Hazardous	Hazardous	—	—	—
Dielectric constant	260 ± 0.1	2.2–2.3	2.1	1.93	2.4 ± 0.1	2.3 ± 0.1	2.0–2.1
Electric breakdown	2 × 10^7 V/mil	5 × 10^7 V/mil	5 × 10^7 V/mil	5 × 10^7 V/mil	3 × 10^7 V/mil	5 × 10^7 V/mil (best)	NA
Hardness (Gpa)	0.2	—	—	0.5	3.1 (R. #1) 1.6 (R. #2)	3.8 (R. #1)	—
Dissociation T(°C)	430°C in N$_2$	530°C in N$_2$	320°C in air	360°C in air	570°C in N$_2$	590°C in N$_2$	—
Thickness vs. annealing temperature	No change to 350°C in N$_2$	No change to 500°C in N$_2$	—	—	No change to 530°C in N$_2$	—	—
Structure as deposited	Crystalline	Crystalline/ amorphous	Crystalline/ amorphous	Amorphous	Microcrystalline/ amorphous	Mcirocrystalline/ amorphous	Cross-linked
Cracks	Yes (annealing at 300°C in N$_2$)	No (annealing at 510°C in N$_2$)	—	Yes −15°C in air 250°C in N$_2$	No (annealing at 570°C in N$_2$)	No (annealing at 600°C in N$_2$)	No

groups could have resulted from reactions with H_2O or O_2 during the deposition or upon exposure of the films to air. A compelling explanation of how the $C-O$ defects were incorporated into the PNT-F film is not clear at present, but these oxygen-based defects will cause a higher dielectric constant and a lower thermal stability.

In a subsequent experiment, the quality of the deposited PNT-F film was greatly improved by taking pains to prevent atmospheric contamination, yielding a carbon atom ratio (calculated from the area of the binding energy peaks), C ($C-C$ and $C-H$) : C ($C-F$) = 59 : 41, where the theoretical ratio is 60 : 40 (Figure 18.19). The deposited PNT-F film had a very low dielectric constant (2.1–2.2) and high thermal stability (590°C in nitrogen). Table 18.10 shows the XPS results for the improved film.

18.3. CONCLUSIONS

An overview of the properties of the materials we are studying is presented in Table 18.11. The objective of this work was to find new approaches to the problem of generating new media with low dielectric constants and high thermal stabilities for use as interlayer dielectrics in microelectronic interconnection applications. We have been partially successful in this quest but there is still much more work to be done. The materials we have been able to deposit remain to be characterized in full detail, which includes not only elucidating their molecular structure but also measuring the panoply of physical properties necessary for practical applications.

VDP also provides a route for the synthesis of thin films without the use of toxic solvents, an opportunity not to be overlooked in this era of increased environmental watchfulness. There remains, however, the challenge of preparing potential source materials useful in this methodology. The amazingly productive field of Organic Chemistry will most certainly be able to meet this challenge.

ACKNOWLEDGMENTS: It is a pleasure to acknowledge the collaboration of my colleagues Prof. Jack McDonald and Prof. Toh-Ming Lu, who have provided insight into their worlds of Electrical Engineering and Condensed Matter Physics while maintaining steadfast good humor in the face of so much Organic Chemistry. The measurement skills of Dr. G.-R. Yang were indispensible in this effort and are greatly appreciated. The hard work, synthetic skill, and smiling perseverance of Dr. Chi-I Lang, currently employed at Applied Materials, Santa Clara, Cal., are the heart and soul of this effort. The gracious willingness of Prof. James Tour to share his hard-won synthetic experience in the preparation of aryl diacetylenes prior to publication is thankfully noted. Financial support, in part, by the IBM Corporation and the Semiconductor Research Corporation is deeply appreciated.

18.4. REFERENCES

1. T.-M. Lu, Personal Communication (1994).
2. R. R. Tummala and E. J. Rymaszewski, *Microelectronics Packaging Handbook*, Van Nostrand Reinhold, New York (1989), pp. 673–725.
3. C. C. Ku and R. Liepins, *Electrical Properties of Polymers*, Hanser, Munich (1987).
4. D. S. Soane and Z. Martynenko, *Polymers in Microelectronics*, Elsevier Science, New York (1989), p. 10.
5. S. D. Senturia, R. A. Miller, D. D. Denton, F. W. Smith, III, and H. J. Neuhaus, in *Recent Advances in Polyimide Science and Technology* (W. D. Weber and M. R. Gupta, eds.), SPE, Poughkeepsie, NY (1987), pp. 351–361.
6. G.-R. Yang, S. Dabral, L. You, J. F. McDonald, T. -M. Lu, and H. Bakhru, *J. Electronic. Mat. 20*, 571–576 (1991).
7. Chi-I Lang, Synthesis of new, vapor-depositable low dielectric constant materials for use as on-chip dielectrics, Doctoral Dissertation, Rensselaer Polytechnic Institute (1995).
8. M. L. Hitchman and K. F. Jensen. *Chemical Vapor Deposition: Principles and Applications*, Academic Press, New York (1993).
9. A. Kruse, C. Thuemmler, A. Killinger, W. Meyer, and M. Grunze, *J. Electron. Spectrosc. Relat. Phenom. 60*, 193–209 (1992).
10. N. Than-Trong, P. Y. Timbrell, and R. N. Lamb, *Chem. Phys. Lett. 205*, 219–224 (1993).
11. R. F. Saraf, C. Dimitrakopoulos, M. F. Toney, and S. P. Kowalczyk, *Langmuir 12*, 2802–2806 (1996).
12. M. J. Szwarc, *Discussions Faraday Soc. 2*, 46–49 (1947).
13. W. F. Gorham, *J. Polym. Sci. A-1 4*, 3027–3039 (1966).
14. W. F. Gorham, U.S. Patent 3,342,754 [*CA 68*, P3320u (1967)].
15. B. L. Joesten, *J. Appl. Polym. Sci. 18*, 439–488 (1974).
16. W. R. Hertler, *J. Org. Chem. 28*, 2877–2879 (1963).
17. S. A. Fuqua, R. M. Parkhurst, and R. M. Silverstein, *Tetrahedron 20*, 1625–1632 (1964).
18. S. W. Chow, L. A. Pilato, and W. L. Wheelwright, *J. Org. Chem. 35*, 20–22 (1970).
19. S. W. Chow, W. E. Loeb, and C. E. White, *J. Appl. Polym. Sci. 13*, 2325–2332 (1969).
20. W. R. Dolbier, Jr., M. A. Asghar, H. -Q. Pan, and L. Celewicz, *J. Org. Chem. 58*, 1827–1830 (1993).
21. W. R. Dolbier, R., A. Asghar, and H. Q. Pan, U.S. Patent 5,210,341 [*CA 119*, P116991d (1993)].
22. L. You, G.-R. Yang, C.-I. Lang, P. Wu, T.-M. Lu, J. A. Moore and J. F. McDonald, *J. Vac. Sci. Technol. A 11* (6), 3047–3052 (1993).
23. L. You, G.-R. Yang, C.-I, Lang, J. A. Moore, J. F. McDonald, and T.-M. Lu, U.S. Patent 5,268,202 [*CA 120*], P285893z (1994)].
24. G. Odian, *Principles of Polymerization*, John Wiley and Sons, New York (1981), pp. 126–140.
25. S. Iwatsuki, *Adv. Polym. Sci. 58*, 93–120 (1984).
26. L. S. Tan and F. E. Arnold, *J. Polym. Sci. Pt. A: Polym. Chem. Ed. 26*, 1819–1834 (1988).
27. R. A. Kirchoff, U.S. Patent 4,540,763 [*CA 104*, P34502w (1985)].
28. T. S. Tan and F. E. Arnold, U.S. Patent 4,711,964 [*CA 108*, P132467r (1988)].
29. K. A. Walker, L. J. Markoski, and J. S. Moore, *Macromolecules 26*, 3713–3716 (1993).
30. D. C. Burdeaux, P. H. Townsend, J. N. Carr, and P. E. Garrou, *J. Electronic Mat. 19*, 1357–1366 (1990).
31. R. A. Kirchoff, C. J. Carriere, K. J. Bruza, N. G. Rondan, and R. L. Sammler, *J. Macromol. Sci.— Chem. A28*, 1079–1113 (1991).
32. S. F. Hahn, S. J. Martin, and M. L. McKelvy, *Macromolecules 25*, 1539–1545 (1992).
33. R. A. Kirchoff, and K. J. Bruza, *Chemtech 23*, 22–25 (1993).
34. T. Tuschka, K. Naito, and B. Rickborn, *J. Org. Chem. 48*, 70–76 (1983).

35. R. A. Kirchhoff, C. J. Carriere, K. J. Bruza, N. G. Rondan, and R. I. Sammler, *J. Macromol. Sci.— Chem. A28*, 1709–1113 (1991).

36. J. G. Eden (ed.), *Photochemical Vapor Deposition*, John Wiley and Sons, New York (1992), pp 5–8.

37. L. J. Bellamy, *The Infrared Spectra of Complex Molecules*, John Wiley and Sons, New York (1975).

38. D. Lien-Vien, N. B. Colthup, W. G. Fatley, and J. G. Grasseli, *The Handbook of Infrared and Raman Characteristic Frequencies of Organic Molecules*, Academic Press, New York (1991).

39. H. Kudo, R. Shinohara, and M. Yamada, Materials Research Society, Spring Meeting, San Francisco, Personal Communication with R. Shinohara (1995).

40. J. G. Speight, P. Kovacic, and F. W. Koch, *J. Macromol. Sci. Revs.—Macromol. Chem C5* (2), 295–386 (1971).

41. P. Kovacic and F. W. Koch, *J. Org. Chem. 30*, 3176–3181 (1965).

42. M. Sato, K. Kaeriyama, and K. Someno, *Makromol. Chem. 184*, 2241–2249 (1983).

43. J. H. Banning and M. B. Jones, *Polym. Preprints 28*, 223–224 (1987).

44. S. K. Taylor, S. G. Bennett, I. Khoury, and P. Kovacic, *J. Polym. Sci.: Polym. Lett. Ed. 19*, 85–87 (1986).

45. M. Satoh, F. Uesugi, M. Tabata, K. Kaneto, and K. Yoshino, *J. Chem. Soc., Chem. Commun. 1986*, 550–551.

46. S. Zecchin, R. Tomat, G. Schiavon, and G. Zotti, *Synth. Met. 25*, 393–399 (1988).

47. S. Hara and N. Toshima, *Chem. Lett. 1990*, 269–272.

48. R. C. Bergman, *Acc. Chem. Res. 6*, 25–31 (1973).

49. T. P. Lockhart, P. B. Cornita, and R. G. Berman, *J. Am. Chem. Soc. 103*, 4082–4090 (1981).

50. J. A. John and J. M. Tour, *J. Am. Chem. Soc. 116*, 5011–5012 (1994).

51. C. C. Ku and R. Liepins, Electrical Properties of Polymer, Harvel, Munich (1987), p. 173.

52. D. T. Clark, and A. Dilks, *J. Polym. Sci.: Polym. Chem. Ed. 17*, 957–976 (1979).

19

Ultrathin PTFE, PVDF, and FEP Coatings Deposited Using Plasma-Assisted Physical Vapor Deposition

K. J. LAWSON and J. R. NICHOLLS

19.1. INTRODUCTION

The unique properties of fluorinated polymers have led to their wide use as surface protection coatings and as components within protective coating systems.[1-4] Applications are diverse, from cookware, through environmental protection coatings for architectural panel work, antigraffitti and soil-resistant coatings in the automotive industry, to low-friction coatings[5] within the precision tool industry. This diversity reflects the chemical and thermal stability, hydrophobicity, and low surface energies associated with fluoropolymers.

Traditionally these coatings have been deposited by solvent-spraying, spin-coating, or powder-spraying technologies, resulting in coating thicknesses in excess of 10 μm. The increase in possible applications of ultrathin fluoropolymer coatings particularly for environmental protection and sensors[6,7] has led to increased interest in plasma polymerization[8,9] and sputtering[10-19] techniques to produce such films.

This chapter examines the deposition of fluorinated polymers using plasma-assisted physical vapor deposition. Ultrathin coatings, between 20 and 5000 nm have been produced, using RF magnetron sputtering. The method of coating, fabrication, and deposition conditions are described.

K. J. LAWSON and J. R. NICHOLLS · Cranfield University, Cranfield, Bedford MK43 0AL, United Kingdom.

Fluoropolymers 1: Synthesis, edited by Hougham *et al.*, Plenum Press, New York, 1999.

313

19.2. RF MAGNETRON SPUTTERING OF FLUOROPOLYMER FILMS

Over the last decade, selected papers[11-14] have examined the deposition of fluoropolymers, using RF magnetron sputtering. All of these papers have examined the deposition of PTFE, with some of them[12,14] also studying the deposition of polyimide (PI) films. This chapter extends these studies and will report on the sputter deposition behavior of PTFE (polytetrafluoroethylene), PVDF (polyvinylidenefluoride), and FEP (fluorinated ethylene propylene copolymer) films.

19.2.1. The Process

The sputtering process shown in Figure 19.1, utilizes the phenomena associated with a low-pressure gas discharge. The system comprises an anode and cathode; generally the low-pressure chamber is earthed and forms an "infinite" area anode; the small cathode surface is the "target" material from which gas ion-etching occurs, resulting in condensation of the material onto workpieces within the vacuum chamber.

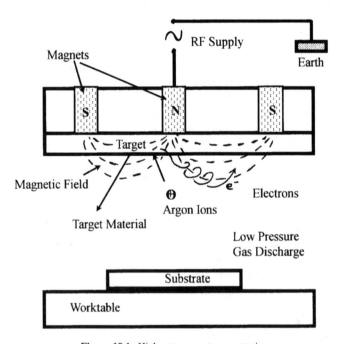

Figure 19.1. High-rate magnetron sputtering.

In order to be able to sputter dielectric materials, an RF glow discharge must be established. An RF voltage is coupled through the inductive/capacitive impedance presented by the electrode, dielectric target, and plasma. The ionization is maintained by electrons within the system. Normally the ion species is an inert gas (e.g., argon). The ion-etching of the cathode occurs owing to a net negative dc bias being established at the insulating surface of the target material. This is due to the high mobility of the electrons in the plasma in comparison with the ions. This net negative bias then accelerates ions to the target surface, resulting in ion-etching.

The magnetron arrangement incorporates magnets, behind the target electrode system, which trap further electrons within the discharge volume, resulting in enhanced ionization. This leads to an increase in etching rate but also traps electrons that might otherwise strike the workpieces to be coated, hence reducing heating that might be caused by these secondary electrons. This is a distinct advantage when depositing onto heat-sensitive materials or, indeed, sputtering heat-sensitive coatings.

As the source of material (sputtering target) is a solid, a variety of electrode configurations can be used; e.g., depositing material downward, upward or sideways. It is possible to electrically bias workpieces being coated. This can encourage a level of ion bombardment that can modify the surface and structure of a coating.

19.2.2. Sputter Deposition of PTFE, PVDF, and FEP

The sputtering equipment was used with sputter-up and sputter-down configurations. 150-mm-diameter disks of the materials were used, with the PTFE and PVDF being 5 mm thick and the FEP 2.3 mm thick. The targets were loosely clamped to the sputtering electrodes to allow for thermal expansion. High-purity argon gas was introduced into the chamber as the glow discharge gas.

The vacuum chambers were pumped down by means of an oil diffusion pump backed by a rotary vane vacuum pump. The base pressure achieved was 1×10^{-5} Torr (1.33×10^{-3} Pa). High-purity argon gas was bled into the chamber, the high-vacuum valve throttled, and the chamber pressure maintained as close as possible to 2×10^{-2} Torr (2.66 Pa). For some of the experiments, the dc self-bias on the magnetron electrode was also measured.

The parameters that were varied were: (a) change from electrically earthed to electrically isolated work table; (b) argon gas mass flow over the range 12–27 sccm; and (c) RF power density over the range 0.18–1.13 W/cm². The detailed process parameters for each deposition are summarized in Table 19.1. A range of substrate materials have been used, primarily soda lime glass slides, but also nickel, aluminum, silicon, and gold- and silver-coated glass slides.

Table 19.1. Deposition Parameters for Fluoropolymer Films

Material	RF power density (W/cm^2)	Work table	Substrate	Rate of deposition (nm/h)
PTFE	0.3	Electrically isolated	Glass	126
PTFE	0.57	Electrically isolated	Glass	4140
PTFE	1.13	Electrically isolated	Glass	9000
PVDF	0.18	Electrically isolated	Glass	126
PVDF	0.3	Electrically isolated	Glass	300
FEP	0.18	Electrically isolated	Glass	552

19.3. RESULTS AND DISCUSSION

19.3.1. Deposition of PTFE

This work has shown that PTFE films can be deposited using RF magnetron sputtering over a range of deposition conditions, with power densities varying from 0.3 to 1.13 W/cm^2. In all cases adherent films were produced showing no delamination even after extended exposure times (out to 16,000 h). The rate of deposition was found to increase in a near linear manner with deposition power, as can be seen in Figure 19.2.

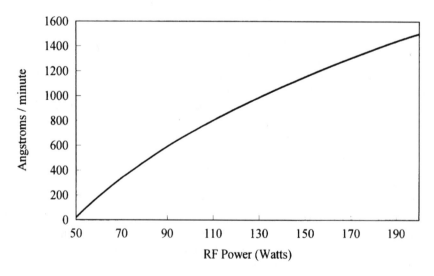

Figure 19.2. Sputter deposition rate of PTFE vs. RF power.

Measurements of the IR absorption spectrum of the as-deposited film was similar to that for the target material, implying similar bonding to that in target material. Figure 19.3 shows the IR reflectance spectrum obtained from a 10-μm PTFE coating sputtered onto a polished silicon slice. The RF power density to the target was 0.57 W/cm^2. The periodicity at the higher wave number values is due to optical interference. The spectra of films deposited at higher power densities were similar, with absorption peaks through 1110 to 1360 cm^{-1} and 610 to 735 cm^{-1}. Previous studies[12-14] had similarly found that RF-sputtered PTFE films possessed attributes similar to the bulk material and that they also closely resembled those deposited by plasma polymerization.[14]

In this current work, films between 20 nm and 15 μm have been produced, the latter at relatively high rates of deposition (3 μm h). The previously reported problems[9] associated with depositing thicker films were not observed in this study.

Equally in this study, changes in the RF power level did not appear to result in significant optical degradation of the coating. In previous studies[13,14] the inability to deposit at high rates, and hence thickness, was thought to reflect damage to the films owing to argon-ion bombardment. Such damage was not observed in these studies, as this would have been evident in the color of the films.

Although physical, chemical, and mechanical properties have not been evaluated so far for these films, the implication from the works of Yamada *et al.*,[12] Ochiai *et al.*,[13] and Hishmeh *et al.*[14] is that these ultrathin films are excellent insulating materials, with good hydrophobicity and superior chemical resistance. Ochiai also observed that the contact angle was 100° at room temperature, similar

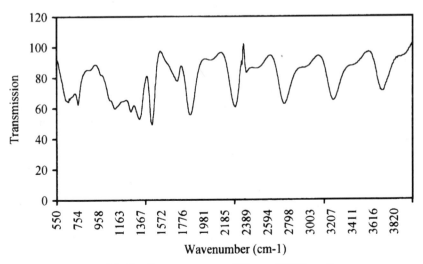

Figure 19.3. IR reflectance spectrum of 10 μm of PTFE on silicon.

to conventional PTFE, which further confirms that the morphology of this coating would be similar to conventional PTFE.

19.3.2. Deposition of PVDF and FEP Films

The sputter deposition of PVDF and FEP films has not been widely reported in the open literature. In this work, both materials have been studied with films deposited in the thickness range 30 nm to 4 μm. The processing conditions that have been used are summarized in Table 19.1. Generally the films that were formed were adherent, although more prone to degradation during deposition than the PTFE films.

For PVDF it was observed that small changes in power density can lead to a rapid increase in deposition rate. From Table 19.1 it can be seen that increasing the power from 0.18 to 0.3 W/cm^2 increased the rate of deposition by a factor of 2.4 (0.125 to 0.3 μm/h). By comparison, the deposition rate for PTFE varied in a near linear manner with sputtering power density, as can be seen in Figure 19.2, and rates typically eight times higher than for PVDF could be achieved.

The sputter deposition of FEP showed that high rates of deposition are possible at low power 0.55 μm/h at 0.18 W/cm^2). Thus FEP can be deposited at approximately four times the rate of PVDF at the same power density. High rates of deposition for PVDF onto glass resulted in films with poor adhesion, although similar films deposited onto a metal substrate gave good adhesion. Conversely, for the FEP films low rates of deposition resulted in optical degradation of the films. Clearly, the morphology and structure of the deposited films are a result of a careful balance between the rate of film deposition and growth, the degree of argon-ion bombardment, and the extent of backsputtering that can occur. For PVDF, high rates of deposition can result in insufficient ion bombardment during deposition to ensure good adhesion, while for FEP films at low deposition rates, the increase in argon-ion bombardment, relative to the deposition rate gives rise to color centers and therefore optical degradation of the films.

Hishmeh et al.[14] in their study of RF magnetron sputtering of PTFE films reported that the films they produced could be readily damaged by argon-ion bombardment, being reduced to CF$^-$ containing groups. Although not observed for PTFE films in this study, such damage could account for the color centers observed in the FEP-deposited films obtained at low deposition rates.

19.4. CONCLUSIONS

The deposition rate of PTFE varied nearly linearly with RF power density. High rates of deposition could be achieved owing to its temperature tolerance allowing higher power densities. The coatings retained PTFE-like properties with good adhesion. PVDF sputtered at rates that were nonlinear with RF power density, resulting in coatings with variable adhesion. However, at equivalent power densities, PVDF sputtered at about twice the rate of PTFE. FEP sputtered at high rates for low power densities, about four times that of PVDF for the same power density. However, low rates resulted in film degradation as evidenced by film discoloration.

This work has shown that it is possible to deposit a number of fluoropolymer films by RF magnetron sputtering by carefully controlling the rate of deposition to degree of ion bombardment during film formation. Films from 20 nm to 15 μm have been produced that are of high integrity and adherent. The fluoropolymers deposited in this study included PTFE, PVDF, and FEP; of these only PTFE had been extensively studied by RF magnetron sputtering prior to this work.

ACKNOWLEDGMENTS: The authors wish to thank Mr. G. Jefferies and Mr. A. Green of the Molecular Electronics Group at Cranfield for their assistance with the IR spectroscopy. They are also grateful to the Paint Research Association for agreeing to allow publication of this chapter, which is based on a paper presented at the Second Fluorine Conference in Munich, Germany, 1997.

19.5. REFERENCES

1. *Proceedings of Fluorine in Coatings*, Salford, UK, Paint Research Association, Teddington, Middlesex, UK (1994).
2. *Proceedings of Fluorine in Coatings II*, Munich, Germany, Paint Research Association, Teddington, Middlesex, UK (1997).
3. A. Bruce Banks, J. Mirtich, S. K. Rutledge, and D. M. Swec, *Thin Sol. Films 127* (1–2), 107–114 (1985).
4. K. A. Ryden, *Proceedings of the Royal Aeronautical Society Conference on Small Satellites* (1991), pp. 4.1–4.11.
5. I. Sugimoto and S. Miyake, *Thin Sol. Films 158* (1) 51–60 (1988).
6. N. Nakano and S. Ogawa, *Sensors and Actuators B:Chemical B21* (1), 51–55, (1994).
7. J. A. McLaughlin, D. Macken, B. J. Meenan, E. T. McAdams, and P. D. Maguire, *Key Eng. Mat. 99–100*, 331–338 (1995).
8. A. Tressuad, F. Moguet, and L. Lozano, in *Proceedings of Fluorine in Coatings II*, Munich, Germany, Paint Research Association, Teddington, Middlesex, UK (1997).
9. S. Kurosawa, D. Radloff, N. Minoura, and N. Inagaki, in *Proceedings of Fluorine in Coatings II*, Munich, Germany, Paint Research Association, Teddington, Middlesex, UK (1997).
10. H. Biederman, *International Seminar on Film Preparation and Etchings by Plasma Technology*, Brighton, Great Britain, (1981).
11. A. Cavaleiro and M. T. Vieira, *Solar Energy Mat. 20* (3), 245–256 (1990).
12. Y. Yamada, K. Tanaka, and K. Saito, *Surf. Coat. Tech. 44* (1–3), 618–628 (1990).
13. S. Ochiai, T. Katao, A. Maeda, M. Ieda, and T. Mizutani, *Proceedings of the IEEE International Conference on Properties and Applications of Dielectric Materials, Vol. 1*, IEEE, Piscataway, N.J., 94CH3311-8, (1994), pp. 215–218.

14. G. A. Hishmeh, T. L. Barr, A. Sklyarov, and S. Hardcastle, *J. Vac. Sci. Tech. A: Vac. Surf. Films 14* (3, pt. 2) 1330–1338 (1996).
15. I. Sugimoti, M. Nakamura, and H. Kuwano, *Thin Sol. Films 249*, 118–125 (1994).
16. J. P. Badey, E. Urbaczewski-Espuche, Y. Jugnet, D. Sage, Duc. Tran Minh, and B. Chabert, *Polymer 35* (12), 2472–2479 (1994).
17. M. J. O'Keefe and J. M. Rigsbee, *J. Appl. Polym. Sci. 53* (12), 1631–1638 (1994).
18. H. V. Jansen, J. G. E. Gardeniers, J. Elders, H. A. C. Tilmans, and M. Elwenspoek, *Sensors and Actuators A: Physical 41*, (1–3, pt. 3) 136–140 (1994).
19. J. Perrin, B. Despax, V. Hanchett, and E. Kay, *J. Vac. Sci. Tech. A: Vac., Surf. Films 4* (1), 46–51 (1986).

Index

Acid end group, 197, 199, 201
Acrylates, 4, 170
  fluorinated, 3–9, 107, 167–178
Acrylo nitrile, 268
Acrylics, 168, 181
Adhesion, 21, 230, 232, 241, 243, 249–250,
  262–266
Adhesives, 250, 252, 258
Adsorbents, 99–100
Aerosol fluorination, 58
Aeropak process, 219, 230
Alkali metal carbonates, 55
Amphiphilicity, 151–153, 156, 164–165
Anisotropy, 161
Annealing, 299
Argon-ion bombardment, 317–318
Aromatic condensation fluoropolymers, 127–128
  poly(azoles), 143–147
    properties of, 146–147
    synthesis of, 143–146
  poly(azomethines), 140–143
    properties of, 141–143
    synthesis of, 140–141
  poly(carbonates), 128–132
    properties of, 131–132
    synthesis of, 128–130
  poly(formals), 132–137
    properties of, 133–135
    synthesis of, 132–133
    thermal degradation of, 135–137
  poly(ketones), 137–140
    properties of, 139–140
    synthesis of, 137–139
  poly(siloxane), 148–149
    properties of, 148–149
    synthesis of, 148

Aromatic condensation fluoropolymers (*cont.*)
  role of hexafluoroisopropylidene in, 128–
    147, 150
Aromatic cyanate esters, 13
Aromatic polyimides, 111
Arrhenius relationship, 45
Aryl ether polymers, 29
Azoles, 143–147
Azomethines, 140–143

Balz–Schiemann reaction, 231
Base-catalyzed homopolyerization, 52–54
Benzoyl peroxide, 8
Bergman's cycloaromatization, 294–295
Biocompatibility, 127, 268–269
  of dialysis membranes, 262
Birefringence, 161
Bis(triphenylphosphoranylidene) ammonium
  chloride, 55
Bisphenol, 128, 187–188
  in synthesis of aromatic condensation
    fluoropolymers, 127–151
Block copolymers
  amphiphilicity of, 151–153
  applications of, 151, 159–163
  modular synthesis concept for, 152
  overview of, 151–153
  phase morphology of, 161–162
  polarity of, 152
  properties of fluorinated, 156–164
    mechanical, 161
    mesophase formation, 161–163
    micelle formation, 156
      role of fluorinated solvents, 156–159
    surface energies, 159–161
    solid state phase behavior, 165

Block copolymers (*cont.*)
  structure of, 159–161
  surface behavior of films, 161, 164–165
  synthesis of fluorinated, 153–156, 164–165
2-bromotetrafluoroethyl aryl ether, 31–34, 41
BS6464 test, 253–254
β-scission, 201, 203–204
Bulk resistivity, 272

Capillary rheometry, 87–88
Carbonates, 128–132
Carbon black, 235
  fluorination of, 235
Carbon dioxide, 191, 203
  in fluoropolymer synthesis, 191–204
  solvent properties of, 192–193, 203–204
Carbonylation agent, 128
Catalysts, 15, 54–55
Chain transfer agents, 84
Chemical vapor deposition, 275, 300
Chlorine-alkali electrolysis, 96
Chlorotrifluoroethylene, 69, 71
Cladding, 249–251
CO₂/aqueous hybrid system, 191, 203
  in fluoropolymer synthesis, 203–204
Coagulation, 81
Coaxial push-out test, 253–254
Cohesion agent, 264
Cohesion energy, 159, 164
Complement activation, 268–269
Complex permittivity, 19
Concrete, 255
  role of surface-fluorinated fibers in, 255–257
Condensation polymerization, 55, 61
Contact angle, 131–132, 134, 140, 142, 159, 161
Copolymerization, 6–7
Critical surface tension, 18
Cross propagation, 201
Cross-linked polymers, 282
Crown ethers, 216–217
Cryptands, 216
Cuprophan/SPAN, 262, 268
Cure skrinkage, 22
Cyanate ester resins, 11–23
  curing reaction, 15
  fluoromethylene on structure of, 11
  functional groups in, 12, 15
  monomer synthesis of, 12–16
  overview of, 11–12
  processing and applications of, 21–22

Cyanate ester resins (*cont.*)
  properties of, 17–19
  dielectric, 19–21
Cyclic oligomers, 55–56, 58
Cycloaromatization
  Bergman's, 294–295
  electrochemical, 294–295
Cyclodimerization, 25–27, 44; *see also* Thermal cyclodimerization
  kinetics of, 26
  of olefins, 25–48
Cyclodimers, 26

Daikin polymers, 86
Defoaming agents, 67
Demnum, 213
Dialkylsiloxanes, 74
Diallyls, 72
Dialysis membranes, 262
  biocompatibility of, 268–269
Diblock copolymers, 161, 165
Dichlorohydrin, 188
Dielectric constant, 19–21, 23, 43, 167–169, 172–175, 178–179, 273, 275
Differential scanning calorimetry thermograms, 13–14
Diglycidyl ether 18, 185, 188
Dihalogenosilanes, 74
Diimidediol, 184–185
Diiodoperfluoroalkanes, 70–71
Diolen, 262, 264–267
Dipole effect, 21
Dipole–dipole interactions, 299
Direct fluorination, 209–220
Disilanol, 74
Dispersion polymerization, 158–160, 164
Drop dissipation test, 266–267
Drop penetration test, 266–267

Elastomers, 51, 67, 151
Electrochemical cycloaromatization, 294–295
Electronegativity, 127
Emulsion polymerization, 81, 86
Entropy, 177
Epichlorohydrin, 216
Epoxy adhesives, 250, 252, 258
Epoxy networks, 181–188
Esters, 97–98
  aromatic cyanate, 11–23
  fluoroaliphatic cyanate, 12, 17
Etherdiacrylate, 170

Etherdiol, 170
Ethers
    aryl, 29
    crown, 216–217
    diglycidyl, 18, 185, 188
    fluorinated, 28–30, 31–38, 41, 43, 48, 52–53,
        59–61, 91
    Ullmann synthesis of, 115

Fiberglass resin, 254
Fluorinated ethylene propylene copolymers, 314
Fluorinated polyacrylates
    applications of, 178
    dielectric constant of, 167–169, 172–175,
        178–179
        structure–property relationships of, 177–179
    processability of, 178
    synthesis of, 169–172
Fluorination, 248, 251
    aerosol, 58
    carbon black, 235
    direct, 209–220
    hollow fiber, 263
    Linde, 219
    radical, 56, 58
    stepwise, 60
    surface, 219–220
        effect on polymer properties, 223–224,
            231–236, 239, 241
        mechanism of, 229
        overview, 223–224
        polymer applications, 241
            benefits for flexible plastics, 245–247
            cementation mixtures, 255–257
            HDPE products, 241–243
            liquid containment, 254
            permeation-based products, 242–247
            perspectives on, 257–258
            pipe products, 251
            products based on improved adhesion,
                247–257
            sheet cladding, 249
        role on polymer films, 232–233
        use of xenon difluoride, 224, 231–236, 239
            industrial application in rubber, 236–239
Fluorine, 168, 224
    chemical reactivity of, 225–231
    electronegativity of, 261
    industrial use of, 261
    isolation of, 224
    properties of, 224–230

Fluorine (*cont.*)
    uses of, 225
Fluoroacrylate polymers, 3; *see also*
        Polyacrylates
    application of, 3
    properties of, 4, 8
        measurement of, 5–6
    synthesis of, 4–8
        monomers, 4–5
        polymers, 5
    role of hexafluoroisopropylidene, 3, 9
Fluoroaliphatic cyanate esters, 12, 17
Fluoroalkyl acrylates, 107
Fluoroalkylation, 40–41
Fluorodiepoxide, 187–188
Fluorodiimidediol, 182
    preparation of epoxy networks, 183–188
    synthesis of, 184–185, 188
Fluorokote, 219
Fluoromethylene cyanate ester resins, 11–23
Fluoroolefins, 91–93, 202
Fluoropolyethers, 52, 59–61
Fluoropolymers, 91–108, 127, 191, 202, 228;
        see also Fluoroacrylate polymers;
        Polymers
    applications of, 96–102
        chlorine-alkali electrolysis, 96
        fuel cells, 96
        ion-selective electrodes, 100–101
        in microelectronics, 271–275, 307
        permselective membranes, 99
        polymer catalysts, 96–97
            alkylation, 98–99
            esterification, 97–98
            oligomerization, 97
        properties of materials, 273
        selective adsorbents, 99–100
        water electrolysis, 96
    aromatic condensation, 127–128
    conductivity of, 102, 104–106
    direct fluorination in synthesis of, 209–220
        overview of, 209–211
        structure–property relationships, 209–211
    functional groups in, 51–53, 59–61, 91–95
    Langmuir–Blodgett films, 102–108
    mechanical testing of plaques, 38–39
    melt-processable, 196–197
    overview of, 91–92, 195–196
    properties of, 95, 102–108, 228
    structure of, 56–62, 92–93, 95, 103–107
    surface fluorination of, 218–220

Fluoropolymers (*cont.*)
    synthesis in carbon dioxide, 191–240
        applications of amphiphilic copolymers, 195
        fluoroalkyl acrylate polymerization, 193–
            195
        fluoroolefin polymerization, 195–202
        homopolymers, 193–194
        random copolymers, 194–195
        solvent systems, 191–204
    synthesis of, 91–95, 211–220, 229
        copolymerization, 92–93
        paracyanogen, 217
        perfluoropolyethers, 212–217
        poly(carbon monofluoride), 211–212
        polyacrylonitrile, 217
        polymerization, 92–93
        radiation grafting, 93–95
    thermal cyclodimerization of olefins, 25–48
    thermal stability testing of films, 39, 44–46
    thin-film deposition, 275–307
        defect structures, 305–306
        dielectric constants, 280–281, 291
        improvement of, 306–307
        infrared spectroscopy, 287–288, 298–302
        parylene F, 277–282
        parylene N, 275–277
        polynaphthalenes, 292–307
            properties, 298–307
        solid state NMR spectra, 302–303
        thermal stability, 291, 298–300
        x-ray photoelectron spectroscopy, 288–
            291, 298, 303–307
    ultrathin film deposition, 313–318
    use in protective coating systems, 313
    use of catalysts in, 54–55
    use of xenon difluoride in, 231–236
Fluororubbers, 228–230, 233, 236–239
Fluorosilicones
    homopolymerization 75–78
    introduction of fluorinated groups, 72–78
        hydrosilylation, 73–74
        organometallic derivatives, 72–73
        polycondensation, 75
    overview of, 67
    properties of, 67–68, 78
    synthesis of, 68, 74–78
        precursors, 68–72
        telechelic oligomers, 70–72
        telomerization, 69–70, 76
        viton cotelomers, 70
    uses of, 78

Fluorosiloxanes, 67
Fluorinated containers, 242
Fluorotelomers, 75
Fomblin Z, 213
Formals, 132–137
Fragmentation, 59
Free volume, 19–21
Free-radical polymerization, 7, 193–194
Freon, 113, 169–170, 172–173
Friction coefficient, 233–234
    in rubbers, 234
Friedel–Crafts reaction, 115–116, 124
Fuel cells, 96
Fullerenes, 225–226

Gap fill, 21–22
Gas permeability, 219, 233
Glass transition temperature, 8, 17, 162
Gorham method, 278
Graft copolymerization, 91, 94–95, 97
Grigart–Putter reaction, 12

Heat-transfer agents, 51
Heterophase polymerization, (158
Hexafluoro-2-hydroxy-2-(4-fluorophenyl)pro-
    pane, 4
2-[hexafluoro-2-(4-fluorophenyl)-2-
    propoxy]acetic acid, 5
Hexafluoro-2-(4-flourophenyl)-2-propyl
    acrylate, 4
Hexafluoro-2-(4-fluorophenyl)-2-propyl
    methacrylate, 4
Hexafluoroacetone, 169
Hexafluoroisopropylidene, 3
    functional group in fluoroacrylate polymers,
        3, 9
Hexafluoropropene, 69, 71, 76
Hexafluoropropylene, 91
High-density polyethylene (HDPE), 241–
    243
    product applications of, 241–245
Homopolymerizartion, 6–7, 75–78, 172
    base-catalyzed, 52–54
Hydrocarbon permeation, 230
Hydrogen fluoride, 212
Hydrogenosilanes, 73
Hydrolytic susceptability, 15
Hydrophobicity, 182
Hydrosilylation, 73–74
2-hydroxylethylmethacrylate, 4
3-hydroxylpropylmethacrylate, 4

Inherent viscosity, 8
Initiator, 8, 85–86
Inorganic dielectrics, 273
Interlayer dielectric, 272–273, 307
Ion etching, 315
Ion-selective electrodes, 100–101
Ion-exchange resins, 197–201

Kel–F polymer, 94
Ketones, 137–140
Kirkwood correlation plots, 194
Krytox, 58, 213

Langmuir–Blodgett films, 102–103
Lap shear strength, 251–253
    test, 252
Lap shear test, 252–254
Linde fluorination, 219
Linear polymers, 53–55
Lithium aluminum hydride, 52
Lubricants, 51, 67, 212–213, 219

Man-made fibers, 261; see also Yarns
Melt-flow index, 199
Mesophase formation, 161–163
Molecular order, 19–21

Nafion, 96–101, 197
Neutron activation, 246
Nitrogen permeability, 246
Nitrogen-containing ladder polymers, 217–219
Nucleophiles, 53

Olefins, 25–27, 53
    cyclodimerization of, 25–27
O-quinodimethane, 283
Organic peroxides, 85–86
Organolithians, 72–73
Organomagnesians, 72–73
Oxidative cationic polymerization, 293
Oxidative stability, 277
Oxyfluorination, 248–249, 251, 253
Oxygen permeability, 246–247

Packing efficiency, 20
Paracyanogen, 217
Parylene-N deposition process, 276–277
Parylenes, 274–282
Peel strength, 250
    test, 250
Peel test, 264

Pendant group, 3, 9
Perfluorinated alkyl vinyl ethers, 91
Perfluoroalkyl iodides, 68
Perfluorocyclobutane aryl ether polymers, 28–30
Perfluoroethers, 51, 59–61
Perfluorooctanoyl peroxide, 82
Perfluoropolyesters, 202, 212–217
Perfluoropolyethers, 212–217
Permeability, 127, 163
Permeation, 242–247, 258
Permittivity, 172
Permselective membranes, 99
Petrol permeation, 244–246
Phase-transfer catalysis, 128–129, 132
Phosgene, 128
Photooxidation, 202
Piezoelectricity, 127
Pipe products, 251
    HDPE, 244
    polyolefin, 251
Planarization, 21, 48, 274
Plasma-assisted physical vapor deposition, 313–319
    ultrathin fluoropolymer films, 313–318
Polarizability, 19, 127
Polyacrylates, 3–9; see also Fluoroacrylate polymers
Polyacrylonitrile, 217
Polyarylate fluoropolymers, 31, 34, 36
    synthesis by cyclodimerization, 31, 34, 36, 47–48
Poly(aryl ether ketones), 111–112
    applications of, 111–112, 123–124
    characterization of, 112
    films, 114
        property measurements, 114–115
    fluorinated, 115–123
    overview of, 111–112
    properties of, 112, 115–124.
    structural relationships, 115–117
    synthesis of, 111–116
        hexafluoroisopropylidine in, 112, 115, 120
        homopolymerization, 116–118
        polycondensation, 119
        polymerization, 113
        via naphthalene moieties, 111, 114, 122–124
Poly(benzocyclobutenes), 274, 282–292
    chemistry of, 283–284
    use in polymers, 284

Poly(carbonates), 128
Poly(carbon monofluoride), 211–212
  structure of, 211
Polychlorotrifluoroethylene, 81, 94
Polydimethyl siloxanes, 68
Polyester adhesives, 258
Polyester resins, 253
Poly(ethylene oxide), 214
Poly(formals), 132
Polyimides, 181, 273–274, 314
Polymerization, 92–93, 113
  aromatic condensation in, 127–151
    poly(carbonates), 128–132
      properties of, 131–132
      synthesis of, 128–130
    poly(azoles), 143–147
      properties of, 146–147
      synthesis of, 143–146
    poly(azomethines), 140–143
      properties of, 141–143
      synthesis of, 140–141
    poly(formals), 132–137
      properties of, 133–135
      synthesis of, 132–133
      thermal degradation of, 135–137
    poly(ketones), 137–140
      properties of, 139–140
      synthesis of, 137–139
    poly(siloxane), 148–149
      properties of, 148–149
      synthesis of, 148
    role of hexafluoroisopropylidene in, 128–
      147, 150
  base-catalyzed, 52–54
  catalysis in, 54–55, 96–99
  condensation, 55, 61
  copolymerization in, 6–7, 92–93, 161, 165
  cross linking in, 282
  cycloaromatization in, 294–295
    Bergman's, 294–295
    electrochemical, 294–295
  cyclodimerization in, 25–27, 44
    kinetics of, 26
    of olefins, 25–48
  dispersion, 158–160, 164
  emulsion, 81, 86
  fluorination in, 58, 128–147, 150, 153–165,
      168, 209–220, 224, 248–251, 314
    direct, 209–220
    Linde, 219
    radical, 56, 58

Polymerization (*cont.*)
  fluorination in (*cont.*)
    stepwise, 60
    surface, 219–220
  free-radical, 7, 193–194
  functional groups in, 12, 15, 51–53, 59–61,
      72–78, 91–95
  graft, 91, 94–95, 97
  heterophase, 158
  homopolymerization in, 6–7, 75–78, 116–
      118, 172
  oxidative cationic, 293
  radical copolymerization in, 92, 97, 172
  random copolymerization in, 51, 130
  role of CO_2 in, 191–204
  role of nitrogen in, 217–219
  solution, 295–296, 300
  suspension, 88
    effect of reaction variables, 83–84
    gel permeation chromatography, 86–87
    initiators, 85–86
    internal viscosity, 84
    overview of, 81
    polymer characterization in, 82–83
      molecular weight measurements, 82–83
      property testing, 83–84, 87–88
    role of organic peroxides in, 85–86, 88
    thermal stability, 88–90
    viscosity-shear rate dependence, 88
    zero strength time, 84
  vapor deposition, 275–307
Polymers, *see also* Fluoropolymers
  adhesion, 230
  aromatic condensation, 127–151
    poly(azoles), 143–147
    poly(azomethines), 140–143
    poly(carbonates), 128–132
    poly(formals), 132–137
    poly(ketones), 137–140
    poly(siloxanes), 148–149
  applications of, 3, 96–102, 111–112, 123–
      124, 151, 159–163, 178, 236–239,
      241–247, 251, 255, 313–318,
  aryl ether, 29
  block copolymers, 151–153
    amphiphilicity of, 151–153
    applications of, 151, 159–163
    modular synthesis concept for, 152
    overview of, 151–153
    phase morphology of, 161–162
    polarity of, 152

Polymers (*cont.*)
  block copolymers (*cont.*)
    properties of fluorinated, 156–164
      mechanical, 161
      mesophase formation, 161–163
      micelle formation, 156
        role of fluorinated solvents, 156–159
      surface energies, 159–161
      solid state phase behavior, 165
      structure of, 159–161
      surface behavior of films, 161, 164–165
      synthesis of fluorinated, 153–156, 164–165
    catalysis, 15, 54–55, 96–97
      phase-transfer, 128–129, 132
    characterization of, 82–83
      molecular weight measurements, 82–83
      property testing, 83–84, 87–88
    cross linking in, 282
    Daikin, 86, 87
    diblock copolymers, 161, 165
    fluoroacrylate, 3; *see also* Polyacrylates
    Kel-F, 94
    linear, 53–55
    nitrogen-containing ladder, 217–219
    silicon, 149
    structure–property relationships of, 115–117,
      177–179, 209–211
    surface fluorination in, 219–220
      effect on polymer properties, 223–224,
        231–236, 239, 241
      mechanism of, 229
      overview, 223–224
      polymer applications, 241
        benefits for flexible plastics, 245–247
        cementation mixtures, 255–257
        HDPE products, 241–243
        liquid containment, 254
        permeation-based products, 242–247
        perspectives on, 257–258
        pipe products, 251
        products based on improved adhesion,
          247–257
        sheet cladding, 249
      role on polymer films, 232–233
      use of xenon difluoride, 224, 231–236,
        239
        industrial application in rubber, 236–
          239
    thermoplastic, 43–44, 47
    thermoset, 28, 45–47
    use of CO_2 in synthesis of, 191–204

Poly(octafluorobisbenzocyclobutene), 285–292
Polyolefins, 247
Poly(propylene oxide), 214
Polystyrene-*b*-polybutadiene, 153–156, 161–
  164
Polystyrene-*b*-poly(vinylperfluorooctanic es-
  ter), 153–155
Polystyrene-*b*-polydimethylsiloxane, 159
Poly(tetrafluoroethylene), 11, 25, 91, 159, 168–
  169, 203, 228, 314
Poly(tetrafluoronaphthalene), 295–297
Poly(tetrafluoro-*p*-xylylene, 278–282
Poly(trifluoropropyl methyl siloxane), 68, 76
Polyvinylidenefluoride, 314
Precursor method, 280
Proton chemical shift, 56
Pull-off test, 21
Pull-out tests, 255
Pulse propagation, 271
PVC-coated fabrics, 264

Radiation grafting, 93–95
Radical copolymerization, 92, 97
Radical fluorination, 56, 58
Radical polymerization, 172
Radio frequency magnetron sputtering, 313–
  316
  fluoropolymer film deposition, 315–318
  process description, 314–315
Random copolymerization, 51, 130
Refractive index, 19–21, 178–179
Reinforcement resin 252
Resin bleeding, 22
Rubbers, 67
  adhesion of yarns to, 264–266
  friction coefficients of, 233–234
  Russian-made, 233–234
  use of xenon difluoride in, 236–239

Scholl reaction, 116–118
Semiconductors, 107–108
Signal coupling, 272
Silanes, 73
Silicon polymers, 149
Silicones, 74
Siloxanes, 68, 76, 148–149
Sodium peroxide, 85
Solid state phase behavior, 165
Solution polymerization, 295–296, 300
Spin coating, 274
Spirodimer, 284

Spontaneous birefringence, 161
Stepwise fluorination, 60
Steric hindrance, 72
Steric repulsion, 210
Steric stabilizers, 158, 164
Styrene, 158–159
Sulfonyl fluoride, 92–93
Surface energy, 159–161, 164, 248
Surface fluorination, 219–220
    effect on polymer properties, 223–224, 231–
        236, 239, 241
    mechanism of, 229
    overview, 223–224
    polymer applications, 241
        benefits for flexible plastics, 245–247
        cementation mixtures, 255–257
        HDPE products, 241–243
        liquid containment, 254
        permeation-based products, 242–247
        perspectives on, 257–258
        pipe products, 251
        products based on improved adhesion,
            247–257
        sheet cladding, 249
    role on polymer films, 232–233
    use of xenon difluoride, 224, 231–236, 239
        industrial application in rubber, 236–239
Surface pressure–area diagrams, 103–104
Surface tension, 248
    critical, 18
Surfactants, 67, 81, 151
Suspension polymerization, 88
    effect of reaction variables, 83–84
    gel permeation chromatography, 86–87
    initiators, 85–86
    internal viscosity, 84
    overview of, 81
    polymer characterization in, 82–83
        molecular weight measurements, 82–83
        property testing, 83–84, 87–88
    role of organic peroxides in, 85–86, 88
    thermal stability, 88–90
    viscosity–shear rate dependence, 88
    zero strength time, 84
Swelling effect, 200
Symmetrical cyanurate heterocycle, 12
Synthetic fibers, 255

Tan delta, 19–20
Taxogen, 69
Teflon, 21, 168–169, 177, 179, 274

Telechelic oligomers, 70–72, 78
Telogen, 69–70
Telomerization, 69–70, 78
Telomers, 69–72, 216
Terminal resonance, 15
Tetraalkylammonium chloride, 55
Tetrafluoroalkyl silanes, 72–74, 78
Tetrafluoroethylene, 91, 195–202
Tetramethylammonium bromide, 185–188
Thermal curing , 15
Thermal cyclodimerization
    kinetics of, 26, 44
    of fluorinated olefins, 25–29
Thermal expansion coefficient, 15, 17
Thermal stability, 8, 18, 67, 80–90, 274–275,
        277, 298–300
    testing of films, 39, 44–46
    thin-film deposition, 291, 298–300
        defect structures of, 305–306
        dielectric constants, 280–281, 291
        improvement of, 306–307
        infrared spectroscopy, 287–288, 298–302
        parylene F, 277–282
        parylene N, 275–277
        polynaphthalenes, 292–307
            properties of, 298–307
        solid state NMR spectra, 302–303
        ultrathin, 313–318
        x-ray photoelectron spectroscopy, 288–291,
            298, 303–307
Thermogram, 13–14
Thermooxidative stability, 131, 1347, 140, 147
Thermoplastic elastomer, 151, 163
Thermoplastic polymers, 43–44, 47
Thermoplastic resins, 151
Thermoset polymers, 28, 45–47
Thin-film deposition, 275–307, 313–319
    defect structures, 305–306
    dielectric constants, 280–281, 291
    improvement of, 306–307
    infrared spectroscopy, 287–288, 298–302
    parylene F, 277–282
    parylene N, 275–277
    polynaphthalenes, 292–307
        properties, 298–307
    solid state NMR spectra, 302–303
    thermal stability, 291, 298–300
    ultrathin, 313–318
    x-ray photoelectron spectroscopy, 288–291,
        298, 303–307
Thin-film processing, 46–47

Thin-film uniformity, 274
T-peel test, 250
Triacrylate, 171, 176
Trichlorotrifluoroethane, 169
Triethylamine, 169
Trifluoroethylene, 68, 71
Trifluorovinyl ethers, 53
Trifluorovinyloxy aryl ether, 34–38, 43, 48
Triol, 171
Twaron, 262, 264–265

Ullman ether synthesis, 115
UV absorption, 287–289

Vapor deposition polymerization, 275–307
    apparatus for thin-film deposition, 296
Vinylidene fluoride, 68, 70–71, 76, 94
Viton elastomers, 68, 70
Vulcabond, 264

Water absorption, 8, 18, 21
Water electrolysis, 96
Water repellency, 262, 266

Wettability, 232
Wetting behavior, 266–267

Xenon difluoride, 224, 239
    industrial application in rubber, 236–239
    use in fluoropolymers, 231–236
X-ray diffractogram, 298
Xylylene, 277–278

Yarns
    adhesion to PVC, 264–266
    aramid, 264
    adhesion to rubber, 264–266
    overview of fluorination on properties of, 262
        methods of application, 262–263
            batch process, 262
            continuous process, 263
            hollow fiber fluorination, 263
    tire, 261, 264
    viscose, 264
    wetting behavior, 266–267

Zero strength time, 81, 84